The Economics of Livestock Disease Insurance

Concepts, Issues and International Case Studies

The Economics of Livestock Disease Insurance

Concepts, Issues and International Case Studies

Edited by

S.R. Koontz, D.L. Hoag and D.D. Thilmany

Department of Agricultural and Resource Economics
Colorado State University
USA

and

J.W. Green and J.L. Grannis

Animal and Plant Health Inspection Service
US Department of Agriculture
Fort Collins, Colorado
USA

CABI Publishing

CABI Publishing is a division of CAB International

CABI Publishing
CAB International
Wallingford
Oxfordshire OX10 8DE
UK

Tel: +44 (0)1491 832111
Fax: +44 (0)1491 833508
E-mail: cabi@cabi.org
Website: www.cabi-publishing.org

CABI Publishing
875 Massachusetts Avenue
7th Floor
Cambridge, MA 02139
USA

Tel: +1 617 395 4056
Fax: +1 617 354 6875
E-mail: cabi-nao@cabi.org

A catalogue record for this book is available from the British Library, London, UK.

A catalogue record for this book is available from the Library of Congress, Washington, D.C.

The economics of livestock disease insurance: concepts, issues and internaitonal case studies / edited by S.R. Koontz ... [et al.].
p. cm.
Includes bibliographical references and index.
ISBN-13: 978-0-85199-077-4 (alk. paper)
ISBN-10: 0-85199-077-0 (alk. paper)
1. Insurance, Agricultural–Livestock. 2. Insurance, Agricultural–Livestock.–Case studies. 3. Livestock–Diseases. 4. Livestock–Diseases–Case studies. I. Koontz, Stephen R. II. Title.

HG9968.6.E25 2005
368.56–dc22

2005012394

ISBN-10: 0-85199-077-0
ISBN-13: 978 0-85199-077-4

Printed and bound in the UK from copy supplied by the editors by Cromwell Press, Trowbridge.

Contents

Contributors

Barnett, Barry J., Department of Agricultural and Applied Economics, University of Georgia, 314-A Conner Hall, Athens, GA 30602-7509, USA, bbarnett@agecon.uga.edu.

Bätza, Hans-Joachim, Federal Ministry of Consumer Protection, Food and Agriculture, Rochusstraße 1, D-53123, Bonn, Germany, hans-joachim.baetza@bmvel.bund.de.

Bervejillo, José E., Agricultural Issues Center, University of California-Davis, One Shields Avenue, Davis, CA 95616, USA, jeberve@ucdavis.edu.

Black, J. Roy, Department of Agricultural Economics, Michigan State University, 305 Ag Hall, East Lansing, MI 48824, USA, blackj@msu.edu.

Bruch, Megan L., University of Tennessee, Center for Profitable Agriculture, P.O. Box 1819, Spring Hill, TN 37174, USA, mlbruch@utk.edu.

Coble, Keith H., Department of Agricultural Economics, Mississippi State University, 321 Lloyd Ricks, Box 5187, Mississippi State, MS 39762, USA, coble@agecon.msstate.edu.

Driscoll, James L., Watts & Associates Crop Insurance Division, 7027 W. 129th Place, Overland Park, KS 66209, USA, jdriscoll@wattsandassociates.com.

Epps, Terri, Agriculture & Agri-Food Canada, 689 SJCB, 930 Carling Avenue, Ottawa, Ontario, K1A 0C5, Canada, eppster@agr.gc.ca.

Gramig, Ben M., Department of Agricultural Economics, Michigan State University, 105 Cook Hall, East Lansing MI 48824-1039, USA, gramigbe@msu.edu.

Grannis, Jennifer L., Veterinary Services, Animal and Plant Health Inspection Service, US Department of Agriculture, 2150 Centre Avenue, Bldg. B, Mail Stop 2E5, Fort Collins, CO 80526-8117, USA, jennifer.l.grannis@aphis.usda.gov.

Green, John W., Veterinary Services, Animal and Plant Health Inspection Service, US Department of Agriculture, 2150 Centre Avenue, Bldg. B, Mail Stop 2W4, Fort Collins, CO 80526-8117, USA, john.w.green@aphis.usda.gov.

Hanson, Terry, Department of Agricultural Economics, Mississippi State University, 312 Lloyd Ricks, Box 5187, Mississippi State, MS 39762,USA, hanson@agecon.msstate.edu.

Hart, Chad, Food and Agricultural Policy Research Institute and Center for Agricultural and Rural Development, Iowa State University, 568E Heady Hall, Ames, IA 50011-1070, USA, chart@iastate.edu.

Hoag, Dana L., Department of Agricultural and Resource Economics, Colorado State University, Fort Collins, CO 80523-1172, USA, dhoag@lamar.colostate.edu.

Huirne, Ruud, Institute of Risk Management in Agriculture, Business Economics, Wageningen University and Research Centre, Hollandseweg 1, 6706 KN Wageningen, The Netherlands, ruud.huirne@wur.nl.

Jansson, Torbjörn, Institute for Agricultural Policy, Market Research and Economic Sociology, University of Bonn, Germany, jansson@agp.uni-bonn.de.

Jarvis, Lovell, Department of Agricultural and Resource Economics, University of California-Davis, One Shields Avenue, Davis, CA 95616, USA, lsjarvis@ucdavis.edu.

Johnson, Kamina, Department of Agricultural and Resource Economics, Colorado State University, Fort Collins, CO 80523-1172, USA, kaminakjohnson@yahoo.com.

Keogh, Rob C., Animal Health Australia, Suite 15, 26 - 28 Napier Close, Deakin, 2600, ACT, Australia, rob.keogh@aahc.com.au.

Koontz, Stephen R., Department of Agricultural and Resource Economics, Colorado State University, Fort Collins, CO 80523-1172, USA, stephen.koontz@colostate.edu.

Lee, John G., Department of Agricultural Economics, College of Agriculture, Purdue University, West Lafayette, IN 47907-2056, USA, jlee1@purdue.edu.

Meuwissen, Miranda, Institute for Risk Management in Agriculture, Business Economics, Wageningen University and Research Centre, Hollandseweg 1, 6706 KN Wageningen, The Netherlands, miranda.meuwissen@wur.nl.

Miller, J. Corey, Department of Agricultural Economics, Mississippi State University, 319 Lloyd Ricks, Mississippi State, MS 39762, USA, miller@agecon.msstate.edu.

Neumann, Geoff B., Geoff Neumann & Associates Pty Ltd, P.O. Box 562, Brighton, SA 5048, Australia, geoff.neumann@bigpond.com.

Norell, Bo, Swedish Board of Agriculture, Jönköping, Sweden, bo.norell@sjv.se.

Ott, Stephen, Veterinary Services, Animal and Plant Health Inspection Service, US Department of Agriculture, 2150 Centre Avenue, Bldg. B, Mail Stop 2E5, Fort Collins, CO 80526-8117, USA, stephen.l.ott@aphis.usda.gov.

Paarlberg, Philip L., Department of Agricultural Economics, College of Agriculture, Purdue University, West Lafayette, IN 47907-2056, USA, paarlbep@purdue.edu.

Pritchett, James, Department of Agricultural and Resource Economics, Colorado State University, Fort Collins, CO 80523-1172, USA, james.pritchett@colostate.edu.

Rabinowicz, Ewa, Swedish Institute for Food and Agricultural Economics, Lund, Sweden, rabinowicz@sli.lu.se.

Seitzinger, Ann Hillberg, US Department of Agriculture, Animal and Plant Health Inspection Service, Veterinary Services, 2150 Centre Ave, Bldg B, Mail Stop 2E7, Fort Collins, CO 80526-8117, USA, ann.h.seitzinger@aphis.usda.gov.

Sempier, Stephen H., Department of Agricultural Economics, Mississippi State University, Box 5187, Mississippi State, MS 39762, USA, sempier@agecon.msstate.edu.

Shaik, Saleem, Department of Agricultural Economics, Mississippi State University, 215E Lloyd Ricks, Mississippi State, MS 39762, USA, shaik@agecon.msstate.edu.

Skees, Jerry R., The H.B. Price Professor of Policy and Risk, Department of Agricultural Economics, University of Kentucky, Lexington, KY 40546-0276, USA, jskees@uky.edu.

Stephen, Bruce, Agriculture & Agri-Food Canada, 689 SJC, 930 Carling Avenue, Ottawa, Ontario, K1A 0C5, Canada, bstephe@agr.gc.ca.

Sumner, Daniel A., UC Agricultural Issues Center and Department of Agricultural and Resource Economics, University of California-Davis, One Shields Avenue, Davis, CA 95616, USA, dasumner@ucdavis.edu.

Thilmany, Dawn, Department of Agricultural and Resource Economics, Colorado State University, Fort Collins, CO 80523-1172, USA, dawn.thilmany@colostate.edu.

Turvey, Calum, Department of Applied Economics and Management, 356 Warren Hall, Cornell University, Ithaca, NY 14583, and the Food Policy Institute, Rutgers University, New Brunswick, NJ. cgt6@cornell.edu.

van Asseldonk, Marcel, Institute for Risk Management in Agriculture, Business Economics, Wageningen University and Research Centre, Hollandseweg 1, 6706 KN Wageningen, The Netherlands, marcel.vanAsseldonk@wur.nl.

Wilkens, Eckhardt, R+V versicherung, Hanover, Germany and Agricultural Committee of Comité Européan des Assurances, Brussels, Belgium, wilkens-burgwedel@t-online.de.

Wolf, Christopher A., Department of Agricultural Economics, Michigan State University, 317B Ag Hall, East Lansing, MI 48824, USA, wolfch@msu.edu.

Acknowledgments

This work is the result of two conferences, the willingness of conference participants to write and contribute to the effort that followed, and the willingness of other professionals to fill in the gaps that were uncovered in the conference discussions.

The Livestock Insurance Products International Conference and Forum: Discovery of Ideas and Issues was held November 2002 at the Natural Resources Research Center in Fort Collins, Colorado. The event was sponsored by Colorado State University and the US Department of Agriculture (USDA) Animal and Plant Health Inspection Service (APHIS), Veterinary Services (VS) and Centers for Epidemiology and Animal Health (CEAH). Over 35 academic researchers, insurance industry representatives and government officials from Australia, Canada, Germany, the Netherlands and the USA participated. The editors would like to recognize the financial support of USDA-APHIS.

The book's contents are also a result of a conference focusing on the Economics of Animal Health, held in July 2003 during the Western Agricultural Economics Association meeting in Denver, Colorado, and sponsored by the Farm Foundation, with additional support from a USDA Economic Research Service (ERS) Cooperative Agreement. In addition to financial support, the Farm Foundation's leadership in bringing industry, government and academic participants to the conference, and discussion among conference participants improved the content and relevance of several of the book's chapters.

A number of other professionals with interests and experience in animal disease management, insurance, and public policy contributed to this work following the conferences. Their experience and expertise certainly help round out and improve the breadth of this book.

Finally, the editors would like to thank CABI Publishing for their interest in our efforts and Tim Hardwick in particular for his patience in working with us.

Chapter 1

Economics of Livestock Disease Insurance – Principles, Issues and Worldwide Cases

Dana L. Hoag, Dawn D. Thilmany and Stephen R. Koontz

Introduction

The potential for animal disease outbreaks in the United States of America, Canada, Europe, Australia and other developed countries is an important concern for livestock producers, livestock allied industries, consumers, citizens and governments. These concerns were validated by recent discoveries of Bovine Spongiform Encephalopathy (BSE) in the EU, Canada, the USA and Japan, outbreaks of Foot and Mouth Disease (FMD) in the EU, and outbreaks of Avian Influenza in Asia. Animal disease insurance has been a key response in some nations, and opportunities are expanding to provide products for livestock producers throughout the world. However, government and industry leaders will need to understand individual private incentives, market impacts, and public policy perspectives on regional, national and international levels if they are to be an active voice in the formation of livestock insurance products and complementary risk management programmes. This book provides a balanced and diverse overview about the economics of livestock disease insurance by looking at what can be learned from research and programme experiences in many different parts of the world.

The motivation for this book comes from two conferences. One conference was organized by the US Department of Agriculture (USDA) Animal and Plant Health Inspection Service (APHIS) and Colorado State University. The second conference was organized by Colorado State University and the Farm Foundation of Oakbrook, Illinois. The USA is relatively inexperienced in livestock insurance, and any livestock insurance scheme would be strongly influenced by APHIS policy. Professionals from countries with more experience were invited to the first conference in late 2002 to discuss what is known and how different countries manage diseases and livestock insurance programmes. Participants represented the USA, Canada, Europe and Australia. The second conference was held in mid-2003 and focused primarily on the USA. Participants from the conferences, and others that could address issues raised at the meetings, were subsequently invited to write the chapters that you will find here. This book will remain topical for some time. Livestock industries will not likely eliminate these diseases in the near future, nor will there be structural changes that will diminish the importance of managing risk associated with livestock disease. It is also unlikely that a spectrum of insurance products will emerge that quickly and easily address disease issues raised herein.

Interest in developing animal industry insurance products has specifically increased in the USA for several reasons. Recent international experiences with disease losses have raised awareness of potential risks. Hard lessons were learned, especially in Canada and Europe, about how devastating an outbreak could be to markets and trade. By animal disease, we and the chapter authors are largely referring to the OIE's (Office International Des Epizooties or World Organization for Animal Health) List A Diseases and List B Diseases. See Table 1.1. Since the value of trade is so important to livestock industries, and the risk of disease transmission has increased, politicians and industry groups in the USA and elsewhere are pressing for development of risk management products. Furthermore, insurance is viewed as one way to lessen the potential burden on the US Treasury should major disease outbreaks occur in the USA.

Despite the fact that the availability of livestock industry insurance products is increasing worldwide, our knowledge and experience is limited; therefore these products are in the state of development. Following the passage of the 2000 Agricultural Risk Protection Act, the USDA Risk Management Agency (RMA) solicited work to summarize alterative insurance issues, schemes and proposals. A review of existing literature and discussion with experts revealed little work that explained, synthesized or discussed details regarding different policy approaches and different insurance programmes used around the world. The main purpose of this book therefore is to present discussion of current policy and insurance institutions and to encourage thinking about new policy and insurance schemes. This is achieved through the presentation of insurance principles, policy perspectives and assumptions, and examples communicated through international case studies. Within the following chapters, the reader will find information from academic experts in agricultural insurance, insurance industry leaders with expertise in agricultural insurance, and government agency personnel that are responsible for addressing animal disease management-related programmes and policies that may influence the effectiveness of insurance products. Many chapters are written by international experts from countries where livestock disease insurance and livestock business revenue insurance are currently available or are being developed.

Intended Uses

The intended readership of this book is a combination of individuals and associations of government agency personnel, insurance and broader industry readers, as well as academics. USDA personnel are responsible for developing and facilitating the offering of animal industry insurance products in the USA while the insurance industry will actually offer the animal industry products. Academics will be involved in the research process that is, in part, used by the insurance industry, and will be involved in training personnel that work within government agencies and insurance firms. Academics have also played an important role in past research on the development of insurance products. Thus, this work will be a resource for all these interested parties to inform the discussion of research and provide worldwide examples related to different products.

Because of the topical and case study nature of the text, the book likely has the greatest potential to be a supplementary text. The specific courses which might use the book would be the same as discussed above – agricultural and food policy, trade and regulation – food safety, food security, biosecurity and risk management – and contemporary research topics courses.

Table 1.1. OIE Classification of Diseases.

List A Diseases	List B Diseases			
	Multiple Species Diseases	Cattle Diseases	Swine Diseases	Avian Diseases
Foot and mouth disease	Anthrax	Bovine anaplasmosis	Atrophic rhinitis of swine	Avian chlamydiosis
Swine vesicular disease	Aujeszky's disease	Bovine babesiosis	Enterovirus encephalomyelitis	Avian infectious bronchitis
Peste des petits ruminants	Echinococcosis/ hydatidosis	Bovine brucellosis	Porcine brucellosis	Avian infectious laryngotracheitis
Lumpy skin disease	Heartwater	Bovine cysticercosis	Porcine cysticercosis	Avian mycoplasmosis (*M. gallisepticum*)
Bluetongue	Leptospirosis	Bovine genital campylobacteriosis	Porcine reproductive and respiratory syndrome	Avian tuberculosis
African horse sickness	New world screwworm (*Cochliomyia hominivorax*)	Bovine spongiform encephalopathy	Transmissible gastroenteritis	Duck virus enteritis
Classical swine fever	Old world screwworm (*Chrysomya bezziana*)	Bovine tuberculosis		Duck virus hepatitis
Newcastle disease	Paratuberculosis	Dermatophilosis		Fowl cholera
Vesicular stomatitis	Q fever	Enzootic bovine leukosis		Fowl pox
Rinderpest	Rabies	Haemorrhagic septicaemia		Fowl typhoid
Contagious bovine pleuropneumonia	Trichinellosis	Infectious bovine rhinotracheitis/infectious pustular vulvovaginitis		Infectious bursal disease (Gumboro disease)
Rift Valley fever		Malignant catarrhal fever		Marek's disease
Sheep pox and goat pox		Theileriosis		Pullorum disease
African swine fever		Trichomonosis		
Highly pathogenic avian influenza		Trypanosomosis (tsetse-borne)		

List A diseases: Transmissible diseases that have the potential for very serious and rapid spread, irrespective of national borders, that are of serious socio-economic or public health consequence and that are of major importance in the international trade of animals and animal products.

List B diseases: Transmissible diseases that are considered to be of socio-economic and/or public health importance within countries and that are significant in the international trade of animals and animal products.

Source: http://www.oie.int/eng/maladies/en_classification.htm. [Accessed 2004].

Other targeted readers include individuals and professional associations in the area of veterinary medicine, industry groups, and other arenas that provide input and commentary on animal disease management. Veterinarians and members of industry organizations frequently work more closely with members of livestock industries and interface with politicians and political organizations more frequently than academics. These groups of professionals provide important input about livestock and meat industries to the political process so it is important that they understand the breadth and complexity of disease insurance issues.

Lessons and Recommendations

One of the most effective methods to summarize the message and contribution of this book's chapters is to outline common themes, lessons learned from case studies, and recommendations for future research and policy work. In this section, messages common to a number of the chapters are quickly surveyed, together with discussion of how they may influence the development of livestock insurance products and complementary animal health management research, programmes and policy instruments.

The first and perhaps most important lesson is that the evidence here supports a joint solution between the private and public sectors. The government has strong and definitive reasons to intervene when there are market failures, as is the case with livestock diseases. However, perverse production incentives will result if there is not a shared responsibility with industry. That is, the decisions of government and industry are interlinked. This interdependence needs to be taken into consideration by both public and private parties in order to understand the choices faced by affected producers. Specifically, consideration of the importance of private incentives, including the moral hazard created when producers perceive economic opportunities in securing public funds during outbreaks, is imperative in the development process. Adverse selection that allows for poor managers to offset underperformance could also be of concern to both private industry (which expects competitive pressure to push inefficient producers out) and public interests (which face public scrutiny for expenditures that are not welfare-enhancing). Many of the chapters herein discuss how producers will adjust their production practices to respond to the structure of insurance programmes, and in contrast, how insurance programmes will change their policies to react to the structure of production.

Policymakers can address how complex interactions between private and public incentives should best be addressed through intensive study. Authors here examine a variety of issues in a number of settings. Some authors investigate how incentives matter as well as the implications for policymakers and managers by modelling a theoretical or actual disease outbreak on farms or the farm sector. Janson, Norrel and Rabinowicz, for example, found that making insurance compulsory for EU livestock producers, rather than being taxpayer financed, would increase net social welfare. Wolf showed why beef and dairy producers would choose different risk management tools to address the current bovine tuberculosis problem in Michigan in the USA and what might be done about it. One of the important lessons from these models is that institutions and market mechanics have the effect of wrapping around insurance programmes and can severely affect their pricing. Other authors addressed the complexity and interaction of public and private incentives another way by looking at insurability conditions, such as how compensation is made, incentives, availability of data, and the feasibility of different premium levels. Shaik *et al.*, for example, develop a disease risk management integration matrix to assess insurability conditions when there are varying conditions like those that exist in livestock disease management. Practitioners will find this matrix most useful.

A second common theme in the book is the importance of effective design, credible probability assumptions and pricing of insurance products. These tasks are made more difficult by the diversity of issues, objectives and business environments related to animal disease management. Many authors here conclude that the dimensionality is much greater for animal disease management than for many other agricultural risks. There may be few similarities, for example, between catfish and cattle. The economic principles of public goods, externalities, moral hazard, adverse selection and incentive compatibility will vary from one situation to another, increasing the effort required to develop viable products. Other insurance principles that might need to be considered include certainty of loss determination, measurability of risks, depth of reinsurance markets, systematic risks and economically feasible premiums.

An important question is whether there is sufficient information to design actuarially sound insurance products. Green, Driscoll and Bruch's chapters presume there are not enough data to create fairly priced livestock insurance products, but Meuswissen *et al.* think there is enough information if we draw from a wide array of research. In reality, the concern about availability of sufficient data may be conditional on the chosen design of insurance products, as discussed and outlined by Shaik *et al.* in their insurability matrix. This insurability matrix is a take-home tool for designers of products. Additionally, data may not be the only limitation. Meuwissen *et al.* and Turvey demonstrate how modelling actuarially sound premiums can be a difficult task in itself.

Related to effective design is a third important theme, incentives. Shaik *et al.,* as well as many other authors, examine the differences in incentives (and likely participation) for coverage of direct relative to consequential losses in compensation plans. Interest in coverage of a wider array of losses exists, and is one of the primary reasons given for why current products are underutilized. Some would argue that the public should only underwrite (or be involved in) development of risk management tools that address direct losses. Yet, it is not clear whether enough producers would participate in some industries (i.e. dairy) if there was not fuller coverage of business interruption losses, and if participation is low then underinvestment in the public good of animal health management may occur.

The fourth lesson is the complexity and the substantive nature of livestock disease insurance. The models needed to price actuarially fair premiums are not simple but are important. Turvey and Meuwissen *et al.* show this. Details within the models – be they drawn from data or from expert opinion – have potentially large ramifications on premium levels. But the devil is not all in the details of the modelling. The incentive structure and the potential for hidden behaviour wrap around these pricing models. Improper incentives will reveal themselves and will defeat a well-designed model and efficiently priced product. This is the concern of Gramig *et al.,* Shaik *et al.,* Coble *et al.* and many other authors. Insurance schemes that are developed and offered have the potential to deeply impact an industry's structure. The work by Jansson, Norell and Rabinowicz clearly shows this impact and the interviews of producers by Grannis, Green and Bruch shows that producers recognize this impact. Insurance schemes and the willingness of governments to address animal disease issues also have potentially far-reaching market impacts. Seitzinger *et al.* provide a positivistic modelling example of the extent of impact. Measurement of these impacts necessitates equilibrium models which measure changes to new prices and quantities – both in production and trade – that may be quite different from where we are now. Jansson, Norell and Rabinowicz, through the use of a normative programming model, show that the change from a taxpayer-financed scheme to a self-financed scheme has the potential to substantively change the location and density of animal agriculture in the EU.

Finally, several lessons can be drawn from a comparative analysis of participating countries. For example, it appears that geographical boundaries matter for both the optimal animal health management scheme, and likely success of risk management products such as insurance. Australia and the USA benefit from having few borders. Effects of broader animal health management strategies will also influence the incentives, perceived risks and optimal design of risk management products. For example, Australia and the USA focus on keeping diseases out of the country while Europe focuses on actively managing those that exist. Therefore, producers, government and the insurance industry in the EU know that outbreaks will occur. The USA and Australia assume that they will not have or that outbreaks will be stopped close to the border.

On a related note, perceptions of risk levels will differ, and these perceptions will influence the perceived demand for risk management products. A certain level of uncertainty about animal health may be acceptable to producers in other areas (such as Europe) because they think of risk as familiar, while the US industry may have irrationally high perceptions of the real impact from a disease since it is something they have not managed in the past. Alternatively, producers in Australia may underestimate the probability of an outbreak because of overconfidence in their ability to keep the disease out. In either case, misperceptions (and the lack of data) may influence a producer's choice to purchase insurance, or for the full livestock industry to support public expenditures on subsidized products.

The obvious lesson from a comparative analysis across countries is that, within the EU and Australia, producers know they will incur some, if not most of the cost of the outbreak. Subsequently, programmes from those regions appear to be financed through both public and private means. Yet, within the USA, recognition of any financial responsibility among industry participants is not apparent, possibly due to overly low perceptions about private risks, and the assumption that public good aspects make the government's incentives to insure more acute. This is interesting. US agricultural producers have long banded together to offer property, life and other insurances. But they do not offer and do not appear to be thinking about offering coverage for livestock diseases, whereas in The Netherlands, Germany and Australia producers do offer livestock disease insurance. Canada is somewhere in the middle since it is offering research and development assistance for the private sector to develop self-help programmes.

Together, one could surmise that there are relevant roles for both the public and private sector in the development of insurance products. In fact, several authors suggest that the development of successful programmes requires leadership and participation from both industry and government. One potential public role is in seeding pilot programmes that allow private participants to gain a greater capacity to assess risk (collect data) and test the market's response to differently designed instruments (following the experiences of other countries). As it does in many other instances, where industry innovation may be below optimal levels, the public sector can overcome gaps or asymmetries in information through encouraging private feasibility assessment of pilot products with partial public funding support.

But, there is no one blanket scheme that will work for a variety of industries covering a diverse set of diseases. The incentive structure, lack of data, possibility to reinsure, disease biology, geographic and trade borders are just too different. Further, disease insurance schemes should not be confused or necessarily mixed with well-known price or revenue products – including futures, options, forward contracts and government products. Once again, the interdependency between the public and private sectors requires careful planning to prevent conflicting incentives.

Not only are market failures possible, they are likely. Insurance products will have to be well-designed, are likely specific to a given industry and disease, possibly to a specific location, and specific diseases or sets of similar diseases. And even then, the lack of reliable actuarial data and the degree of systematic risk may be overwhelming and market-based premiums too high. Multiple overlapping schemes will be the norm. Private companies, producer risk pooling organizations and governments will all be involved – through regulation and subsidization, and even then there will likely remain some degree of self-insurance.

Summary of Chapter Contents

The chapters in this book are organized into three major parts. With this introduction is Chapter 2, by Grannis and Bruch, which discusses how the US Department of Agriculture (USDA) Animal and Plant Health Inspection Service (APHIS) manage disease in the USA. Chapters 3-8 include a variety of discussions about what is known about how to build a livestock insurance programme. The remaining chapters offer a diverse discussion about disease management issues and programmes in Australia, Canada, Europe and the USA. These later chapters include more discussion about how to build economically sound insurance programmes, and observations are based on modelling or observing case studies.

Chapters 3-8 begin with a look at the conceptual basis for government involvement in the management of livestock diseases, including prevention, control, regulation and eradication. The discussion is picked up by looking at incentive compatibility and insurability conditions in the private sector, emphasizing how livestock disease management is unique. Compensation is also examined, including what losses should be compensated, choosing a method to value the losses, determining the portion of the losses to compensate, and outlining a potential role for insurance. Finally, the complexity of risks at the farm level is demonstrated using a model that evaluates revenue insurance.

The remainder of the book reviews international case studies. Public and private livestock insurance schemes are reviewed in Canada, Australia, the USA and the European Union for everything from aquaculture to sheep. The discussions examine crop and livestock insurance, risk mitigation schemes, government funding and philosophies, and cover a wide spectrum of diseases including FMD, BSE, scrapie, bovine tuberculosis, hog cholera and various aquaculture diseases. The methods used to address livestock diseases in the four regions are discussed thoroughly and many examples are presented to estimate how different management schemes might impact producers, consumers and society.

A brief summary of each chapter follows:

The Role of USDA-APHIS in Livestock Disease Management within the USA by Grannis and Bruch

In 2001 the US Department of Agriculture (USDA) Animal and Plant Health Inspection Service (APHIS) funded and co-organized an international conference to explore the use of livestock insurance in the USA. APHIS is responsible for conducting basic actions to protect the health of the US livestock herd and was responding to increasing pressures for the creation of a livestock disease insurance programme in the USA. Discussion at the conference revealed that any livestock disease insurance system must recognize and interact with government agencies that implement policy. However, the function of APHIS is not well known, even by animal industry members within the USA.

APHIS is tasked to protect animal health in the national herd. APHIS takes actions that directly affect the risk of an animal disease entering and spreading within the USA. If diseases are found, responses include depopulation, quarantine or limiting movement. APHIS also indemnifies producers for losses incurred due to animal diseases. The authors discuss how current APHIS activities might delay the development of livestock insurance in the USA because the government indemnity is reduced proportionately to insurance receipts and because US producers have become accustomed to receiving compensation without paying for insurance. Clearly, development of livestock disease insurance in the USA will need to consider APHIS actions and regulations. Participation in insurance programmes could be used to determine eligibility for indemnification since participation would represent adherence to biosecurity measures that protect the health of the USA livestock herd. For example, participation in livestock insurance programmes can provide producers with coverage for disease costs not reimbursed by the government, such as consequential losses. The future of livestock disease insurance in the USA is still uncertain and policy development will need to be coordinated with the actions of APHIS.

The Role of Public Policy in Controlling Animal Disease by Sumner, Bervejillo and Jarvis

This chapter discusses the prevention of animal disease from an economic perspective considering public goods and externalities. Disease is inherent in animal production. Animal diseases cause significant economic losses, though the losses would be much greater if it were not for efforts to manage disease. Private incentives to manage animal disease are clear and strong – reduce animal suffering and prevent the loss of wealth. Nonetheless, private agents do not always provide a socially optimal level of disease management. The authors argue that disease control and eradication policies produce externalities and have public good effects that may lead to private underinvestment in disease management. This underinvestment provides justification for public intervention. Rationale for collective action in animal diseases comes in several forms: 1) infectious disease management as a public good and the closely related idea of externalities related to the costs and benefits of private efforts to control infectious diseases, 2) distributional impacts of disease outbreaks and the idea that major losses may occur at random with little opportunity for mitigating behavior, and 3) potential for animal diseases to be tools of attacks on national security with the related idea that collective action for national security is a core role of government. The role of government in animal disease management may be seen as supporting private activity, providing an underlying infrastructure or filling in where private incentives could be diluted.

They note that the nature of externalities and public good characteristics, particularly whether they are large and thus warrant concern, is determined by geography and biology. Both of these affect the natural habitat for specific diseases and thus the likelihood that a disease will spread from one region to another. The likelihood of spread is a crucial consideration in government intervention, whether regarding efforts to exclude, control or eradicate the disease from a region. These considerations will thus influence the design of policies and whether these policies should conform to a locality or a region, and whether they should include single or multiple political entities. They have also considered the distributional effects of disease outbreaks and how forward pricing, or the use of futures and options markets, may mitigate risks. Compensation for animals destroyed during eradication or control efforts may also reduce direct losses and if industry assessments are used to fund such compensation, costs may be aligned with benefits. Finally, they explore the issue of biosecurity. Biosecurity is closely linked to national security, which spreads the costs of management and generates a greater potential benefit.

Incentive Compatibility in Risk Management of Contagious Livestock Diseases by Grammig, Barnett, Skees and Black

This chapter addresses the importance of the incentives present in the livestock production and animal health systems. The major challenge in designing disease insurance mechanisms is recognizing the conflict between encouraging producer herd health management and biosecurity measures while maintaining incentives for early disclosure of problems. This is an incentive compatibility problem. A key to the success of designing public and private mechanisms must involve full recognition of the sometimes competing incentives that influence individual producer behavior.

This chapter clearly emphasizes the importance of incentives when designing animal health policy or risk management instruments. An underlying theme and concern is that improperly designed livestock insurance solutions could increase the disease risk problem for the entire sector. Livestock disease has great potential to cause widespread economic damage. If the individual producer incentives are not considered when designing regulations, offering government disaster payments, or developing insurance products, the result could be a weakening of the animal health system that worsens the effect of an outbreak.

The authors first explore considerations for insuring livestock diseases, including asymmetric information, external effects, systemic risk and intertemporal effects. Good schemes should encourage individual producers to improve management in ways that facilitate herd health and biosecurity measures, while also providing incentives for early disclosure in the event of a suspected disease occurrence. Attaining this goal is complicated by the presence of information asymmetry. Frequency and severity of disease outbreak are heavily dependent on individual behaviour, which is costly for regulators or insurers to monitor and correct. The authors point out that managing livestock disease may be more important than crops because there are external effects, such as diseases that affect humans. They then explore the similarities between systemic risks that affect crops, like drought, and risks from livestock disease. Many diseases are not systemic and are therefore more difficult to address. Finally, livestock are unique in that there is an intertemporal impact related to the time it takes for producers to report an outbreak and the time it takes to react. The authors extend their findings about incentives to sections about the role of good management and early disclosure on biosecurity, public and private incentives, and investments in food safety.

Insurability Conditions and Livestock Disease Insurance by Shaik, Barnett, Coble, Miller and Hanson

The success of agricultural insurance is closely linked to addressing important conceptual economic issues, posed here as insurability conditions. This chapter is divided into five major sections. In the first section, the authors discuss insurance as a risk management issue. Then, they discuss several insurability conditions. For example, losses should be determinable and measurable. There should be a large number of approximately homogeneous, independent exposure units. Losses should be accidental and unintentional and have no risk of catastrophic losses. The chance of loss should be calculable, and premiums should be economically feasible. In the third major section, they discuss potential policy tools in the context of insurability conditions. They review the APHIS indemnification programme, private disease insurance, government disease insurance, and business interruption insurance among others. They propose a matrix to compare risk management tools with insurability conditions, and in the last section they explore challenges and opportunities for the future.

Risks will vary accordingly with insurability conditions. Therefore, there is no one risk management tool that will work best all of the time. The appropriate tool depends critically on the risk characteristics of the disease. The disease risk management integration matrix compares five market losses and six production losses against risk characteristics and insurability conditions (i.e. sufficient information) to infer which risk management tools might be most appropriate. The authors conclude that many livestock diseases deviate significantly from ideal insurability conditions.

Issues Associated with US Livestock Disease Compensation in the 21st Century by Ott

This chapter discusses the current process that APHIS uses to indemnify livestock producers from losses and outlines issues related to that process given Dr. Ott's APHIS experiences on the frontline with respect to the compensation of livestock losses. Issues include deciding what losses should be compensated, choosing a method to value the losses, determining the portion of the losses to compensate and a potential role for insurance. Traditionally, the compensated loss is the animal. Compensation is a requirement of the US Constitution, detailed in volumes of the US Federal Code of Regulations. Yet, there are still a lot of unanswered questions. Other potential losses the USDA is under pressure to cover include disinfection and clean-up costs, disposal costs, other production expenses, and most importantly, lost revenue. Animal loss, in the past, has been compensated in cases of major diseases and species. Now, the USDA may be more willing to cover less severe diseases and minor species due to political pressure. However, there are interactions between vertical integration, externalities, public goods, moral hazard, and determining fair market values that create difficulties in doing so.

Paying owners compensation for destroying their animals in order to control disease has been an important tool in the eradication of livestock diseases in the USA. Not only does the US Constitution require compensation whenever the government takes an animal away from its owner, economically it makes sense to do so as well. Individually, owners of diseased animals have little incentive to report diseases, which is detrimental to the industry as early detection is important in the control of transmittable diseases. Compensation for animals destroyed becomes the financial incentive for reporting livestock diseases. Determining the right compensation level can be problematic. If too low, owners may decide disease reporting is not worth it. If too high, the industry may become too complacent in disease prevention since the government pays for most or all of the financial losses associated with the disease. Insurance, industry-generated compensation funds and permit-assurance bonds are alternatives to government-financed compensation. These alternative schemes internalize to the individual the potential industry loss from slow- or non-reporting of disease.

Conceptual Issues in Livestock Insurance by Turvey

The purpose of this chapter is to establish some basic principles related to livestock insurance and to illustrate how these principles can be applied in practice using many techniques available today. First, a general model is used to illustrate the complexity of the risks at the farm level, and several possibilities for insuring all risks are discussed in a qualitative way. Second, a general model is developed to illustrate the principles of frequency, intensity and duration for developing livestock disease insurance. Third, a more specific class of net revenue insurance models is presented and empirically evaluated. These models assume certainty in production and feed use, but allow for variability in livestock and feed prices. Monte Carlo approaches to calculating the value of several conventional and path-dependent livestock net revenue insurance possibilities are illustrated assuming the existence of a futures market. Fourth, insuring catastrophic market risks arising from the introduction of a disease that would cause market livestock prices to evaporate is modelled as a jump process with a disease arriving at unknown times, but with known frequency. Subsequently, calculation of a Poisson-induced indemnity as an insurance product could be considered in addition to conventional livestock or revenue insurance, or the revenue insurance should be adjusted to include the probability of catastrophic market risk.

Turvey's chapter shows in well-developed detail the methods for finding actuarially fair premiums. The models and representations should be considered within the context of a new area of study that will evolve over time. This chapter's intent was to provide a set of principles from which livestock insurance can be designed. Disease insurance based on the properties of frequency, intensity and duration is easier to state as a set of mathematical principles than to implement in practice. Research must be undertaken to document these variables for each of the diseases that are endemic in the USA and efforts must be made to identify probabilities for diseases that are not endemic. There is a role for APHIS in this task, but there is also a need for academic scrutiny, farmer buy-in and consideration by the RMA. Consideration must also be given to whether insurance should be publicly or privately provided. Unlike market price risk, which is borne by and within the control of farmers, animal diseases, especially those with large epidemic possibilities, could impose disastrous or catastrophic externalities on the farm sector, with considerable and significant welfare losses for consumers. While one might argue that consumers should bear any of the market losses from animal diseases, an equally sound argument can be made for public intervention to provide stability.

Data Requirements for Domestic Livestock Insurance by Green, Driscoll and Bruch

Insurance companies evaluate all of the relevant information that they can when determining whether a policy will be offered, and so does the government. Insurance underwriting and premium rates depend on the insurance strategy, the heterogeneity of the group or subgroup and the variability in year-to-year outcomes. Developing a premium rate requires data to estimate the frequency and severity of loss. Data relating to the risk characteristics of individuals in the population to be insured are also needed, as well as data describing the effectiveness of government safeguarding programmes. Information requirements will depend on the structure of the comprehensive programme, but will certainly be greater than the data currently available. An indemnification model must be defined for the type of loss to be covered. In order for risk rating to be possible, the frequency and severity of possible losses for each peril must be estimated. Models contain common elements of frequency, severity, number of animals affected and costs. The selected model may need to incorporate a measure of the potential for recurrence.

While the potential market for livestock insurance in the USA is large, there are many unknowns that will affect the success of any livestock disease insurance product. The size of the domestic herd defines the extent of this market. This chapter discusses the size of the market for cattle, sheep, poultry, swine and goats, then looks at data on slaughter and mortality. No matter which insurance approach is used, livestock mortality data are essential in determining potential payouts on insurance policies and, thus, the premiums necessary for insurance carriers to make reasonable profits. The adequacy of currently available death loss data is assessed for the purpose of government indemnification and/or private sector provision of policies. The authors go on to discuss the US Animal Identification Plan as the principal government programme currently being developed. Options other than insurance may need to be explored. Insurance may not achieve the necessary market penetration or achieve other goals. One option might be to push for statutory changes that would allow mandatory spending for indemnification so government officials could take action when necessary without requiring an emergency declaration.

Public and Private Schemes Indemnifying Epidemic Livestock Losses in the European Union: A Review by van Asseldonk, Meuwissen, Hurine and Wilkens

This chapter describes the most recent developments for agricultural insurance programmes in the EU. There are a large variety of systems worldwide, most with some mixture of public and private financial backing. Examples include, crop and livestock insurance in the USA, Canada, Greece, Italy, France, UK and Germany. The chapter reviews compensation for direct and consequential losses among several countries and discusses how they are financed.

The chapter specifically focuses on the development of the German animal disease insurance that includes coverage for BSE and FMD coverage for business interruption, production loss and income loss that was offered by some private insurance companies to livestock breeders. These products were not well received by farmers as the premiums were very high. The high density of livestock breeding in some areas, the presence of very large farms, and risks due to wild livestock populations contributed to the high premium rates. Capacity within the insurance industry was also limited at the time. A "risk partnership" between producers, private insurance and government to provide solutions to cover losses from catastrophic risks was discussed. Only a limited number of EU member states offer free assistance, and there are restrictions on subsidization strategies to consider from the World Trade Organization. They recommend a mandatary system to finance direct losses, but suggest consequential losses might be under voluntary programmes.

Designing Epidemic Livestock Insurance by Meuwissen, van Asseldonk, Skees and Huirne

This chapter deals with how to design epidemic livestock insurance that provides proper incentives for producers to behave in the collective interest, how to design a financing model to deal with systemic risk and diminishing government financing, and rating insurance with expert information and Monte Carlo simulation methods. It focuses on the 1997-98 outbreak of FMD and hog cholera in the Netherlands, a country with some of the most advanced animal disease insurance products, government-private systems and supporting research programmes. The direct costs of disease outbreaks in this case are shared by the EU, the Netherlands and the producers. Producers finance their portion through levies deposited into the Animal Health Fund. Consequential losses are not paid, and insurance is not available to cover them.

The chapter covers incentives for risk management, including prevention, rapid disclosure, and compliance with movement stand-still, dealing with systemic risk, including a proposed alternative financing model, calculating risk premiums and the annual loss distribution for their scenarios in the Netherlands. The factors considered, and resulting simulation, include the number of farms infected, duration of the epidemic and the size of restricted areas. The number of epidemics in a five-year period was estimated based on expectations of experts in the field. Two risk-financing instruments, a levy system funded by producers and guaranteed by a bank, if necessary, and insurance, are evaluated and compared. Average direct losses were higher for insurance due to transaction costs. The minimum direct costs for the insurance were relatively high since premiums would have to be paid each year; however, the maximum was relatively low. In a levy system, assessments may not have to be paid each year if no losses are sustained, so the minimum is relatively low. The maximum direct costs were relatively high for the levy system since costs of an outbreak must be paid in full within five years. Possible roles of the government were given as control and regulation of epidemic diseases, reinsurance, subsidization of premiums and tax incentives. Interest in consequential insurance is low in the Netherlands, possibly due to underestimation of probability risks, perception that risk to self is lower than risk to others, belief that governments will provide disaster assistance, and insurance requirements such as premiums and deductibles.

The authors conclude that the three major difficulties with insurance will be human behaviour in the course of an epidemic, the systemic character of risk and lack of data. However, they conclude that epidemic livestock insurance is feasible.

The German System of Compensating Animal Keepers in Cases of Outbreaks of Animal Diseases by Bätza

This chapter discusses the Animal Disease Act in Germany as updated through April 11, 2001 and the Animal Disease Fund. The act provides compensation for animals that are destroyed or have died from disease. Compensation is based on the market value of the animal. The act does not allow compensation for consequential losses. However, private insurance is available to cover this risk. Compensation may be reduced or denied if animal keepers failed to abide by statutory provisions regarding management practices, reporting or other factors. This work is the German counterpart to the US work by Ott but it does not discuss the vagaries of policy.

There is an Animal Disease Fund in every "land" or state. The fund is managed by a governing board and a board of directors. Funds are raised through producer contributions. The cost of the act is shared equally by the states and producers of horses, cattle, swine, sheep, poultry and freshwater fish. The state covers the costs in other cases, and if the burden on producers is too high, for poultry and freshwater fish. Bätza explains that the German government sees animal disease compensations as a "right granted by legislators on both grounds of equity and as a matter of convenience".

Managing the Risks and Impacts of Animal Diseases in the Australian Livestock Sector by Keogh and Neumann

Australia's annual livestock production is valued at approximately US$8 billion with exports making up 70 percent of the total. The animal health status of the country has always been very good by world standards because of its geographical isolation, low intensity livestock production methods and a climate that is unfavourable to many diseases. Therefore, Animal Health Australia's philosophy for risk management is to "maintain access by risk avoidance and minimization". This is done through exclusion by strong border quarantine, strategic eradication, sound biosecurity, and early detection through surveillance, preparedness and/or prompt stamping-out policies.

Animal Health Australia is a not-for-profit company established by government and livestock groups. It was incorporated in 1996 to "facilitate a national approach to enhancing Australia's animal health status". The key activities of the company include planning and coordinating national priorities, coordinating partnership arrangements, brokering special programme funding, managing national programmes and being a custodian of national processes such as the Australian Veterinary Emergency Plan and the Emergency Animal Disease (EAD) Response Agreement. Funding comes from the Commonwealth government, states/territories and livestock industry groups, with a small contribution from other sources. National Performance Standards define the responsibilities of each member group in several areas related to disease control and define current compliance levels. Insurance is not currently used much in livestock production.

Australia has focused its efforts on risk minimization. This may change, though, as experiences in other countries, increased travel and trade, potential bioterrorism, emerging diseases and intensification of livestock industries lead to a perception of increased risks and consequences. Livestock industries might consider insuring against the maximum share of the costs that they could be responsible for according to the EAD Response Agreement or against costs beyond the EAD maximum. Livestock producers may also want to insure against potential consequential losses.

Livestock Industry Insurance: Canada by Stephen and Epps

The chapter describes the Canadian strategy used to find solutions to risk management problems. Essentially, the government recognizes the need for public support and is addressing the problem by helping industry find solutions. When an industry group brings the government a problem, first the risk is assessed by prioritizing the perils and analyzing the selected perils. Then the financial services industry is consulted to develop risk management options. The selected option is promoted and placed with the private sector. Livestock producers have formed the Canadian Animal Health Coalition for promoting a collaborative approach to the health of animals. The group is working on an economic impact study of Foot-and-mouth Disease, a Canadian Animal Health Emergency Management Strategy, a Geographic Information System, animal welfare issues and a strategic plan. A programme called the Business Risk Management programme, Private Sector Risk Management Partnership, seeks to assist agricultural industry in filling gaps by providing funding for research and development for private sector risk management solutions.

The partnership programmes are intended to contribute to Canada's overall policy framework by helping to develop more risk management tools and organization. The approach will create an environment that helps producers understand and actively manage their business risks.

The Current State of US Federally-Supported Livestock Insurance by Hart

In the USA, the Agricultural Risk Protection Act of 2000 allowed for livestock pilot insurance programmes in the USA. Before the passage of this act, livestock was explicitly excluded from coverage under federally-supported agricultural insurance programmes. This chapter discusses the types of programmes that can be considered under the act and the usage and performance of the products that have been tried to date. Six federally-supported insurance products that can cover livestock are: Livestock risk protection (LRP) for hogs, fed cattle and feeder cattle; livestock gross margin (LGM) for hogs; adjusted gross revenue (AGR) and adjusted gross revenue-lite (AGR-Lite). LGM and LRP targets livestock price risks, LGM covers livestock and feed price risk, and AGR covers the whole farm.

This chapter illustrates the state of US federally supported livestock insurance schemes. Most are close substitutes with existing futures and options (or baskets thereof) products. These products are not disease coverage.

Hart finds that LRP-Swine and LGM covered 1.3% of hog sales in 2002 and 5.8% in 2003. Fed cattle producers insured nearly 100,000 head at nearly $20 per head and feeder cattle producers insured 72,000 head. Moral hazard and adverse selection are suspected culprits suppressing participation. These products were not designed for catastrophic losses. In fact, the Risk Management Agency suspended sales of LRP-Fed Cattle and LRP-Feeder cattle after the announcement of the BSE case in the USA in December, 2003. Hart points out that these instruments mainly target price risk and not production risks. Availability of data is a possible reason to explain differences in the type of risk covered by programmes. However, we think that the risk in these products is easily re-hedgable in the futures and options markets. Thus, they are not necessarily substitute products in that the insurers are reinsuring the risk through existing and organized markets.

Livestock Disease Eradication Programmes and Farm Incentives: the Case of Bovine Tuberculosis by Wolf

Disease management is even more urgent when it poses a threat to human health. This chapter looks at the case of bovine tuberculosis in the state of Michigan in the USA. Michigan lost "tuberculosis free" accreditation in June 2000, thereby curtailing marketing and production options for Michigan beef producers. The state has given producers the option to depopulate or to develop a continuous test-and-slaughter protocol. Subsequently, beef producers chose to depopulate and dairy producers chose the testing protocol. Wolf examines the incentives and on-farm decisions that lead to these decisions.

By examining the on-farm costs and returns for different scenarios, Wolf concludes that business interruption losses are the key to understanding farm decisions. Dairy farms have relatively more complex financial implications compared to beef operations. Specifically, the larger business interruption costs make it more difficult for dairies to incur the down time for depopulation. Hence dairy producers prefer very different risk management solutions than beef producers. His findings suggest that government assistance, with business interruption insurance subsidies for example, could help expedite the Michigan industry's ability to regain their tuberculosis-free status.

Economic Impacts of Eradicating Scrapie, Ovine Progressive Pneumonia and Johne's Disease on US Sheep, Lamb, Sheep Meat and Lamb Meat Markets by Seitzinger, Paarlberg and Lee

This chapter explores one of the most important public sector considerations, which influences the public pressure for, and resulting investment choices, related to animal health management schemes: the welfare implications of eradicating disease. It examines the potential market impact of eliminating scrapie, ovine progressive pneumonia (OPP) and ovine Johne's disease (OJD) in the US sheep flock. The USA announced in 2000 that it would spend \$100 million to eradicate scrapie while efforts to control OPP and OJD have been voluntary. The market and economic welfare effects of eliminating these programmes are accomplished by constructing a model of the US lamb, lamb meat, sheep and sheep meat markets and simulating the consequences of eradication on prices, domestic output, domestic use and trade.

Elimination of these three diseases is estimated to help both producers and consumers. Eliminating just scrapie would increase revenue by \$10.8 million on the production side, and the elimination of all three diseases would increase revenue an estimated \$20.5 million. Consumer welfare would increase by \$1.7 million annually. The chapter contains a detailed description of the model and some discussion of the specific lessons from the analysis, as well as what it might suggest for the broader aspects of public support for livestock health management programmes. This model has been used in evaluating the market-related impacts of disease management and eradication.

Understanding Broader Economic Effects of Livestock Insurance and Health Management: Impacts of Disease Outbreak on Allied Industries by Pritchett, Thilmany and Johnson

Understanding how an animal disease event will impact the animal products marketing channel is a complex, multidisciplinary problem, but as Paarlberg *et al.* show, such estimates are often needed when assessing public welfare implications. This chapter provides a typology matrix of research approaches and foci that should be explored to assess the full public and private impacts from disease outbreaks (or avoidance of such outbreaks). To appropriately model a wide variety of animal disease impacts, a system of economic relationships is needed that accounts for the interdependencies and degree of response (elasticities) among the various production, marketing and consumer sectors of the economy. Interdisciplinary work should encourage the merger of sophisticated epidemiological models used to trace the growth and demise of disease outbreaks and economic models that capture the technological and economic relationships linking stages, potential structural change and performance of the marketing channel.

Without considering the ripple effects and persistence of outbreak shocks on upstream and downstream sectors, the public may underinvest in animal health management programmes (including insurance). As policymakers explore strategic responses to animal disease, the distributions of losses, policy costs and programme benefits become particularly important. The spatial dimension of animal disease also deserves additional attention. Too often data limitations prevent analysis of spatial economics when evaluating outbreak scenarios. This article concludes with a discussion of future directions for animal disease studies at various market levels, which can subsequently feed better baseline data to broader sector, regional and national analyses.

US Livestock Industry's Views on Livestock Disease Insurance by Grannis, Green and Bruch

Producer group listening sessions were conducted by APHIS to identify perils and risks in livestock production, producer interest in insurance, the best components for insurance products, and obstacles or issues to consider from the producers' point of view. Listening sessions were held for members of the National Pork Board, National Cattlemen's Beef Association, Delmarva Poultry, Inc., National Milk Producers' Federation and American Sheep Industry Association. Worksheets that asked about the disease perils to livestock production, current practices used to manage perils and perceptions about government and individual responsibilities in managing disease risk were completed by some groups. This chapter summarizes the points raised in those meetings and broader findings from those sessions.

Several conclusions were made from these sessions. US producers, for the most part, are supportive of the livestock disease insurance concept. However, the cost of insurance and ability to complement, rather than substitute, for government indemnities are major concerns. Interestingly, disease is not perceived as the most important peril faced by livestock producers and they do not regularly incorporate disease risk management into their common production practices. In fact, different challenges are faced by producers of each species, segment and operation size so a single insurance product will not fit the needs of every livestock producer. For example, poultry producers preferred revenue insurance to protect against disease, while beef producers were opposed to this type of insurance because of the potential to distort the market.

Modelling the Impact of Compulsory Foot and Mouth Disease Insurance in the European Union by Janson, Norell and Rabinowicz

This chapter employs the authors' knowledge of Sweden, extends it to the EU, and compares two ways of financing programmes to combat FMD: a purely tax-financed system versus a compulsory insurance scheme. The first of these systems is largely a representation of the current EU system. The second recognizes that under the current system, producers lack incentives to prevent diseases. The authors point out that while animal health is a public good, the financing of epidemic cleanups is not. The authors examine the impacts of requiring agricultural producers to self-fund an insurance scheme to cover the cost of FMD outbreaks. There are substantial ramifications for animal location as animal density impacts the disease spread, and subsequently, insurance cost.

The authors use the Common Agricultural Policy Regional Impact Analysis (CAPRI) model to estimate implications for production agriculture. Results show that the quantitative effects of compulsory insurance on production are small, but net welfare is increased except in Denmark, Greece, Ireland and the Netherlands (because of high-density production in those countries).

Currently, insurance costs are paid partly by the member state and partly by the EU. The compulsory insurance programme provides incentives to reduce the total costs of FMD by shifting responsibility to industry members and the member state where the outbreak occurs.

While premiums are shifted from taxpayers to producers, there are net welfare gains. But, this shift to producers, and redistribution among EU members, make the political implications more complex.

Investigating the Feasibility of Livestock Disease Insurance: a Case Study in US Aquaculture by Coble, Hanson, Sempier, Shaik and Miller

Aquaculture presents unique challenges for insurance. This chapter discusses a large-scale study about the feasibility of developing and implementing risk management programmes for US aquaculture species (catfish, salmon, trout and baitfish) with the greatest economic value. If risk management programmes are found to be feasible, then the programme will supply products, policies and other information to the Federal Crop Insurance Corporation. By way of the aquaculture example, the authors describe an approach for developing an animal disease insurance product. The steps involve determining risks, evaluating the nature of the risks, developing draft underwriting language to define coverage, collecting actuarial data and assessing producer willingness to pay.

Aquaculture is uniquely different, even among livestock insurance diseases. The nature of a pond for example does not lend itself to identification of disease among individuals. The authors determine that every situation in aquaculture is sufficiently unique that they need to examine insurability conditions on a case-by-case basis. They use an insurability condition matrix and discuss how to elicit risk for these unique settings. Then they classify the risks and discuss implications and potential for aquaculture insurance. They conclude that they can establish guidelines that will enable the development of insurance programmes for aquaculture.

Chapter 2

The Role of USDA-APHIS in Livestock Disease Management within the USA

Jennifer L. Grannis and Megan L. Bruch

Introduction

In 2001, a joint research project conducted by the US Department of Agriculture (USDA) Animal and Plant Health Inspection Service (APHIS) and Colorado State University investigated the applicability of insurance as a tool for supporting the goals of US disease risk mitigation strategies. A conference held in November 2002 brought industry and government experts from Europe, Canada, Australia and the USA together to discuss a variety of topics related to livestock disease insurance. Some of these topics included livestock disease economics, the insurability of livestock for mortality resulting from disease, public-private indemnity-insurance programmes, US livestock industry interest in livestock disease insurance and US activities to prevent disease introduction and spread.

During the conference it became clear that an important part of livestock disease risk, and therefore the insurability of livestock disease risk, was government action to mitigate risk of disease introduction and spread. This chapter discusses USDA-APHIS and the role the agency plays in mitigating livestock disease risk within the USA. Domestic US conference participants, as well as international participants, were unaware of many of the actions conducted by USDA-APHIS and other US federal government agencies to protect the health of the US livestock herd.

European, Australian and Canadian chapters in this volume describe livestock disease management strategies that include industry participation and livestock disease insurance coverage. At the time of the conference, the USA was just beginning to consider livestock disease insurance as a possible tool for livestock disease risk management. Understanding the US government disease management system relates directly to the development of livestock disease insurance in the USA. Since the USA does not currently have a significant livestock disease insurance industry, this discussion of animal disease risk management is focused on existing government actions to mitigate disease risk.

Role of Government in the Supply of Animal Health

Actions the USDA takes to minimize the risk of disease introduction are driven by the awareness that animal health is a public good. A livestock disease outbreak can represent a market failure – the undersupply of animal health – that can be mitigated by government action. A healthy livestock herd provides sufficient production of food to satisfy demand and ensures that zoonotic diseases are not transmitted to humans. Optimal animal health provides benefits to the entire population, yet each person's benefit does not exclude or diminish the benefit to the rest of the population. Because the individual producer receives little return from actions to maintain superior animal health, the optimal production level of animal health may not be achieved without government intervention.

Risks contributing to animal disease outbreaks occur at many levels of the livestock production chain. While individuals' actions on the farm contribute to the national supply of animal health, risks associated with trade in livestock and livestock products cannot be easily managed by individual producers. To protect animal health in the national herd, the government minimizes the risk of disease introduction from outside the USA by protecting livestock from foreign animal diseases (FADs). APHIS, in cooperation with state governments and animal industries, has successfully eradicated a number of diseases in the USA. Ongoing disease eradication efforts of other production diseases and the exclusion of exotic diseases have resulted in a healthy livestock herd that supplies meat for consumers.

Government involvement in maintaining animal health has implications for the insurability of losses due to livestock diseases. Risks faced by insurance companies are related to the effectiveness of government programmes that keep disease out of a country and eliminate a disease outbreak when it does occur. Government control measures, such as depopulation of affected animals, quarantines and movement restrictions, may cause more losses than the disease itself. Federal and state governments work together to provide compensation to livestock producers when animal disease control strategies result in animal deaths. However, those programmes usually only cover the value of the animal. Once animals leave the farm and enter the food chain, agencies such as the USDA Food Safety Inspection Service (FSIS), the Department of Health and Human Services (DHHS), and the Food and the Drug Administration (FDA) play vital roles in supplying safe food to US consumers. The role of APHIS is discussed below and federal livestock health management influence on livestock disease insurance is discussed in the conclusions.

Role of USDA-APHIS in Disease Prevention and Response

Bovine spongiform encephalopathy (BSE, commonly referred to as mad-cow disease) was confirmed in a cow in Washington State on December 25, 2003. This event, covered extensively by the media, demonstrated the role and actions APHIS plays in the event of a foreign disease event within the USA. Immediately upon the preliminary diagnosis on December 23, federal veterinarians with expertise in investigating FADs were dispatched to support the Washington State veterinary staff (USDA, 2003). The FAD experts began the process of identifying the farm of origin for the infected cow, her age, her birth farm, and traced forward her progeny and backward her herd-mates. This response is typical of the way animal disease diagnosis, quarantine and eradication is completed by interacting federal and state agencies. However, eradicating foreign animal diseases is only one part of the strategy APHIS employs to maintain the health of the US livestock herd.

APHIS Organization and Role in Disease Prevention

APHIS is the USDA agency tasked with safeguarding US agriculture. APHIS keeps foreign diseases and pests out of the USA and reacts to prevent outbreaks when diseases or pests are introduced. The five major programme areas of APHIS which protect American agriculture are:

1. Plant Protection and Quarantine (PPQ)
2. International Services (IS)
3. Veterinary Services (VS)
4. Animal Care (AC)
5. Wildlife Services (WS)
6. Biotechnology Regulatory Services (BRS)

Each of these programmes has a role in excluding disease and pests from the USA. APHIS also has a strong role to play in ensuring access to export markets for plant and animal products, addressing sanitary and phytosanitary (SPS) issues in trade, and monitoring some animal welfare issues. The APHIS approach is to provide an umbrella of services that safeguard animal, plant and ecosystem health to restrict agricultural disease and pest loss and to secure trade markets for agricultural products (VS, 2003).

Within the APHIS programme areas, five objectives were established to achieve the goal of safeguarding livestock production (APHIS, 2001c). The Veterinary Services (VS) programme area works to achieve these objectives:

1. Keeping foreign animal diseases from entering the country.
2. Providing an emergency response when livestock diseases slip past US borders.
3. Controlling or eradicating major domestic livestock diseases.
4. Preventing the interstate spread of diseases.
5. Facilitating exports by attesting to the health status of outgoing animals.

Keeping FADs from Entering the USA

APHIS has the authority and responsibility to exclude foreign animal diseases by prohibiting imports of animals, animal products and other materials that pose a risk of introducing such diseases. APHIS bases its foreign animal disease exclusion activities on an evaluation of the animal health status of countries or regions. Animals and animal products may be imported into the USA if they originate from countries or regions that APHIS has evaluated and recognized as free from diseases, like FMD, or for which measures exist which can mitigate the risk of the animal or animal product carrying the disease (i.e. required testing of the animal prior to import, required quarantine period before or after importation, treatment of the product by cooking or other processing). Animals and animal products are prohibited or restricted from countries or regions that APHIS has not evaluated (APHIS, 2001a).

The evaluation of country or region disease status and identification of appropriate mitigations for imported animals and animal products is conducted according to criteria defined in APHIS's 1997 regionalization rule (US-CFR, 1997). The goal of the regionalization rule was to create a mechanism to establish regionalized, risk-based import requirements that were consistent with APHIS's obligations under the World Trade Organization's Sanitary and Phytosanitary (WTO-SPS) Agreement. APHIS regulations require that the request for recognition of the animal health status of a country or region be accompanied by information on the following topics:

1. Authority, organization and infrastructure of the veterinary services organization in the region.
2. Disease status of the region.
3. Status of adjacent regions with respect to the agent.
4. Extent of an active disease control programme.
5. Vaccination status of the region.
6. Degree to which the region is separated from adjacent regions of higher risk through physical or other barriers.
7. Extent to which movement of animals and animal products is controlled from regions of higher risk and the level of biosecurity regarding such movements.
8. Livestock demographics and marketing practices in the region.
9. Type and extent of disease surveillance in the region.
10. Diagnostic laboratory capabilities.
11. Policies and infrastructure for animal disease control in the region, i.e. emergency response capacity.

APHIS evaluates the information provided by the petitioning country or region and conducts a site visit to verify the information. A risk assessment is then carried out and a decision is made regarding whether, and under what conditions, importations can be safely allowed (APHIS, 2001a).

APHIS actively monitors disease status worldwide using information disseminated by the World Organization for Animal Health (OIE), information received from APHIS personnel stationed overseas, news reports, and information received from other government agencies such as the Department of Homeland Security (DHS) and the Department of Defense. When a new outbreak occurs in a country or region from which APHIS has allowed importations, APHIS evaluates the need to impose a ban on imports. If a ban is imposed, prior to releasing the ban, APHIS re-evaluates the disease status of the country or region.

APHIS import requirements are enforced through inspections overseas and at domestic ports of entry. In 2003, a portion of the border monitoring activities formerly administered by APHIS were transferred to the DHS. DHS officers enforce APHIS import requirements and monitor for contraband in commercial shipments and baggage or individual passengers. DHS also monitors mail parcels for prohibited products.

Responding to FADs

APHIS encourages reporting of unusual or suspicious symptoms suggestive of foreign or emerging animal disease by producers, private practitioners, state and private laboratory officials, and state animal health officials. Diagnosticians, trained in foreign animal diseases, investigate reports received. Testing for foreign animal diseases occurs at APHIS laboratories. APHIS is currently building a National Animal Health Laboratory Network and foreign animal disease testing will be available at state-level laboratories participating in this network.

In the case of an animal disease outbreak, Veterinary Services (VS) tests, identifies and eradicates the disease. Past success in eradicating diseases from the USA includes screwworm, fowl plague, glanders, vesicular exanthema and FMD (APHIS, 2001c). Recent FAD emergencies have included Exotic Newcastle Disease (END) in commercial layer poultry in southern California in 2002-2003, High-Pathogen Avian Influenza (HPAI) in Texas in 2004, and BSE in Washington State in 2003-2004. Also, in 2002, VS combated Low-Pathogen Avian Influenza (LPAI) in live bird markets in eastern states and commercial poultry operations in Virginia.

A variety of tools are available for VS to employ when eradicating a livestock disease. These tools include certification, depopulation, movement controls, quarantine and vaccination. Depending on the disease's virulence, the species affected, the risk to the human population, industry impacts and trade implications, a disease management strategy appropriate to control a particular disease outbreak is determined. For END, a fast-spreading avian disease, quarantine, movement restrictions and depopulation were all used in eradicating the disease from southern California. The current BSE disease management strategy matches the characteristics of BSE. There is surveillance-based strategy to find cases of the disease, depopulate infected cattle, and exclude beef products that could potentially transmit the disease from the human food chain. In 2003, in the case of spring viremia of carp (SVC), the infected ponds were drained and infected fish were depopulated. Ponds were emptied, the pH was adjusted to a level that destroyed the virus, and restocking was allowed using only SVC certified-free stock. Each disease management strategy is designed to eradicate the disease, return the affected industry to a disease-free status, and recover trade markets as quickly as possible.

Controlling or Eradicating Major Domestic Livestock Diseases

APHIS also collaborates with state animal health officials and industry to combat domestic (endemic) diseases that have significant economic and trade impacts such as brucellosis, pseudorabies, scrapie and tuberculosis. Surveillance on farms, at slaughter facilities and at points of sale is combined with traceback and removal to identify and eliminate animals infected with these diseases. A certification system may also be utilized to identify disease-free herds from which producers are encouraged to purchase stock for some disease management programmes.

Eradication of endemic diseases within the USA relies on testing, depopulation and certification of disease-free status following internationally approved scientific guidelines. Specifics differ depending on the specific eradication programme, but in general, animals that test positive for a programme disease are depopulated. Endemic diseases within the USA for which eradication programmes have been established are often referred to as programme diseases. Depending on the programme, the herd from which the infected animal is from may also be depopulated. Once a herd tests free of the disease, a disease-free certification may be conferred depending on the requirements of the different programmes. Individual states can be declared disease-free once herds within the state are free of the disease. Programmatic depopulation and indemnification for three disease eradication programmes is summarized for the National Scrapie Program, the Accelerated Pseudorabies Eradication Program (APEP) and the Brucellosis Eradication Program.

Identification and eradication of Scrapie, a neurological disease affecting sheep, has been the focus of several USDA programmes since 1952. The Scrapie Flock Certification Program was established in 1992 for producers to register their flocks and allow checks on the disease status of flocks. For the fiscal year ending September 30, 2002, 1,533 flocks were participating in the programme with 641 newly enrolled during the period. Infected cases confirmed by the National Veterinary Services Laboratories totalled 259 sheep and 5 goats; 11,751 animals were tested for the disease.

APEP was initiated in 1999 to prevent the spread of pseudorabies in swine. Pseudorabies, also known as Aujeszky's disease, is a herpesvirus that can affect cattle, horses, dogs, cats, sheep, and goats (Taft, 1999; VS, 2001). Due to very low swine prices during that time, some producers were not vaccinating their animals. The emergency programme allowed APHIS to purchase and depopulate infected herds from owners who volunteered for the programme. The greatest activity was in FY1999 when 854 thousand swine from 857 herds were depopulated. Indemnity paid in FY1999 was $59 million with total cost of the programme reaching $77 million. In FY2001, depopulated animals fell to 124 thousand from 152 herds at a total cost of $13 million. Indemnity payments in FY2001 for depopulated animals were $6 million, almost half of the total cost of the programme. Total depopulations fell again in 2002 to 28 thousand from 21 herds. Indemnity paid for depopulated swine in 2002 was $671 thousand. As of May 2004, only four states, Florida, Iowa, Pennsylvania and Texas, had herds remaining under surveillance for pseudorabies (VS, 2004).

The first Brucellosis programme was established in 1934 but the eradication option was added in 1954 and continues to be the focus today. If the programme were to be discontinued, it is estimated that in less than ten years the production costs of beef and milk would increase by $80 million annually (APHIS, 2003a). In FY 1999 1,161 swine were depopulated at a cost of $80 thousand (APHIS, 2001b). A total of 409,880 bovines were destroyed between September 1996 and August 2001 with $3 million paid in indemnities. Ninety-nine percent of the depopulated animals were located in Texas where 73% of total indemnities were paid. As of May 2004, only Texas and Wyoming were not classified as free of bovine brucellosis.

Other diseases that could impact livestock production but which are concentrated in wildlife, such as rabies, are monitored by WS and state veterinarians. Chronic Wasting Disease (CWD) affects both wild and farmed cervids (deer and elk). Government management and eradication of CWD involves VS, WS, state veterinarians and the US Department of the Interior National Park Service as many of the affected animals are found in national parks. Other diseases, such as Johne's disease and LPAI, also have control programmes designed to reduce the incidence of these diseases and to strategically manage these diseases to reduce economic impacts.

Other APHIS Activities

Animal identification could play an important role in managing the spread of livestock diseases. Disease investigations, which trace backwards and forwards the herds and individual animals that came in contact with a diseased animal, are slow as they rely on individual non-standardized producer records. Animal identification has the potential to reduce the amount of time it will take APHIS to conduct animal disease investigations and benefit individual producers by facilitating the management of market characteristics of animals. While animal identification in itself will not reduce animal disease risk, the potential applications for disease eradication and management make it an important component of the future animal health strategy for APHIS. APHIS is working with state and industry personnel to implement the National Animal Identification System. Existing state and industry programmes and new approaches are being evaluated for their appropriateness for the national system. Implementation of this system will begin in 2004. Under this voluntary system a unique identifier will be created for each animal production premise in the USA. Details of how unique identification numbers will be assigned for premises, individual animals, or groups of animals are being developed as of 2005.

APHIS facilitates the exportation of livestock by completing animal health certifications as required by trading partners. APHIS participates in the WTO and OIE to ensure that export markets for US livestock remain open to US livestock and products. APHIS maintains databases that track imports and exports of animals and import and export certificates for animal products. APHIS is also developing the capacity to track interstate movement of animals and animal products.

Indemnification

US disease management strategy is driven by the epidemiological goal to eradicate diseases quickly once identified, especially for exotic diseases. Indemnity payments to livestock owners are used to encourage early participation in control strategies. Recent animal disease outbreaks such as BSE, END, infectious salmon anaemia, and LPAI and HPAI resulted in millions of dollars of indemnity payments to producers across the USA. Indemnities were paid for the value of the animal and, occasionally, for cleaning and disinfection costs.

US law guarantees indemnification for producers whose animals are depopulated for diseases determined by the Secretary of Agriculture to be of significant impact to the affected industry to warrant eradication. Indemnity payments may also be contributed to by the State where the disease occurred but can be pro-rated or reduced based on budgetary constraints. The Animal Health Protection Act (AHPA) provides guidance on indemnity payments. USDA-APHIS regulations specifying indemnity programmes are found in the US Code of Federal Regulations (US-CFR, Chapter 9 contains APHIS-specific regulations). However, the amount of money that can be paid for the animal is limited. The AHPA states, "Compensation paid any owner under this subsection shall not exceed the difference between i) the fair market value of the destroyed animal article facility, or means of conveyance; and ii) any compensation received by the owner from a State or other source for the destroyed animal, article, facility or means of conveyance".

The AHPA language implies that the value of the indemnity paid will equal the fair market value less any salvage value. Thus, if a disease were to occur and animals could be slaughtered at a plant within the same quarantine zone where the livestock resided, the owner would then receive the fair market value less whatever the amount received for slaughter. Otherwise, when animals were sold to slaughter at a value similar to the expected value prior to the disease outbreak producers would then receive the slaughter value and no additional compensation. Also, the language in the AHPA seems to suggest that if owners of condemned livestock had insurance then they would not be eligible for additional indemnity payments that would exceed the determined fair market value of the animal or herd. This will obviously curb multiple payments for the value of depopulated livestock, as occurred during the 2001 FMD outbreak in the UK, but may limit the development of insurance products where the payouts are based on the value of the livestock.

APHIS currently reimburses producers for the value of animals depopulated using market value, often determined by visual appraisal. A federal indemnity payment reflects market value before the disease outbreak and occasionally other specified costs like cleaning, disinfection or disposal. The portion of the market value and any other costs which are paid to the owner are determined on an outbreak-by-outbreak schedule for exotic diseases. As a risk management tool for the producer, this can be a highly uncertain process.

State governments may also provide indemnity payments to producers during disease outbreaks. Federal and State payments may be combined to provide full compensation to the livestock owner. A total of 128 state and federal animal disease indemnity programmes were in place during 2003. Of those, nine were federal programmes including the previously mentioned APEP, brucellosis, scrapie and tuberculosis programmes. States may have a programme that matches the federal disease eradication programmes, or may not. Each state, and the territories of Puerto Rico and the Virgin Islands, have at least one piece of legislation governing indemnity payments for the destruction of diseased livestock. These programmes cover the major livestock species, including poultry and aquaculture, and may also cover deer, elk, bison and other non-traditional livestock. A programme may cover a single disease, more than one disease, single species, multiple species or combinations of all of these. None of the State programmes compensate owners for losses caused by business interruption or lost production time.

The indemnity payment paid by a programme may be calculated by a variety of methods. For approximately 77% of the State programmes, the indemnity payment is a portion of an appraised or market value, adjusted by salvage value or other payments received by the owner. Depending on the programme, appraisals may be conducted by certified appraisers, a team of appraisers chosen by the owner and administering agency, or by the owner and agency representatives themselves. The payment decision may be based on funds available, the indemnity paid by a federal programme, or specific regulations related to the situation. A maximum payment or payments that depend on the type of animal (commercial grade or registered) may also determine indemnity values.

APHIS and State indemnities generally only compensate the owner for the value of the animal destroyed. Other losses – consequential losses – are experienced both by the owner of the diseased animals and members of allied industries. These consequential losses can include business interruption costs, cost of feeding animals waiting depopulation, lost market access both domestically and internationally, decreased reproductive rates, lost genetic stock and increased surveillance after the disease is eradicated. Allied industries, such as feed suppliers and trucking, are deprived of markets as well. Consequential losses can be expected to exceed the value of the destroyed animals. These losses will likely not be compensated by government indemnities.

While APHIS pays indemnities for animals depopulated for animal disease eradication, other federal agencies may offer affected producers low-cost loans, grants or other emergency compensation. These other compensation programmes are most likely to compensate livestock producers when livestock mortalities, resulting from disease or disease eradication, are the result of a natural or man-made disaster. Most relevant is the USDA Farm Service Agency Emergency Loan Program (EP) that provides emergency loans to livestock owners when they suffer losses as a result of an animal disease quarantine. These loans are low cost, must be repaid, must be backed by collateral, and depend on the producer having taken appropriate actions to mitigate risk. The EP risk mitigation requirement is usually satisfied if a producer has insurance on farm assets. Since livestock disease insurance is not readily available, livestock owners would be excused from the risk mitigation requirement to be eligible for a loan. Other disaster relief programmes may pay indemnity when a disease outbreak is the result of a natural disaster, and the Federal Emergency Management Agency has the ability to offer support depending on the specific disaster event.

As can be seen from these regulations, while indemnity is an important component of disease management strategies, indemnification is not readily available for costs that are incurred beyond the value of the livestock depopulated. Livestock disease insurance could provide an opportunity for producers to reduce the risk they face from a disease outbreak. The US private insurance industry is just beginning to develop insurance products related to animal diseases, and government partnerships to foster that development are also in the early development stage.

Risk planning for disease outbreaks does not appear to be regularly practiced by US livestock producers. Indemnity programmes offset the most understandable risk from disease – the lost value of the animal – but consequential losses resulting from livestock diseases have the potential to be substantial and are often poorly planned for by producers. Federal and State programmes exist for indemnification, potentially lulling producers into not taking a larger role in managing livestock disease risk. Expected costs of disease outbreaks are underestimated by individual producers and demand for risk management tools for livestock disease may be less than is necessary to supply optimal livestock herd health. Developing livestock disease insurance could benefit producers by providing a new tool for livestock disease risk management and also improve the supply of livestock health to the marketplace.

Conclusions

The USDA-APHIS takes actions that directly affect the risk of an animal disease entering and spreading within the USA. APHIS also indemnifies producers for losses incurred due to animal diseases. The development of livestock disease insurance in the USA will need to be coordinated with the APHIS actions and regulations. Livestock disease insurance could support APHIS' role in maintaining a healthy livestock herd. Livestock disease insurance policies can require biosecurity and production practices that benefit producers, achieve APHIS goals and help promote the supply of animal health. Participation in insurance programmes could be used to determine eligibility for indemnification since participation would represent adherence to biosecurity measures that protect the health of the US livestock herd. Participation in livestock insurance programmes can provide producers with coverage for disease costs not reimbursed by the government, such as consequential losses.

The future of livestock disease insurance in the USA is still uncertain and policy development will need to be coordinated with the actions of APHIS. The two most important potential benefits of livestock disease insurance development in the USA include the ability for producers to develop appropriate risk management strategies for livestock diseases and improved on-farm biosecurity. On-farm risk management has been traditionally used by livestock producers and biosecurity is often difficult to implement. The availability of livestock disease insurance can support the actions of APHIS in safeguarding animal health while also facilitating the improvement of on-farm risk management and biosecurity.

References

APHIS (2001a) Draft Interim Report: Animal Disease Risk Assessment, Prevention and Control Act of 2001 (PL 107-9). US Department of Agriculture, Animal and Plant Health Inspection Service, Washington, D.C.

APHIS (2001b) Pest and Disease Management: Brucellosis. Marketing and Regulatory Programs Business Service Fiscal Year 2001 Reference Book. US Department of Agriculture, Animal and Plant Health Inspection Service, Washington, D.C.

APHIS (2001c) *Veterinary Services: Protecting Animal Health and Promoting Trade.* Miscellaneous Publication No. 1550. US Department of Agriculture, Animal and Plant Health Inspection Service, Washington, D.C.

APHIS (2003a) *Questions and Answers About Brucellosis.* USDA-APHIS Factsheet. US Department of Agriculture, Animal and Plant Health Inspection Service, Washington, D.C.

Taft, A. C., The Eradication of Aujeszky's Disease in the United States. US Department of Agriculture, Animal and Plant Health Inspection Service. [Presented 1999. Accessed 2004.] Available from http://www.aphis.usda.gov/vs/nahps/pseudorabies/aujeszkys.pdf.

US-CFR (1997) Importation of Animals and Animal Products, US Code of Federal Regulations, Chapter 9, Parts 92, 93, 94, 95, 96, 97, 98, and 130, Docket No. 94-106-9, RIN 0579-AA71. Final Rule, Tuesday, October 28, 1997, pp. 56,000-56,026.

USDA (2003) Case of BSE in the United States: Chronology of Events. US Department of Agriculture. [Accessed 2004.] Available from http://www.usda.gov/news/releases/2003/12/bsechronology.htm

VS (2001) Pseudorabies Questions and Answers. US Department of Agriculture, Animal and Plant Health Inspection Service, Veterinary Services. [Accessed 2004.] Available from www.aphis.usda.gov/vs/nahps/pseudorabies/q-a.html.

VS (2003) Introduction to USDA, APHIS and VS: Orientation Part 1 Compact Disk for the VS Veterinary Services Career Program. US Department of Agriculture, Animal and Plant Health Inspection Service, Veterinary Services, Washington, D.C.

VS (2004) Pseudorabies Eradication Program Report, Update Report May 1, 2004. US Department of Agriculture, Animal and Plant Health Inspection Service, Veterinary Services. [Accessed 2004.] Available from www.aphis.usda.gov/vs/nahps/pseudorabies/update.html.

Chapter 3

The Role of Public Policy in Controlling Animal Disease

Daniel A. Sumner, José E. Bervejillo and Lovell Jarvis

Introduction

Diseases reduce animal productivity and may cause animal death. Diseased animals can also threaten humans with disease, whether because diseased animals yield contaminated products or because diseased animals produce viruses and bacteria that can infect humans even if they are not consuming animal products. Animal diseases cause significant economic loss in nearly all countries every year, though the losses would be much greater if it were not for efforts to manage disease. Animal disease management activities include efforts to keep a disease out of a region, protect animals against disease through vaccination, control the severity of symptoms in infected animals, and deal with animal deaths. Eradication of diseases that have infected a region may occur via vaccination and/or slaughter of infected animals and cleanup of infected premises.

Farmers, ranchers and other private agents fund and conduct most animal disease control activities in the United States of America and elsewhere (Bicknell, Wilen and Howitt, 1999). Private incentives to manage animal disease are clear and strong. For private firms (farms, ranches or others), such incentives are the potential loss of profit or asset values. For individuals or groups who own or control animals for recreation or companionship, the incentives are equally strong to maintain the services the animals provide, to reduce potential animal suffering and to prevent the loss of wealth.

Nonetheless, private agents do not always provide a socially adequate level of disease management. Disease control and eradication produce externalities and/or have public good effects that often lead to private underinvestment in disease management and that provide scope for public intervention. The role of government in animal disease management may be seen as supporting private activity, providing an underlying infrastructure or filling in where private incentives have been thought to be diluted. Government expenditures in the USA on agricultural pest and disease management amount to less than one billion dollars annually, though such expenditures generally benefit consumers and producers to a much greater degree.

This chapter will review the conceptual basis for the role of government in animal disease prevention, control, regulation and eradication. We will mention along the way some of the main governmental activities related to animal disease policy and compare the ideal conceptual role of government to the actual activities of government. Examples and case studies for the management of invasive species, which has similar conceptual characteristics to those for management of animal disease, are discussed more fully in Anderson, McRae and Wilson (2001); Bervejillo and Jarvis (2003); Ekboir, Jarvis and Bervejillo (2003); Sumner (2003a); and Sumner, Bervejillo and Jarvis (2005). Illustrations of the concepts discussed here are also presented in other chapters in this book, such as that by Wolf. Most of the focus is on animal agriculture, but companion animals, recreation animals and wildlife may be affected by the same diseases that affect agriculture, so the issues are broader than those raised in agriculture alone.

Rationales for collective action in animal diseases come in several forms:

1. Infectious disease management as a public good and the closely related idea of externalities related to the costs and benefits of private efforts to control infectious diseases,
2. Distributional impacts of disease outbreaks and the idea that major losses may occur at random with little opportunity for mitigating behaviour, and
3. Potential for animal diseases to be tools of attacks on national security with the related idea that collective action for national security is a core role of government.

Public Goods, Externalities and Infectious Animal Disease

The general economic concepts that apply to public policy in other areas apply as well to animal diseases. The most important of these is the idea that for some goods or services, private firms will provide socially insufficient quantities due to insufficient private economic incentives. This lack of sufficient private incentives may be attributed to "public good" characteristics or to the occurrence of external costs or benefits. The economic definition of a "public good" relates to the characteristics of demand and of cost of a good (or service), not whether it is actually provided by government or through other collective action. The concept of externality is related to how private agents other than those directly producing or consuming a good are affected by its production or consumption and whether this effect is reflected in the incentives given to the decision-making agent.

Under the standard definition, a "public good" has two distinguishing characteristics. First, consumption of the good or use of the service by one agent does not preclude or diminish the consumption by another. This characteristic of non-rivalry means that when some level or degree or quantity of the good or service is provided to anyone it can just as readily be made available to other consumers at little or no additional cost (Viscusi, Vernon and Harrington, 2000). One commonly used example of non-rivalry is the provision of bus services in an area where the bus is never expected to be full. Since it costs only a trivial amount more to operate a bus carrying 21 riders than one with 20, the use of the service by an additional rider does not diminish the availability of service for other customers, nor does it raise the cost of providing the service. Second, by definition, it is very costly or impossible to exclude consumption by any member of the public when a public good is provided. For example, when radio signals are broadcast, anyone with a radio tuned to the right frequency may listen in even if they have not paid for the service. In agriculture, the public good most often discussed relates to research and development activities since the benefits of the knowledge obtained often cannot be fully captured by the investor (Alston, Norton and Pardey, 1995).

The other reason that private agents may have insufficient incentive to invest in the supply of a good or service relates to the concept of external costs or benefits. An activity gives rise to external benefits or costs if the consequences of its production or consumption are felt outside the calculus of the decision-maker considering the activity. For example, if by removing certain noxious weeds from his pastures a rancher reduces the likelihood of infestation on a neighbour's pasture, the removal of the weeds would have an external benefit on the neighbouring ranch that does not directly affect the rancher considering the action. The neighbour should thus be willing to help finance the removal of the noxious weeds; without such help, the initial rancher may not undertake removal or may invest too little relative to the total economic benefit that such investment would achieve.

Some services associated with animal diseases seem to have characteristics that meet the criteria of public goods and provide external costs or benefits. For example, eradicating a disease from a region may benefit all owners of susceptible animals in the region with no additional costs. In addition, benefits of eradication would be difficult to withhold from those who did not help pay for the eradication. If a disease infects some animals in a region, it raises the likelihood that nearby animals will also become infected. That said, economists have found that private market suppliers have in fact provided some "obvious" public goods and goods with externalities. As noted below, public goods will be provided in cases where the business environment presents opportunities for firms to capture enough benefits to cover their costs. In some cases, private agents may not provide the good, in other cases they may provide the good, but in a socially insufficient amount, and in still other (rare) cases, they may provide the optimal amount.

For example, Coase (1974) discussed the classical example of a public good that was for many years given as services provided by lighthouses. The idea was that the signal for a nearby coastal hazard would be available to all ships that passed in the darkness and it would cost the lighthouse no more if ten ships passed and were warned of the danger than if only nine ships passed and were warned. Furthermore, the idea was that lighthouse operators would not know when ships were passing, or whether those that passed had paid for the lighthouse service or not. Therefore, it would be practically impossible to exclude service from those who had not paid for the service. It turns out that many lighthouses were, in fact, privately owned and operated and had solved the problem of non-excludability by turning on the light only when a paying customer was scheduled to pass. If so, of course, once excludability had been achieved, the lighthouse was no longer a public good. Alternatively, of course, one might find that it was profitable for one shipping company to maintain a permanent light for its own ships were traffic sufficiently frequent, and that investment could have provided an optimal amount of lighthouse for all ships. Likewise, Cheung (1973) and Muth *et al.* (2003) consider facts surrounding data of one classical example of positive externalities, the provision of pollination services for orchards and the provision of bee habitat for beekeepers. They show that a well-functioning market for pollination services exists in which commercial orchards pay beekeepers to move hives to locations useful for pollination, though this result does not mean that all of the externalities (pollination versus honey production) were captured by the market.

Issues of the same type frequently arise when animal disease management is discussed. Many producers may find that there is sufficient private incentive to vaccinate their animals against a serious contagious disease. Together, they may largely solve the public good problem of vaccinating against a disease. However, if there are still a few individuals for whom the private incentive is insufficient, they may not vaccinate and thus a reservoir of infection may remain that is capable of infecting other animals. That reservoir will require that the vaccinating producers continue to vaccinate, at significant cost, rather than be able to cease vaccinating, as would occur if the disease could be eradicated in the region. There is thus reason for government animal disease control, either by requiring vaccinations or by directly carrying out the vaccinations required to eliminate the reservoir.

The Importance of Geography and Biology

The question of whether public good characteristics are involved in the management of specific animal diseases often depends on the biological characteristics of the disease and the habitat of the disease. The degree to which private incentives for disease control are reduced, whether as a result of non-rivalry in consumption or the inability to cheaply exclude non-payers, applies in different degrees depending on how the disease is likely to spread across the habitat. Thus, biological and geographical/topographical characteristics influence the degree to which government intervention in animal disease is justified. If two politically separate regions are geographically and economically integrated so that disease in one area is highly likely to spread to the other, it makes little sense for one region to consider animal disease control in isolation of the other. If there is justification for public intervention in one region, there is justification for (coordinated) public intervention in both regions. Alternatively, if one region is effectively separated from another by geographical barriers, there may be little reason for the public sector in one region to be concerned about whether private investments in the other region are sufficient – since any public good aspects of their behaviour will be contained in that other region. Similarly, diseases that spread rapidly constitute a greater threat to multiple farms/ranches than those diseases that do not spread rapidly, and thus also are more likely to call forth public intervention.

Consider border inspections and detection services that lower the probability that a new disease will enter a region, and particularly whether these services satisfy the condition that the service not be rival in consumption. The total benefits that accrue from excluding the disease from that region increase with the number and value of susceptible animals, but the benefits per animal do not diminish if more animals benefit. For example, in a competitive industry, the number of other producers that also experience lowered marginal costs (from not treating the disease individually) does not affect how much the per-unit cost declines for any individual producer. Furthermore, the cost of inspections and detection at the border depends on the characteristics of the border or ports of entry and does not rise in proportion to the size of the industry protected. The number of farms or animals that benefit often does not affect the measures required to keep a disease from a region. Of course, the number of animals can be related to the likelihood that the disease will enter, as when contraband traffic is related to the number of animals in the region.

Next, consider whether it is feasible to exclude the benefits of border inspection from some animals in a region. In a typical situation, if a disease is kept out of a region, all susceptible animals are protected from the disease. If the disease never enters a region, no animal will need treatment and all producers and consumers of animal products and services benefit, whether or not they contributed to providing the service.

The degree to which public good characteristics and externalities are associated with keeping animal diseases outside of a region depend on the appropriate definition of the region. If natural regions define the group of animals (and those who own or benefit from them) over which there is non-rivalry in demand, it is infeasible to exclude neighbours from protection and contagion creates a positive externality connected with disease control.

Such regions may not be coincident with national or state administrative or political borders. For example, in the USA diseases are often contained in one habitat region and successfully kept from spreading to other regions. The stronger the natural barriers to disease spread, due to differences in natural habitat, the more those public good and externality features affect local decisions. Such sub-national public goods and externalities are rationales for state and local funding and the operation of animal disease prevention and control measures. The more a region is naturally separate from other habitat regions, the more natural it is to consider sub-national public goods and externalities with respect to animal diseases. By this reasoning, California may be expected to have more disease protection, control or eradication activities operated at the state level than say, Illinois, which is physically close to other states with similar habitats and livestock populations. In Europe, Britain would have more individual efforts for animal disease exclusion and eradication than, say, Belgium, where it would be difficult to control a disease without a coordinated effort with nearby nations.

The nature of the disease also matters. A disease that is highly contagious and spreads rapidly would suggest a regional or national rather than a local effort because the public good and externality features of the disease management services cover a wider area. Therefore, we would expect a broad-based public approach to control of Foot and Mouth Disease (FMD) or Exotic Newcastle Disease (END), while we might expect a more narrowly focused approach to the control of bovine brucellosis.

The criteria of non-rivalry and non-excludability also may be addressed in the context of disease eradication programmes. Eradication of a disease from a region allows animal owners to forego private costs of disease management, increase productivity per animal, reduce production losses, and perhaps achieve higher consumer demand.

The same disease and habitat characteristics that imply that border exclusion measures would lower costs for all producers in a natural disease control region also apply to eradication programmes. Those potentially affected by the disease in that region share the benefits of eradication services. Additional production in the region increases the benefit of eradication services for expanding producers, but does not diminish the benefit of such services for existing producers. If some farmers in a natural habitat region refuse to pay for eradication services, there would be no feasible way to exclude them. Furthermore, disease control costs are lower for neighbors if the disease is controlled in nearby animals. These characteristics suggest broad-based funding of eradication programmes.

However, efforts to eradicate differ in other respects from efforts to exclude the disease from a region. Eradication is likely to be more costly the more animals there are in the infected region. Of course, other factors also affect eradication costs, including the existence of multiple hosts, some of which may be wild species, the nature of disease control regulations that are required for control, and the features of the infested region, such as its topography and whether it includes urban as well as rural areas.

Where eradication costs rise with the number of animals potentially affected, a per-unit assessment may be implemented to tie the funding to the beneficiaries of the programme. When eradication benefits a specific group of producers, a user fee or a per-unit fee ties payment for the cost of the programme directly to those who benefit rather than to the general taxpayer.

Many diseases that raise costs also reduce demand for animal products. A fall in demand may occur because the diseases affect the perceived quality or safety of the product, as with Bovine Spongiform Encephalopathy (BSE). It is common that consumers find it costly to acquire detailed information about product safety. Consumers also may find it difficult to differentiate among products so that safety concerns tend to reduce demand of all products that consumers perceive may be affected. Therefore, even if a disease is not contagious and has no cost externality or production-side public good connection, other firms in an industry may lose from infections in a single firm. Food safety scares associated with isolated events have occurred for many products and make clear that widespread demand shocks are common when safety information (or misinformation) is released. (See Buzby, 2003; or Alston *et al.*, 2005; and the references they cite.) These reputation externalities suggest that industry-wide incentives for disease management exceed the sum of private incentives, perhaps by a large amount. In these cases, consumers gain if product safety is improved and they continue to consume products that they enjoy. Given reputation externalities, regulations or other collective action to improve food safety may benefit both producers and consumers.

Animal disease outbreaks in one country or jurisdiction encourage governments in other jurisdictions to restrict market access to products from the region where the infection occurs. The response from foreign governments differs in its effects from those caused by a reduction in foreign consumers' willingness to pay. For example, a government may restrict access in order to protect its producers from the threat of the disease even if consumers are not directly threatened by the disease, e.g. FMD. Such foreign government reactions do not mean that foreign consumers necessarily value the product less highly. Thus, effective demand for the product of the infected region can decline even if individual consumers have not reduced their demand. Moreover, if some foreign consumers suffer reduced access to a product because of their government's decision, consumers in other countries who remain in the market likely benefit from lower prices since overall demand has decreased.

Alternatively, if foreign governments push for improved disease control measures in the country where a product originates, as an alternative to excluding exports, these measures can increase costs to producers in the infected country, increase costs to all consumers, and benefit producers in the importing country. Thus, unlike responses to broad-based consumer concerns, responses to satisfy governmental demands can have a variety of effects on both consumers and producers in different countries.

The recent North American BSE outbreak provides an example (APHIS, 2004). Japan and South Korea requested that the USA test all slaughtered animals for BSE. The US government declined to do so, arguing that the proposed tests were scientifically inappropriate, though it understood that the refusal to accept the tests meant that Japan would deny access to its market for US beef. The US government concluded that if it imposed new regulations on the processing of US cattle to satisfy demands by the Japanese or Korean governments, the result would likely be higher production costs in the USA, higher prices for US and foreign consumers, and little or no improved food safety for US or foreign consumers (Sumner, Bervejillo and Jarvis, 2005). Alternatively, not to conduct the testing implied not having to further increase costs of production, but a lower total demand, with lower prices for producers and US consumers and no consumption by Japanese and Korean consumers.

The public good and externality characteristics we have been discussing are defined for producers and consumers of the goods or services that the animals provide. Thus, the scope of the "public" served is limited to a part of the total population. The more a disease is specific to a single species or agricultural industry, the more it is natural to consider applying the costs of providing border protection and eradication services to producers and consumers in that industry. Such services are sometimes termed "industry collective goods" to distinguish them from public goods that apply to a broader spectrum of the population. In the case of industry collective goods, product or animal unit assessments are a natural funding mechanism.

Diseases that also affect wildlife or pets or that have other widespread benefits, such as military security, are more natural candidates for general public funding.

Distribution of Costs and Risks of Disease Outbreaks

When cost or demand-side externalities exist, losses spread beyond those with infected animals and benefits of disease management accrue to owners of animals that are not infected. In these situations, governments can create incentives for recognizing external effects by regulation, taxation or subsidy. The same effects on disease management can occur whether governments apply regulation, taxation or subsidy (Coase, 1960). However, the effects on the distribution of net gains and losses may be quite different.

Agricultural policy in the USA (and in many other rich countries) consists largely in regulating agriculture much as any other economic activity is regulated, with some special exemptions or other conditions. In addition, agriculture is provided with exceptional subsidy and trade protection (Alston and James, 2002). In the USA, the meat industries (hogs, beef cattle, poultry) as well as the egg industry, receive little protection or subsidy. The rate of subsidy for these industries is low compared to subsidies for grains, cotton or oilseeds. The dairy industry receives trade protection, direct payments, price supports and a government-sponsored cartel to raise prices and the overall level of dairy subsidy is significant (Sumner, 2003b). For crop agriculture, subsidized yield or revenue insurance is also widespread, with subsidy outlays in the range of $2 billion. No such costly and widespread subsidized insurance is available for livestock producers.

For animal disease policy, the US government has used subsidy, regulations and taxes or fees to achieve policy goals. The basis for choice among instruments seems to depend on the distributional impacts as well as on enforcement and transaction costs. When major outbreaks occur, such as the recent END outbreak in California and the finding of a single case of BSE in Washington State, large government outlays occur for detection and eradication and for enforcement of additional regulation. The END case entailed government outlays of more than $200 million, with a relatively small share of the funds devoted to compensation to poultry owners facing animal destruction in the eradication programme.

The recent BSE case in the USA is instructive. The BSE case entailed substantial new regulation and much less extensive animal destruction. Compensation was likely a small part of government outlays and certainly was small relative to total economic losses associated with the case. The major economic losses were to the US cattle industry, especially to the cattle feeding segment, which faced lower fed cattle prices while the supply of cattle was approximately fixed. These lower prices were attributable to import embargos from major trading partners, especially Japan and Korea. Feedlots incurred losses even though domestic regulators and domestic consumers seem to have readily accepted that the safety of the product was not compromised. Gross revenue losses were in the range of 20 percent, as judged by the magnitude of the price decline. Because most of the affected beef was sold in the domestic market, most loses were incurred by producers because of the price decline, while domestic consumers enjoyed the greatest gains. Thus, the net loss to the USA as a whole from the BSE occurrence was caused by the price decrease associated with the decrease in export market demand, with the benefit to domestic consumers largely offsetting the loss to domestic producers. Foreign buyers who continue to import US beef now obtain it at a lower price. Of course, additional losses may occur from regulatory overreaction to the BSE occurrence, especially if broad responses impose added costs to satisfy Japanese or Korean regulators. In that case, US producers may gain from a higher market price, but US consumers would lose doubly, first from the higher costs and second from higher prices when Japan re-enters the market.

Recent animal disease events in the USA testify to the risk and uncertainty that disease outbreaks may impose on livestock producers. As the BSE event suggests, major losses can occur through price impacts even when very few animals are affected and almost no losses are directly attributable to eradication. Price risk in the cattle markets may be managed in several ways, including forward contracts, futures contracts and options contracts. Use of these tools is not subsidized and has long been available to livestock producers.

Two cost-side risks are associated with animal malady management. One is the chance that a disease outbreak will directly raise costs. Livestock producers face these risks as a normal part of their business just as they face the potential for weather-driven feed cost increases. A second supply-side disease risk relates to the chance that a disease eradication programme requires that animals on a specific farm or ranch be killed. For narrowly focused programmes, no offsetting price gains benefit producers. Compensation of animal losses may act to spread benefits and costs of eradication or control programmes across an industry (Kuchler and Hamm, 2000). If industry assessment funds rather than general tax funds are used, the connection between benefits and costs of such programmes is even more direct. The role of a government-subsidized insurance programme is harder to place in the context of public goods or externalities of animal diseases.

Policy for Animal Disease in the Context of Security Against Intentional Introduction of Disease

For obvious reasons, biosecurity against potential terrorist attacks on the food system, commercial agriculture or wildlife has become a significant part of national security policy in the USA and elsewhere in recent years. The public good rationale for policy action has been long established for national defense in general. It adds little to the cost of homeland security to protect additional residents, and security for one resident does not reduce the security available for others. Furthermore, excluding non-payers from security of the food system would be costly and technically difficult.

Nonetheless, even in the context of biosecurity, it is useful to examine the nature of the public good argument carefully. For efforts to reduce the impact of threats on the security of the livestock product supply, the beneficiaries are producers and consumers of animal products. Thus, to align costs with those who benefit, it is natural to consider assessments to cover added biosecurity costs. However, biosecurity threats are linked broadly to national security, because terrorist attacks have the potential to contribute to widespread panic or a breakdown in social order and the normal functioning of society. The losses from an attack are far larger than the direct economic costs to consumers and producers. Therefore, additional efforts to reduce the probability of attack and mitigate the losses from attack follow.

Intentional attacks on animal agriculture may be of several sorts, all of which may warrant particular efforts for prevention and mitigation. First, consider intentional introduction of diseases that have large economic consequences, but little threat to human health. For example, introduction of FMD would not affect human health and would not, by itself, cause significant animal death. However, as Ekboir, Jarvis and Bervejillo (2003) show, the expected economic losses and animal destruction associated with an outbreak and subsequent eradication of the disease are very large. Furthermore, if the disease was introduced in several locations at once and in a way that allowed it to spread before detection, eradication may not be economically feasible except over a long period. Such an introduction would also cause major international trade losses and likely cause disruption of tourism and international commerce more broadly.

A second kind of attack is one that affects the food system by introducing a disease that affects food safety. One might consider introduction of BSE, but this is problematic. BSE is much less contagious than many other diseases, progresses slowly and may go undetected. Use of BSE is likely to be ineffective as a terror agent unless some special efforts were undertaken. Ideas along these lines are best left unexplored in publicly accessible documents.

A third approach to using animal disease as an agent of terrorist attack would be to explicitly target wildlife or other animals outside the farm or food system. Using intentional disease introductions to attack important or symbolic endangered or threatened species could cause widespread unease and significant losses to the sense of homeland security.

In all these cases, governments have a clear role in attempting to exclude intentional disease introduction and to reduce losses. The amount of resources and how to marshal those resources must be considered in the same way other national security issues are analysed and is beyond the scope of this paper, except to note that standard public good considerations and benefit and cost calculations apply.

Summary and Conclusions

Private agents have adequate incentive to carry out most of the expenditures that are made for the management of animal disease. Nonetheless, because animal disease management frequently has some externalities and/or public good characteristics, government often has a role to play in disease management. We have reviewed briefly the concepts of public goods and externalities in the context of animal disease. We note that the nature of externalities and public good characteristics, particularly whether they are large and thus warrant concern, is determined by geography and biology. Both of these affect the natural habitat for specific diseases and thus the likelihood that a disease will spread from one region to another. The likelihood of spread is a crucial consideration in government intervention, whether regarding efforts to exclude, control or eradicate the disease from a region. These considerations will thus influence the design of policies and whether these policies should conform to a locality or a region, and whether they should include a single or multiple political entities. We have also considered the distributional effects of disease outbreaks and how forward pricing, or the use of futures and options markets, may mitigate risks. Compensation for animals destroyed during eradication or control efforts may also reduce direct losses and if industry assessments are used to fund such compensation, costs may be aligned with benefits. Finally, we considered the public good characteristics of protection and mitigation of intentional threats to biosecurity. Terrorist threats to animal industries, the food system, or wildlife all raise relatively new roles for public policy in animal disease management.

References

Alston, J.M. and James, J.S. (2002) The incidence of agricultural policy. In: Gardner, B.L. and Rausser, G.C. (eds) *The Handbook on Agricultural Economics,* Volume II(a). Elsevier, Amsterdam, pp. 1869-1929.

Alston, J.M., Norton, G.W. and Pardey, P.G. (1995) *Science Under Scarcity: Principles and Practice for Agricultural Research Evaluation and Priority Setting.* Cornell University Press, Ithaca, New York.

Alston, J.M., Brunke, H., Gray, R.S. and Sumner, D.A. (2005) Demand enhancement through food-safety regulation: a case study of the marketing order for California pistachios. In: Alston, J.M. *et al.* (eds) *The Economics of Commodity Promotion Programs: Lessons from California.* Peter Lang Publishing, Inc., New York.

Anderson, K., McRae, C. and Wilson, D. (2001) *The Economics of Quarantine and SPS Agreement.* Centre for International Economic Studies, Adelaide, and AFFA Biosecurity, Canberra, Australia.

APHIS (2004) Report on measures relative to BSE in the U.S. US Department of Agriculture, Animal and Plant Health Inspection Service. [Accessed 2004.] Available from http://www.aphis.usda.gov/lpa/issues/bse/US_BSE_Report.pdf

Bervejillo, J. and Jarvis, L. (2003) Bovine spongiform encephalopathy: lessons from the UK. In: Sumner, D. (ed.) *Exotic Pests and Diseases: Biology and Economics for Biosecurity.* Iowa State Press, Ames, Iowa, pp. 71-84.

Bicknell, K.B., Wilen, J.E. and Howitt, R.E. (1999) Public policy and private incentives for livestock disease control. *The Australian Journal of Agricultural and Resource Economics* 43, 501-521.

Buzby, J.C. (2003) International trade and food safety: economic theory and case studies. US Department of Agriculture, Economic Research Service, Agricultural Economic Report Number 828.

Cheung, S.N.S. (1973) The fable of the bees: an economic investigation. *Journal of Law and Economics* 16, 11-33.

Coase, R. (1960) The problem of social cost. *Journal of Law and Economics* 3, 1-44.

Coase, R. (1974) The lighthouse in economics. *Journal of Law and Economics* 17, 357-76.

Ekboir, J., Jarvis, L.S. and Bervejillo, J.E. (2003) Evaluating the potential impact of a foot-and-mouth disease outbreak. In: Sumner, D.A. (ed.) *Exotic Pests and Diseases: Biology and Economics for Biosecurity.* Iowa State University Press, Ames, Iowa, pp. 85-98.

Kuchler, F. and Hamm, S. (2000) Animal disease incidence and indemnity eradication programs. *Agricultural Economics* 22, 299-308.

Muth M.K., Rucker, R.R., Thurman, W.N. and Chuang, C.T. (2003) The fable of the bees revisited: causes and consequences of the US honey program. *Journal of Law and Economics* 46, 479-516.

Sumner, D.A. (2003a) Economics of policy for exotic pests and diseases: principles and issues. In: Sumner, D.A. (ed.) *Exotic Pests and Diseases: Biology and Economics for Biosecurity.* Iowa State University Press, Ames, Iowa, pp. 3-18.

Sumner, D.A. (2003b) Implications of the USA farm bill of 2002 for agricultural trade and trade negotiations. *Australian Journal of Agricultural and Resource Economics* 47, 117-140.

Sumner, D.A., Bervejillo, J.E. and Jarvis, L. (2005) Public policy, invasive species and animal disease management. *International Food and Agribusiness Management Review* 8, 78-97.

Viscusi, W.K., Vernon, J.M. and Harrington, Jr., J.E. (2000) *Economics of Regulation and Antitrust.* MIT Press, Cambridge, Massachusetts.

Chapter 4

Incentive Compatibility in Risk Management of Contagious Livestock Diseases

Ben M. Gramig, Barry J. Barnett, Jerry R. Skees and J. Roy Black

Introduction

One of the most difficult challenges in designing mechanisms that address the risks posed by contagious livestock disease is the potential conflict between encouraging producer herd health management and biosecurity measures while maintaining incentives for early disclosure of suspected health problems. This chapter addresses the importance of the incentives present in the livestock production and animal health systems. When considering the design of policy and market instruments, the focus must be on incentive compatibility as a major determinant of the potential success or failure. Producers face multiple objectives in the course of managing their farms or ranches and must strike a balance between production, financial obligations and animal health. While this book is about systems that involve both the public and private sector for improving risk management and livestock disease management, a key to the success of designing such mechanisms must involve full recognition of the sometimes competing incentives that influence individual producer behaviour.

The intent of this chapter is not to propose specific solutions to address incentive compatibility in livestock disease management, but rather to clearly emphasize the importance of incentives when designing animal health policy or risk management instruments. An underlying theme and concern is that improperly designed livestock insurance solutions could increase the disease risk problem for the entire sector. Livestock disease has great potential to cause widespread economic damages. If the individual producer incentives are not considered when designing regulations, offering government disaster payments or developing insurance products, the result could be a weakening of the animal health system that worsens the effect of an outbreak.

This chapter focuses specifically on those diseases with the greatest potential to cause widespread or systemic losses. Animal diseases with the potential to cause a range of socio-economic problems or public health consequences are classified by the Office International des Epizooties (OIE is also known as the World Organization for Animal Health) as either List A or List B diseases. Unless otherwise noted, all reference to animal disease in this chapter refers to List A diseases, where risk of transmission is irrespective of national borders. Included in this category are diseases such as Foot and Mouth Disease (FMD), Classical Swine Fever (CSF), Newcastle disease and highly-pathogenic avian influenza.

Recent events have clearly demonstrated the devastating impact that a livestock disease can have on the entire sector within a country – trade restrictions follow and market prices are severely impacted. Consequently, the benefits of disease-free status are shared by all participants in the market. These benefits are non-exclusive in the sense that any individual engaged in livestock production benefits from an unimpeded market for their products. With the emergence of a highly contagious disease (and some less contagious ones), export demand often decreases as trade partners ban imports from the infected country in an attempt to prevent human illness or the spread of the disease to their domestic herds. Depending on the nature of the disease, demand from domestic consumers may decrease as well. All domestic producers in the market, not just those with infected animals, suffer from the resulting lower market prices. In this way, a single producer's actions that cause a disease outbreak can jeopardize the value of production as well as investments made in production and processing for all market participants. The non-exclusive nature of an unimpeded market for livestock products has characteristics similar to those of a common property resource where access and consumption of the resource are not restricted by property rights. Disease-free status maintains consumer confidence and keeps export markets open, but is heavily dependent on individual behaviour of livestock producers.

The focus of government agencies such as the Animal and Plant Health Inspection Service (APHIS) in the USA has largely been to protect against incursion of disease at national borders through inspection, quarantine and other methods, as well as to identify, contain and eradicate animal health threats if they emerge from within the country. These are vital activities in maintaining disease-free status and are provided by the government based on their "public good" dimensions (see Grannis and Bruch, this volume). Much attention has been paid to government compensation to producers when their animals are slaughtered in response to an outbreak. However, herd depopulation represents only one cost to producers and does not account for economic losses incurred by other market participants. If a major objective of public or private risk management programs is to mitigate and prevent a potential disease outbreak, policy makers must understand the full set of consequences and the individual incentives of producers. The old adage that it is best to "shoot and bury an infected animal" still haunts those who wish to design effective discovery programs in the name of early detection.

Considerations for Insuring Contagious Livestock Disease

Livestock disease insurance is typically available only for companion or sporting animals, breeding stock and other high-valued animals. High cost typically precludes the purchase of livestock disease insurance for production animals. The limited availability of insurance for production animals has typically been only for named peril coverage (i.e. specific diseases). The fact that only limited insurance for production livestock has been offered by insurers may point to the complexity of disease-based coverage or even the lack of demand for such coverage from producers. While it is not the goal of this chapter to determine why private sector insurance for livestock disease has not been available, it is useful to reflect on this question when considering public policy and insurance designs that will mitigate and prevent contagious disease outbreaks in the livestock sector.

Asymmetric Information

Insurance purchasers will almost always know more about their risk exposure than the insurer. In addition, they will know more about their efforts (or lack of efforts) to mitigate their exposure to loss risk. This asymmetric information creates important challenges for those designing insurance products.

Hidden information will lead to adverse selection in insurance. If insurers are unable to properly classify potential policyholders according to their risk exposure, those who have been classified as having low risk when they actually have high risk will buy the insurance. The converse is also true – those who have low risk but are classified as having high risk will not purchase. As indicated by Shaik *et al.* (this volume) livestock disease insurance poses difficult risk classification challenges. If someone has insurance that is underpriced relative to their risk, they may also expand production to generate further expected benefits from the insurance product. Thus, in addition to inequities that follow hidden action and adverse selection, the availability of an insurance product with misclassification problems can actually increase the risk-taking behaviour of individuals. As indicated previously, increased risk-taking by the individual can put the entire sector at increased risk given the nature of catastrophic animal diseases and the link back to markets.

Hidden action (also called "moral hazard") is another problem caused by asymmetric information. It occurs when, subsequent to purchasing insurance coverage, policyholders change practices so as to increase their risk exposure. Hidden action is most problematic when loss risk is highly conditioned on management or production decisions rather than on random events that cannot be influenced by the policyholder. Such is the case with many contagious livestock diseases where loss risk is often dependent on sanitary practices and other disease control measures (see Shaik *et al.*, this volume). Once again, if hidden action cannot be controlled, actions of the individual will put the entire system at increased risk.

External Effects

For contagious livestock disease insurance, the implications of asymmetric information problems are more significant than they are for crop insurance. This is because of the external effects associated with livestock disease, especially those diseases that have human health implications.

Unlike crop insurance for most risk events, livestock producers with contagious disease in their herds are not the only individuals affected by an outbreak. With the exception of contagious crop disease problems, when crop insurance policyholders act in ways that increase expected losses, the production losses (reduced yields) incurred as a result of the change in behaviour are generally confined to the policyholders' farms or ranches. Once again, the hidden actions taken by a producer can increase the risk of a disease outbreak where both economic losses and risk to humans extend well beyond the individual producers.

The individual production losses from animals quarantined and slaughtered to contain and eradicate diseases represent only a fraction of the economic cost of an outbreak. A crude numerical example taken from the December 2003 case of Bovine Spongiform Encephalopathy (BSE) in the USA illustrates the widespread economic effect of the presence of a disease classified as being a socio-economic or public health threat within countries (World Organization for Animal Health, 2004). While BSE was found only in a single herd of cattle in one state, all farmers, ranchers, feedlots and processors in the entire country were affected. When the BSE case was confirmed, US cattle prices decreased as several countries, most notably Japan and South Korea, ceased importing US beef. By comparing the December 2003 live cattle futures prices for the first quarter of 2004 with the observed cash prices (fed cattle) for those months, a loss in market value of $566 million is revealed (calculations based on market data from the Livestock Marketing Information Center).[1] The loss in market value indicates how the individual loss of animals to eradication may pale in comparison to the magnitude of the market losses across all producers.

Systemic Risk

Livestock disease epidemics and natural disasters, such as drought or flood, represent what are known as "systemic risks". Distinct from independent losses that are well-suited to insurance applications, systemic risks result in correlated losses and pose a significant challenge to financing insurance. The distinctions between the systemic risks in agriculture created by natural disasters as compared to livestock disease are found in the respective income and inter-temporal effects.

When a drought occurs it affects a large number of producers in an entire region who suffer economic losses because of reduced yields. If the drought affects a large enough quantity of production, the supply of a commodity or commodities may be affected and producers in areas of the country unaffected by the drought may receive higher prices for their crops. If the drought is localized, affecting the yields of a relatively small number of producers, commodity prices will remain unaffected and the impact of the drought will be confined to those who suffer production losses.

There are both similarities and differences when you consider the income effects from livestock disease. These similarities and differences arise largely from the characteristics of the particular disease in question. Diseases such as rabies, brucellosis or tuberculosis are classified as List B diseases (World Organization for Animal Health, 2004) and are not deemed a threat beyond the borders of the country in which they are present. These types of diseases are generally considered acute risks and as such are a localized threat similar to a drought or flood that affects only a small share of production and does not have implications for widespread income effects because of trade restrictions or widespread quarantine. A small number of List B diseases, such as BSE, are considered systemic risks because of associated human health concerns. Because this chapter focuses on those diseases that pose threats irrespective of national borders and because these diseases present policymakers with the greatest risk management challenges, emphasis remains on List A diseases here.

---

[1] Based on December 23, 2003 live cattle futures prices and the actual cash prices received (Nebraska only for comparison with USDA projections) in the first quarter of 2004, multiplied by the live weight of cattle (heifers and steers only) actually slaughtered during that time period, a loss in value of $566 million or 8.11% has occurred in the aftermath of BSE (Livestock Marketing Information Center, 2004). All losses are not attributed to BSE alone. It should be noted that because the market had been experiencing strong prices for the past year or more, the futures prices prior to the confirmed case of BSE already indicated an expected decline in prices starting in early 2004. USDA market estimates prior to the confirmed case compared to observed cash prices indicate a loss of over $1 billion in value based on actual slaughter over the same period (USDA-ERS, 2003-4).

As described in a previous section, in many cases no producers in the specific livestock sector affected by a disease epidemic, regardless of the disease status of their individual herd, will be able to avoid economic losses.[2] Income losses can result from downward price movements from export losses or a change in consumer confidence that affects domestic demand for longer time periods. Livestock producers in other sectors may benefit from a disease that emerges in another species as export and/or domestic demand for other meat products may increase in response to an outbreak. The full magnitude of demand and price effects depends on the characteristics of the particular disease and the environmental conditions present when and where an outbreak occurs, which may affect the rate of transmission.

Inter-temporal Effects

Drought is generally a single-year event, but even if a multi-year event occurs, drought does not affect the openness of export markets for grains, oilseeds or other commodities. Livestock diseases on the other hand may have substantial inter-temporal effects. There are two dimensions to the duration of impact to livestock markets: the market level and the farm or ranch level.

At the market level, the duration of the impact is determined by the length of time it takes to eradicate the disease and regain the confidence of trade partners and domestic consumers. Determinants of this duration include how quickly a problem is disclosed or discovered and the traceback mechanisms for disease and food-borne illness in the livestock industry. Early discovery and traceback will be discussed in greater detail. Without early discovery the disease spreads, enlarging the task of containment and eradication. The length of time it takes to locate the source of the disease once it is identified in the food system may also impact the confidence of domestic and foreign markets regarding the safety of a country's livestock products.

At the farm or ranch level, the inter-temporal impacts are experienced not only through depressed prices, but also through the time it takes to re-establish herds when depopulation is required for eradication. Investments made over time by farmers and ranchers in herd genetics are real costs that an indemnity based on the market value of the animal does not reflect. It often takes years to recover from depopulation and re-establish a biosecure herd of the same quality as before a disease episode.

The case has been made that for livestock disease risk there are unique ramifications of hidden action and the potential for sustained market effects complicate public policy and insurance compared to other agricultural risks. Also unique to livestock disease as a correlated agricultural risk, is the fact that, unlike drought, which is totally uncontrollable, diligent management can reduce the probability and severity of a disease outbreak in many cases. Bioterrorism (Sumner, Bervejillo and Jarvis, this volume) is one exception to the premise that producer behaviour is a major determinant not only in whether a disease outbreak occurs (management) but also the extent of the outbreak (early disclosure).

---

[2] This is not the case for diseases that can be contained within a country and are classified as not being a socio-economic threat "irrespective of national borders" (World Organization for Animal Health, 2004). Wolf (this volume) mentions tuberculosis as an example of this type of disease which is present in some US states. This type of disease has a more geographically limited impact, many times affecting only producers located in areas within a country where the infection is present.

Table 4.1. Summary of Recent Animal Disease Outbreaks.

Disease	BSE UK 1996/97	FMD Taiwan 1997	CSF Netherlands 1997/98	FMD UK 2001	FMD Rep. of Korea 2000	FMD Japan 2000
Disease confirmed	1996	20 Mar 1997	4 Feb 1997	20 Feb 2001	2 Apr 2000	4 Apr 2000
Duration of disease	Ongoing	4.5 months	18 months	7.5 months	1 month	1 month
Number of outbreaks	6,271	6,147	429	2,033	15	3
Control policy	Stamping out	Stamping out and mass vaccination	Stamping out	Stamping out	Stamping out and ring vaccination	Stamping out
Animals slaughtered						
- Infected	6,271	4.03 million	0.7 million	1.30 million		
- Pre-emptive			1.1 million	3.10 million		
- Welfare			9.2 million	5.43 million		
Total	**6,271**	**4.03 million**	**11.0 million**	**9.83 million**	**2,216**	**740**
Direct costs [a]						
- Compensation	2,433	188	1,183	2,223	377	0.5
- Control measures		66	138	1,335	66	14.5
Sub-total	**2,433**	**254**	**1,321**	**3,558**	**433**	**15**
Indirect costs [a]						
- Agricultural sector		2,202	423	489		
- Related industries		3,212	596	267		
- Other		949		4,890		
Sub-total	**1,395**	**6,363**	**1,019**	**5,646**	**n.a.**	**n.a.**
Total costs [a]	**3,828**	**6,617**	**2,340**	**9,204**	**433**	**15**
Impact on GDP	–0.4%	–0.64%	–0.75%	–0.2%	n.a.	n.a.
Cost to public sector	63.5%	3.8%	43.5%	38.6%	n.a.	n.a.
Cost to private sector	36.5%	96.2%	56.5%	61.4%		

[a] $US million

n.a. = not available.

Source: United Nations Food and Agriculture Organization (2002).

Good Management, Biosecurity, and Early Disclosure

The role of good livestock health management is fundamental to preventing the emergence of contagious disease and involves not only vaccination of animals and proper feeding practices, but also farm- or ranch-level biosecurity measures to prevent the transmission of disease. Biosecurity measures play an important role not only in protecting the integrity of a producer's herd health from incursion, but also in controlling the spread of disease from the production unit.

The importance of early disclosure of a disease outbreak or a suspected problem by a producer cannot be overstated. The recent experiences of countries around the world in identifying, containing and eradicating contagious livestock disease is instructive. The data contained in Table 4.1 indicate the country where the outbreak occurred, the details that define the magnitude of the epidemic, the number of animals slaughtered and the direct and indirect costs incurred. This information demonstrates the finding of the United Nations Food and Agriculture Organization that "early detection and appropriate reaction to a disease outbreak is paramount in minimizing the consequential losses" (UN-FAO, 2002).

There is great disparity in the duration and cost of contagious disease events reported in Table 4.1. The FAO has found that the major determinants of both duration and cost are the length of time it takes to identify (confirm) an outbreak and the deployment of appropriate resources to control the spread of the disease and eradicate it. The outbreaks of FMD in Japan and the Republic of Korea in 2000 provide the best examples of how early detection and timely, effective action to contain and eradicate disease greatly reduce the total cost. BSE in the UK is included in this table to demonstrate how a non-contagious disease can also have enormous financial impacts and how BSE is unique as a livestock disease because it has an incubation period of many years, so that by the time it was confirmed in the UK in 1986 thousands of animals had already been infected (UN-FAO, 2002).

The FMD outbreak in Taiwan had infected 28 farms before the disease was confirmed, in the UK 92 herds were infected before the source of the FMD outbreak was identified, and in the Netherlands CSF spread for over a month and infected 39 herds before being detected. These outbreaks in Taiwan, the UK and the Netherlands cost billions of dollars compared to millions of dollars in Japan and Korea. Complications that affect the duration of the disease have included insufficient slaughter capacity to keep pace with the spread of disease and, in the case of Taiwan, the decision to use mass vaccination rather than pre-emptive slaughter which enabled the disease to persist in diseased vaccinated animals lacking clinical signs of illness.

Management and early disclosure represent two of the biggest challenges to designing effective instruments for risk mitigation and transfer because they are a result of individual behaviour. Verifying that individual behaviour is consistent with either government regulations or contingent claims contracts (such as an insurance policy) is very costly given the number of farms or ranches involved. Structuring incentives so that a producer's actions are consistent both with their individual objectives and those of policy-makers is a difficult task, but this is the essence of incentive compatibility. To add to the complexity of this issue, the desire to hold bad managers accountable is likely to run directly counter to the need for incentives that increase the likelihood of early disclosure.

Compensation for slaughtered animals infected with disease has been a common component of eradication programs with costs into the billions of dollars (see Direct costs – Compensation in Table 4.1), but if early disclosure is an important objective then the incentive effects of compensation need closer scrutiny. Consider a producer who anticipates that disclosure of disease or suspected disease exposes him/her to greater regulatory scrutiny while also having a realistic expectation that if his/her animals are depopulated in the event of an outbreak he/she will be compensated. Equal compensation for lost animals, whether early disclosure occurs or not, fails to create an incentive to act consistent with policy objectives. The incentives for good management (punishment/accountability for bad management) and early disclosure do not appear to be compatible as they currently stand in the USA. As long as producers can expect the same level of compensation for animals, whether disclosure occurs or not, there is little incentive to subject oneself to the scrutiny or other recourse that accompanies disclosure (the management incentive). In an attempt to create incentives for early reporting, Belgium and the Netherlands have instituted compensation programs that no longer pay producers for dead animals and only partial compensation is given for diseased animals (Horst *et al.*, 1999). The motivation is that once the animal is dead, too much time has passed to reward the producers. Creating incentives for early detection is a paramount goal.

The issue of compensation is at the core of incentive compatibility for either government or market risk management mechanisms. Insurance that creates incentives that are at odds with animal health policy may actually increase disease risk. Consistency in insurance design and public policy is only half of the incentive compatibility problem, however. For incentive compatibility to be achieved, either individual behaviour must be observable or the producer is expected to perfectly comply with regulatory and insurance contract requirements *with certainty.*

Public and Private Incentives in Livestock Management

In both the public and private sectors there are incentives in place that influence individual behaviour. Whether designing public or private market mechanisms for risk management, these incentives are important. To illustrate this point, first consider the incentives faced by a farmer or rancher in the private market.

Private incentives likely stem from production, marketing and financial obligations that the producer may have. Under production, consider that the farmer or rancher must produce a certain amount at or below a certain cost per unit of production to be profitable and that net income is likely to be a major incentive in making production and management decisions. Based on the marketing arrangement of the particular operation, producers may be subject to a production contract in an integrated industry or may be responsible for directly marketing what they produce. Under contract production, prices may be known and certain inputs provided by the integrator, but there may also be environmental or food safety provisions in the production contract that represent additional costs of production. If these aspects of the contract are violated, it is likely to cost the producer even more in lost revenue. For an independent producer, there may be greater price risk in the market and the additional costs involved in direct marketing must be absorbed. Any financial debt on equipment or production facilities as well as operating lines of credit further strengthen production incentives to ensure that loan payments are not interrupted.

While none of these private incentives deals directly with animal health, there are clearly private incentives to manage disease and biosecurity when you consider that, in and of itself, the occurrence of disease translates into lost income. How the presence of publicly provided animal health services or indemnities for losses due to disease affects these private incentives warrants consideration.

There are also public incentives that influence producer behaviour. The first class of public incentives takes the form of management guidelines or regulations for animal health and biosecurity. If animal health regulations require measures over and above what producers would do to prevent disease in the absence of regulation, such management requirements translate into additional costs of production and lower profitability. Enforcement of some kind would have to accompany this type of regulation to provide adequate incentives for implementation to occur. A second class of public incentive that is very different from regulation is the desire of all market participants to maintain disease-free status because of the potential cost of a disease outbreak. This is the public extension of the aforementioned private incentives that exist for disease management.

Coase (1988, p. 118) argues that for many public policy problems there is no *prima facie* reason to expect that either government or the market would necessarily achieve a more efficient result (see also Samuelson, 1947, who preceded Coase in making this point). The management of livestock disease risk is no exception. While market failure is often cited as a justification for government intervention in the market, it must be noted that both government and the market are susceptible to failure. Wolf (1988) has provided a theory of government failure, whereby the unintended consequences of government actions result in what he refers to as derived external effects. Two examples of government actions taken that have apparently had such unintended effects on livestock disease management are provided.

Private and public incentives that exist at a given point in time must be taken into account as a whole when new public policies or private market products are introduced. While private management incentives to avoid disease may seem clear, the introduction of a new policy may actually work counter to the desired result. By providing free or heavily subsidized animal health services as a public good designed to facilitate disease control, the government may simultaneously provide a disincentive for private investment in animal health (Ramsay, Philip and Riethmuller, 1999). An example of this is the finding of Bicknell, Wilen and Howitt (1999) that New Zealand's government policies to control bovine tuberculosis removed some private incentives to control the disease. By providing these types of services, the government may also create an incentive for poorly motivated producers to enter the livestock sector based on an expectation that government will resolve livestock disease problems if they arise (Ramsay, Philip and Riethmuller, 1999).

Another example of the unintended consequences of public policy is a provision in the 2002 Animal Health Protection Act (AHPA) (discussed by Grannis and Bruch, this volume) that limits federal indemnity payments to the market value of the animals slaughtered and requires that any indemnity amount received from another source be subtracted from the government payment issued. The intent of this provision in the law was meant to avoid problems experienced in other countries where producers were indemnified by state or provincial authorities as well as the federal government for the same animal. This aspect of the law was well intentioned and should minimize government indemnity costs in the event of an outbreak by preventing "double dipping".

The APHA, however, may have the unintended effect of diminishing the demand for private insurance products to indemnify producers in the future, as well as affecting some private insurance products already available to cover a capped, fixed amount over and above government indemnity levels for animal loss. In addition to affecting livestock indemnity, there is uncertainty about whether the reduction in federal indemnity applies to private coverage for consequential losses from livestock disease. By reducing federal indemnity if private coverage for the value of the animal is purchased, the value of an insurance policy is reduced because the producer has to forgo government indemnity to receive full compensation under private insurance. The government is in effect providing free insurance to producers, albeit that coverage levels won't match the full market value of the animal and the level of coverage is uncertain.

There would be reduced incentive to purchase animal insurance coverage in the presence of government indemnity. In contrast to insurance coverage there is typically no underwriting with a government indemnity program. That is, there is no effort to classify producers according to their risk characteristics and increase premiums or reduce benefits (in the case of a government indemnity program) accordingly. Effective underwriting provides incentives for insured producers to engage in risk mitigation. If government indemnity programs crowd-out insurance purchasing, the result may be reduced incentives for risk mitigation and thus increased disease risk throughout the animal production system. The question of whether coverage for consequential losses reduces federal indemnity for herd depopulation is another source of uncertainty for the producer. It is critical to recognize that there is also potential for market failure in livestock disease insurance.

The costs involved in designing insurance contracts for livestock disease are a result of data availability and monitoring requirements.[3] The development of new insurance products for livestock disease will require an assessment of the nature of the risk and whether insurability conditions are met (see Shaik *et al.*, this volume). Hidden action in livestock production, variability in contagious disease characteristics (transmission, potential socio-economic impacts) and uncertainty associated with the external effects of contagious disease outbreak (implications for correlated losses) all contribute to the transaction costs involved in designing public policy or insurance for risk mitigation. Specific to insurance design is the impact of these costs on premium rates, and if insurance is unaffordable, demand will not be sufficient to cover financing of the catastrophic risks that are present in these types of insurance products.

Because health management and early disclosure are both dependent on individual behaviour, insurers must monitor insureds to limit hidden action problems and discover fraud. Insuring management in any setting is problematic because of the verifiability problem associated with hidden action (Holmstrom, 1979). Insurers are faced with the same hidden action problem as government regulators with limited enforcement resources at their disposal. Monitoring may have major implications for insurance affordability because of its impact on premium loading. The use of third party auditing in tandem with insurance instruments has been suggested as one method for addressing hidden action when risk is highly conditioned on management (Kunreuther, McNulty and Kang, 2002), but this too comes at a cost to the insured. If affordable insurance policies cannot be designed with sufficient incentives for maintaining good management, the provision of insurance under these conditions may actually create incentives for poor management when there is an expectation of compensation. If poor management is the result of insurance with weak monitoring provisions, this would clearly contribute to the risk of outbreak and may inhibit containment efforts even beyond what would occur without the insurance. Investments in the food safety system may reduce information asymmetry and the transaction costs involved in monitoring.

Public and Private Investments in the Food Safety System

Public and private investments in the food safety system provide resources to minimize the impact of disease outbreak. Information systems and risk management planning are important factors that can help reduce the length of time it takes to traceback food-borne illness or animal disease to their origin in the food production system. The experience of other countries in dealing with livestock disease epidemics (see Table 4.1) demonstrates the importance of response time in minimizing the duration and cost of a disease event to society. There are public and private benefits associated with investment and participation in the food safety system.

---

[3] See Green, Driscoll and Bruch (this volume) for detailed discussion regarding data requirements.

Public investments in information systems for traceback represent investments in the public good dimensions of livestock disease control. By instituting a system that may not be provided by the market alone, government can absorb some of the transaction costs and reduce information asymmetry: both obstacles to insurance design. Elements of the information system for traceback include animal identification (premises, group/lot, or individual), animal testing and auditing for *ex ante* mitigation similar to Hazard Analysis of Critical Control Points (HACCP) in livestock slaughter and processing for food-borne illness detection and prevention. Information systems involve all levels of production and processing, and facilitate the discovery and trace-back necessary for a fast response to outbreak.

Information systems like the National Animal Identification System being developed in the USA under the US Animal Identification Program (see http://www.usaip.info/), facilitated by USDA, are partially funded by government and partially funded by the private sector, and can include all stages in the lifespan of a livestock product from the farm or ranch to the consumer. These types of systems provide a vast amount of information that has not been previously available – from producer-level herd health and nutrition management, to transport and movement of animals in finishing and slaughter to domestic or export markets. With traceback ability, a weakness in the food system may be identified anywhere along the path from production to consumption. Auditing similar to what occurs through HACCP provides a method for risk mitigation where technical expertise can be brought to bear to help identify a remedy for any deficiency (Skees, Botts and Zeuli, 2001). In addition to housing information that facilitates the disease identification and eradication objectives of government, private benefits are realized when greater amounts of production information become available through record keeping.

Good managers have firm-level economic incentives to invest time and resources in traceback. There are economic returns to management when improved herd health measures are taken (Buhr *et al.*, 1993; Marsh, 1999; van Schaik *et al.*, 2001). These returns can be achieved through production efficiency improvements identified by utilizing the veterinary, breeding, feeding and animal movement or transport information available through improved record keeping. If farm or ranch profitability can be improved and the desired positive externalities from biosecurity and herd health management are generated, the producer and society can benefit from investments in traceback. Greater availability of information may also help a producer identify a potential or emerging disease problem that may have otherwise gone unnoticed and creates an opportunity for early disclosure to occur.

Public investment in information systems for food safety and traceback is believed to create private incentives for improved health management and early disclosure if public policy and insurance are developed in an incentive-compatible manner. The importance of traceback in the duration of a disease event is as crucial to producers as it is to society as a whole because it determines the response time and, in turn, the ultimate cost of the outbreak to an individual producer.

Financing is Fundamental

Government compensation in the USA for animal loss as a result of mandated depopulation has been *ad hoc* because compensation schedules are developed *ex post*. A government indemnity program that acts as a free insurance policy for producers creates an expectation of compensation if an outbreak occurs – potentially counteracting the desired incentives for good health management and early disclosure.

An established indemnity pool of a known size with pre-determined compensation rules is desirable so that producers have *ex ante* incentives for good management. Part of structuring incentives for management and disclosure includes creating a reliable expectation for the producer that reduces uncertainty about the outcome of a disease outbreak. One mechanism for creating a dynamic that introduces a tension, similar to that created by the presence of prices in the market, is the establishment of a program similar to a check-off where a per head or per hundredweight assessment is paid by the producer into a standing federal indemnity fund for compensation. This type of funding mechanism links the benefits and costs of the proposed assessment, as distinct from the concentrated benefits and diffuse costs that are characteristic of government subsidization of agriculture. It would be important that assessment amounts vary across regions to reflect differences in risk exposure. Higher (lower) risk regions would pay higher (lower) assessments. While a regional indemnity pool would not generate all the incentive compatibility benefits of very stringent insurance policy underwriting, the transactions costs would be significantly lower.

The animal compensation check-off program may not be considered politically feasible, but it is thought that in the absence of insurance it may represent an improvement in incentive compatibility over the present *ad hoc* government indemnity policy. The economic question is whether a check-off program funded by producers will meet the objectives of public policy and whether or not it achieves those objectives more efficiently than the compensation scheme that already exists.

The efficiency of market insurance that is incentive compatible is inhibited by transaction costs in information collection for insurance design and monitoring to overcome hidden action. Given the perceived political and economic obstacles to a producer-funded assessment, the cost to society of an incentive-incompatible *ad hoc* government compensation scheme coupled with insurance may exceed that of a disease outbreak under *status quo* government compensation without insurance. The cost of operating any social arrangement must be taken into account, including the cost of transition to a new system, when devising or choosing among different arrangements (Coase, 1988).

Conclusion

When designing risk management mechanisms for contagious livestock diseases, a goal is to encourage individual producers to improve management in ways that facilitate herd health and biosecurity measures while also providing incentives for early disclosure in the event of a suspected disease occurrence. Attaining this goal is complicated by the presence of information asymmetry. Frequency and severity of disease outbreak are heavily dependent on individual behaviour, which is costly for regulators or insurers to monitor and correct. Disease transmission characteristics and the effectiveness of containment efforts once an outbreak is discovered are sources of uncertainty in determining the extent of a disease event. Livestock disease epidemics represent a unique agricultural risk that poses challenges for public decision makers and the risk management industry alike.

If it is economically feasible to design public policy and market (insurance) mechanisms that are incentive compatible, the negative external effects from an outbreak may be minimized and society is likely to benefit. If, however, insurance is introduced that is incompatible with the incentives provided by public policy (regulation and *ad hoc* disaster payments), the effect of a disease outbreak could be exacerbated and result in a larger problem than would occur in the absence of insurance. In addition to incentive compatibility, society must consider whether the transaction costs involved in operating any social arrangement (not just the availability of indemnification for producers) justify the benefits provided.

This chapter has presented a deliberate approach to mechanism design for public policy and market instruments that is well suited to addressing incentive compatibility issues. The goal of animal health public policy has been to reduce the risk of livestock disease outbreak and to have measures in place that will effectively limit the spread of disease if outbreak does occur. If either policy or insurance is designed without consideration of the full scope of incentives present in the livestock production system, it is possible to increase the risk present in the animal health system. Successful *ex ante* risk mitigation must take into account the incentive effects of public and private mechanisms that influence producer behaviour and could:

1. Reduce incentives for individual producers to improve their health management and biosecurity, and
2. Worsen the effect of an outbreak rather than foster early disclosure for timely containment and eradication of a disease or suspected disease.

The economic risks of contagious livestock disease extend far beyond livestock producers and processors, and could potentially impact a number of seemingly unrelated economic sectors, including tourism. Government and market-based solutions that can improve the effectiveness of risk mitigation are desired, but care must be taken so that the combination of risk management tools available to producers are incentive compatible and feasible in the presence of potentially high transaction costs.

References

Bicknell, K.B., Wilen, J.E. and Howitt, R.E. (1999) Public policy and private incentives for livestock disease control. *Australian Journal of Agricultural and Resource Economics* 43, 501-21.

Buhr, B.L., Walker, K.D., Kliebenstein, J.B. and Johnson, S.R. (1993) An industry-level economic conceptual model of the effects of improved animal health. *Preventive Veterinary Medicine* 16, 3-14.

Coase, R.H. (1988) The problem of social cost. *The Firm, the Market, and the Law.* The University of Chicago Press, Chicago, Illinois.

Holmstrom, B. (1979) Moral hazard and observability. *Bell Journal of Economics* 10, 74-91.

Horst, H.S., deVos, C.J., Tomassen, F.H.M. and Stelwagen, J. (1999) The economic evaluation of control and eradication of epidemic livestock diseases. *Scientific and Technical Review* 18, 367-379.

Kunreuther, H.C., McNulty, P.J. and Kang, Y. (2002) Third-party inspection as an alternative to command and control regulation. *Risk Analysis* 22, 309-318.

Livestock Marketing Information Center. (2004) Database entries for futures prices, cash prices, and live weight of cattle slaughtered. [Accessed 2004.] Available from http://www.lmic.info/.

Marsh, W. (1999) The economics of animal health in farmed livestock at the herd level. *Scientific and Technical Review* 18, 357-66.

Ramsay, G.C., Philip, P. and Riethmuller, P. (1999) The economic implications of animal diseases and disease control at the national level. *Scientific and Technical Review* 18, 343-56.

Samuelson, P.A. (1947) *Foundations of Economic Analysis*. Harvard University Press, Cambridge, Massachusetts.

Skees, J.R., Botts, A. and Zeuli, K. (2001) The potential for recall insurance to improve food safety. *The International Food and Agribusiness Management Review* 4, 99-111.

United Nations Food and Agricultural Organization (2002) Animal diseases: implications for international meat trade. Committee on Commodity Problems, Intergovernmental Group on Meat and Dairy Products, Nineteenth Session, Rome, August 27-29, CCP:ME 02/3.

United States Department of Agriculture, Economic Research Service (2003-4) Livestock and Poultry Marketing Reports. December 2003, January 2004, February 2004 and March 2004.

van Schaik, G., Nielen, M. and Dijkhuizen, A.A. (2001) An economic model for on-farm decision support of management to prevent infectious disease introduction into dairy farms. *Preventive Veterinary Medicine* 51, 289-305.

Wolf, Jr., C. (1988) *Markets or Governments: Choosing Between Imperfect Alternatives.* Rand Corporation Research Study Series, MIT Press, Cambridge, Massachusetts.

World Organization for Animal Health (formerly known as the Office International des Epizooties). (2004) Diseases Notifiable to the OIE. [Accessed 2004.] Available from http://www.oie.int/eng/maladies/en_classification.htm.

Chapter 5

Insurability Conditions and Livestock Disease Insurance

Saleem Shaik, Barry J. Barnett, Keith H. Coble, J. Corey Miller and Terry Hanson

Introduction

The Agricultural Risk Protection Act (ARPA) enacted in June 2000 mandated that the US Department of Agriculture's (USDA) Risk Management Agency (RMA) conduct pilot programmes on risk management tools that would protect livestock producers from price or income fluctuations and production losses. The legislation also authorized RMA to conduct pilot programmes on risk management products for livestock disease perils and various risks faced by aquaculture producers.

To date, RMA has approved two pilot livestock insurance policies: a Livestock Risk Protection (LRP) policy that protects against declining prices for swine, feeder cattle and fed cattle, and a Livestock Gross Margin (LGM) policy that protects the gross margin between the price of fed hogs and the cost of maize and soybean meal. Both products are available only in selected areas. RMA has also funded studies on the feasibility of aquaculture insurance, risk management alternatives for livestock disease risk, non-insurance livestock risk management tools and tools for managing pasture/range and forage production risk.

Risk

Individuals typically become concerned about risk when one or more of the potential outcomes are unfavourable. Thus, when someone says that an individual is "taking a risk", they typically mean that the individual is faced with a situation where there is at least some chance that a loss will occur (Hardaker *et al.*, 2004).

Economic decision-makers are often risk-averse. This implies that given the choice between a less risky alternative and a more risky alternative, they will often choose the less risky alternative even when the expected return is less than the expected return of the more risky alternative. This is why insurance markets exist. Risk-averse individuals are willing to pay an insurance premium to reduce their risk exposure.

From a societal perspective, risk reduces incentives for investment and economic growth (Arrow, 1964; 1996). A simple example demonstrates this point. An individual would be less likely to invest in constructing a building if he/she had to assume the risk of losing the building to a fire. However, if that risk could be shared with a larger pool of investors through the purchase of fire insurance, the individual would be more willing to invest in constructing the building. The point of this example is that economic growth can be stimulated by markets that effectively transfer risk from more risk-averse individuals to less risk-averse individuals. Conversely, the lack of risk transfer mechanisms can retard economic growth.

Of course, entrepreneurship is all about taking risks. This is evidenced by common business maxims like "nothing ventured, nothing gained". Typically, expected returns are positively related to risk. Anyone who has an investment portfolio has been confronted with the risk-return trade-off. Investments that have the highest potential for growth typically are also quite risky. Low-risk investments, such as savings accounts, typically offer lower expected rates of return. Recognizing this trade-off, economic decision-makers do not attempt to eliminate all risk. Instead, they attempt to manage their exposure to risk so as to not take on any more risk than what is required to attain the desired rate of return.

Risk Management and Insurance

In general, economic decision-makers manage risk through mitigation, transfer and/or retention. Mitigation involves activities that reduce the likelihood of loss and/or the magnitude of loss. This may include establishing and following certain best management practices, as well as training employees on the implementation of those practices. It may also include investments in improved production, sanitation or safety technologies. Regardless, risk mitigation typically imposes costs on economic enterprises.

Risk can also be managed through markets that for a fee will transfer risk from those who are less willing to accept it to those who are more willing to accept it. Examples of risk transfer mechanisms include cash forward contracts and hedging using futures or options contracts. Recent years have witnessed growth in markets for innovative risk transfer instruments like catastrophic bonds (Lewis and Davis, 1998; Lewis and Murdock, 1999; Cummins, Lalonde and Phillips, 2000) and weather derivatives (Muller and Grandi, 2000; Martin, Barnett and Coble 2001; Turvey 2001; Richards, Manfredo and Sanders, 2002; Varangis, Skees and Barnett, 2002).

Insurance is probably the most common market-based risk management instrument. Insurance is a legal contract whereby risks are transferred from one party to another in exchange for a premium. Specifically, insurers agree to pay policyholders an indemnity contingent on a measurable index exceeding some threshold. Typically, the index is a measure of losses resulting from specified perils. Insurance purchasers are sufficiently risk averse that they are willing to accept a small loss with certainty (the insurance premium) rather than face the risk of a much larger loss.

Insurance is a unique form of risk transfer in that it typically involves pooling risks. If the loss risks on insurance policies of individuals in the pool are not highly correlated, the statistical law of large numbers implies that pooling can reduce the variance of the insurer's portfolio below the average of the individual policies. Thus, by pooling less than perfectly positively correlated risks, insurers actually reduce society's aggregate exposure to the risk being insured. However, as discussed in the following sections of this chapter, different perils have different risk characteristics and not all perils are insurable.

Retention is a final risk management strategy. Economic decision-makers may choose to simply retain all or part of their risk exposure. Retention may be the result of a deliberate decision based on careful consideration of the likelihood of loss, the potential magnitude of loss, and the costs associated with risk mitigation or risk transfer. However, retention can also occur unknowingly if decision-makers fail to recognize their risk exposure or fail to consider available risk migration and/or risk transfer opportunities. Further, the option to transfer risk is sometimes unavailable because no market exists.

Feasibility of Livestock Disease Insurance

Insurance for livestock disease or death is currently available only on a very limited basis. Most policies are sold for high-valued sporting, companion or breeding animals. Insurance coverage for production animals is generally not available or is available only for losses caused by very specific perils.

As indicated earlier, ARPA directed RMA to investigate the feasibility of insurance coverage for production livestock. When considering the feasibility of new livestock disease insurance products, a number of important questions should be asked. The answers to these questions are critical to determining whether the disease is insurable and, if so, whether sufficient demand will exist for the proposed insurance product. We discuss three important questions in the following sections:

1. What is the peril of concern?
2. What are the risk characteristics of the peril?
3. What risk mitigation or transfer mechanisms currently exist?

What is the Peril of Concern?

One of the first determinations that must be made in product development is exactly what perils will be covered. Failure to include specific language about covered perils will complicate loss adjustment and increase the potential for litigation, as well as make the establishment of accurate premium rates extremely difficult. For example, will the insurance product only cover losses caused by a specific disease, a family of diseases, a set of diseases based on the OIE (Office International des Epizooties or World Organization for Animal Health) Classification of Diseases, or will the product cover losses from all diseases?

What are the Risk Characteristics of the Peril?

In terms of the peril, the insurance developer should know how frequently the disease occurs and the magnitudes of the resulting losses. In other words, what is the probability distribution of insured losses from this disease? Sufficient data must be available to estimate this probability distribution, which may vary across geographic regions, different production systems or even different managers.

Is the disease endemic or exotic? The peril's insurability will be affected if losses incurred by one producer are largely independent of other producers (i.e. the risk is idiosyncratic), or if an outbreak of the disease would likely affect many insured producers simultaneously (i.e. the risk is systemic). The nature of the disease will be determined by the factors that may affect the frequency of occurrence over time (e.g. population changes in animals that serve as disease vectors), how contagious the disease is among production livestock and whether or not the disease can be passed from livestock to humans. Of course, a potentially overriding factor may be the extent to which the disease can be controlled through proper management.

What Risk Mitigation or Risk Transfer Mechanisms Currently Exist?

The potential demand for the proposed insurance product should always be a primary consideration. Demand can be affected by the risk mitigation techniques livestock producers currently employ to protect against the disease, as well as other mechanisms that currently exist to transfer the loss risk to either private markets or the public sector. For example, price risk insurance would likely overlap with the protection afforded by futures markets. All else being equal, the more affordable are available risk mitigation and/or risk transfer mechanisms, the less the potential demand for livestock disease insurance.

Other factors affecting demand include the level of diversification among livestock producers who are the target market for this product, and how much of their household income depends on production from the livestock species at risk from the disease. All other things being equal, the more potential purchasers' incomes are diversified across various sources, the less the potential demand for livestock disease insurance.

Insurability Conditions for Livestock Disease Insurance

Not all perils are insurable. Through the years insurance experts have identified a number of ideal conditions that must be met for a peril to be considered insurable (Rejda, 2001). In the remainder of this section we describe these conditions.

Determinable and Measurable Loss

Can one clearly determine that a loss has occurred? If a loss has occurred, can one clearly determine whether or not the loss was caused by an insured peril? Can the magnitude of the loss be accurately measured? If the answer to any of these questions is "no" it will be very difficult to conduct loss adjustment on insurance claims. Frequent disagreements between the insurer and policyholders will lead to costly litigation regarding whether an indemnity is due or the magnitude of the indemnity. An important insurance principle is "One can't insure what can't be measured".

Accidental and Unintentional Loss

Insurance is designed to pay indemnities only for losses that are accidental and unintentional. However, in some cases, policyholders can engage in hidden actions that increase the probability of loss and/or the magnitude of loss. These hidden actions often involve reduced effort or reduced investment in loss mitigation. For example, an individual who purchases livestock disease insurance may be less careful about following recommended disease-control sanitary practices. These behavioural changes are often called "moral hazard". Fraud is an extreme example, but moral hazard does not necessarily involve illegal or even unethical behaviour. The changed behaviour may simply be a rational economic response given that the individual has purchased an insurance policy.

A problem occurs when insurers cannot effectively monitor these changes in policyholder behaviour and enforce the relevant policy provisions. As a result, indemnities will be greater than expected for reasons that are hidden from the insurers. Assuming they have simply underestimated the risk, insurers will increase premium rates for all policyholders. Due to the higher premium cost, some policyholders will decide to no longer purchase the insurance. All other things being equal, those who are not practising moral hazard will be more likely to cease purchasing insurance while those who are practising moral hazard will be more likely to continue purchasing insurance. Thus, over time, the pool of insurance purchasers will grow smaller and consist more and more of those who are practicing moral hazard.

Deductibles and/or co-payments reduce incentives for moral hazard because the policyholder is forced to absorb a portion of the loss. Insurers may also use good experience discounts or bad experience surcharges to discourage moral hazard. Finally, insurers may attempt to increase monitoring or auditing of policyholder behaviour. This, however, can be quite expensive and will increase the premium charged to policyholders.

The more that policyholders can affect loss outcomes by their management decisions (i.e. the less that losses are accidental and unintentional), the more insurers will be exposed to moral hazard. This is reflected in the insurance principle that says, "You will go broke insuring against poor management".

We finally note an inherent tension exists between precluding moral hazard and encouraging producers subject to quarantine to reveal diseased animals. Various indemnification and insurance programmes have been instituted in part to encourage disclosure. Larger deductibles or co-payments may reduce incentives for moral hazard, but doing so reduces the incentives for early disclosure of suspected disease problems.

Sufficient Information to Conduct Risk Classification

An insurer must be able to accurately classify potential policyholders according to their risk exposure. Those faced with higher (lower) risk exposure will be charged higher (lower) premium rates. If insurers do not have sufficient information to conduct accurate risk classification, adverse selection problems will occur. This is particularly true if potential policyholders have better information about their risk exposure than the insurer. Those who have been misclassified to their benefit (charged premium rates that are too low) will be more likely to purchase insurance, while those who have been misclassified to their detriment (charged premium rates that are too high) will be less likely to purchase insurance. The result will be higher than anticipated indemnities. The insurer, not understanding this was caused by an underlying classification problem, will simply raise premium rates. But this only compounds the problem since those who continue to purchase insurance at the higher premium rates are those whose risk exposure has been most severely underestimated by the insurer.

In the case of livestock disease insurance, accurate risk classification would likely require insurers to make large investments in the collection and analysis of relevant data. This expense would be passed on to policyholders through higher premium rates. Further, it would likely require intrusive inspections of policyholders' livestock production facilities. Mandatory government regulation of livestock management practices could make it easier to conduct risk classification. Similarly, participation in voluntary Best Risk Management Practice (BRMP) certification programmes that require third-party audits could be used in classifying risk exposure.

Sufficient Data to Establish Accurate Premium Rates

Assuming potential policyholders can be accurately classified, an insurer will subsequently establish premium rates *for each risk classification.* To do this the insurer must be able to accurately estimate the frequency and severity of loss. Ideally, an insurer would estimate the statistical probability distribution of loss for each risk classification. And if risk exposure varies across regions, the insurer would estimate the probability distribution of loss by region and risk classification.

The challenge faced by insurers is finding sufficient data to accurately estimate these probability distributions. In many cases, these data simply do not exist. If they exist, they may not be available for the region of interest or the number of observations may be insufficient to allow for accurate estimates of probability distributions. This would certainly be the case for many livestock diseases, particularly rare or exotic diseases.

When considering this insurability condition, categorizing different disease perils according to data availability and the resulting confidence one has in the accuracy of estimated premium rates becomes useful. On a continuum "pure risk" and "uncertainty" are the extremes with "ambiguity" in between (see Knight, 1921).

Pure risk exists when a loss is not certain, but sufficient data are available to accurately estimate the probability distribution of loss. However, in reality we are never able to know the 'true' probability distribution. Uncertainty occurs when both the insurer and potential policyholders recognize some potential for loss exists, but objective data are not available to estimate the probability distribution of loss. In an extreme case of uncertainty, the insurer and potential policyholders cannot even identify all of the potential perils that may cause a loss, much less objectively estimate a probability distribution of losses caused by each of those perils. Ambiguity occurs between pure risk and uncertainty and reflects a situation where relevant data are available, but limited (Ellsberg, 1961). Formal incorporation of risk and uncertainty into economic theory was accomplished by von Nuemann and Morgenstern (1944).

Most livestock disease perils would likely fall somewhere in the ambiguity portion of the continuum. Some data will be available, but likely not sufficient to allow for highly accurate estimates of the probability distribution of loss for each risk classification. Questions may be raised about whether historical data accurately reflect current disease risk; that is, how stable is the probability distribution of loss over time? The available data may be from a different country or different region of the USA, raising questions about the spatial stability of the probability distribution of loss. Extremely rare or exotic diseases will tend more toward the uncertainty end of the continuum while more common diseases (for which more data exist) will tend more toward pure risk.

Insurers typically load premium rates inversely with the amount of data available to generate the rates. This is consistent with the insurer subjectively perceiving there is additional probability of loss not quantified. Perils that tend toward pure risk will be assigned relatively low premium loads since insurers believe they have captured the full probability distribution with their analysis. Those that tend toward the uncertainty end of the continuum will be assigned relatively large premium loads. In extreme cases of uncertainty, insurers may simply refuse to insure the peril or may assign such high premium loads that the insurance becomes unaffordable. Until additional data become available, insurance markets for these perils will simply not exist.

Losses Sufficiently Uncorrelated to Allow for Pooling

Ideal insurability conditions require losses on any one policy be less than perfectly positively correlated with losses on any other policy. When this condition is met, the statistical law of large numbers causes the variability in annual losses on the aggregate portfolio of policies to decrease as the number of policies in the portfolio increases. While insurers cannot accurately estimate whether a loss will occur on any given policy in any given year, they can accurately estimate annual losses on the aggregate portfolio of policies. Automobile and life insurance are examples of insurance lines that come close to meeting this idealized condition.

For most perils, losses for one policyholder will not be completely uncorrelated with losses for other policyholders. As long as losses are not perfectly correlated, pooling will reduce the variability in annual losses for the overall portfolio of policies. However, the more correlated the losses (i.e. the more systemic the risk), the less pooling reduces the variability in annual losses for the overall portfolio. When losses for the overall portfolio are highly variable, the insurer must hold large reserve funds and/or purchase reinsurance to protect against the occurrence of large correlated losses. This increases the cost of providing insurance and drives up premium rates.

In the USA, the RMA provides reinsurance[1] to offset much of the systemic risk faced by private issuers of multiple-peril crop insurance policies. Similar reinsurance arrangements could possibly be used to transfer the systemic risk inherent in many livestock diseases to the federal government.

Economically Feasible Premium

The final insurability condition is that potential policyholders must be willing to pay the required premium. When a peril deviates from the idealized insurability conditions described above, higher insurance premium rates are typically required. For perils that deviate significantly from the idealized conditions insurance markets will not exist because the required premium rate will exceed what potential policyholders are willing to pay.[2]

It is also important to note that willingness to pay for insurance protection depends on the potential buyer's financial portfolio. All other things being equal, the more the portfolio is diversified across alternative agricultural enterprises or off-farm sources of income, the less a potential buyer will be willing to pay for insurance. For example, assuming identical levels of risk exposure to a particular cattle disease, a farmer in the eastern USA who has 50 head of beef cattle, produces ten acres of tobacco and has an off-farm job will likely be willing to pay less for livestock disease insurance than a rancher in the western USA whose income is solely dependent on beef cattle production.

Multiple-peril crop insurance deviates significantly from many of these idealized insurability conditions. To increase purchasing of multiple-peril crop insurance, the US government subsidizes the premiums paid by farmers. It has been widely argued that these

---

[1] Details on the USDA-RMA reassurance agreements are available at http://www.rma.usda.gov/pubs/ra/#overview. [Accessed 2004].

[2] It is important to note that other risk management instruments (i.e. options, bonds, derivatives, etc.) might be used to provide risk protection even when the underlying peril deviates from one or more of the insurability conditions described above. This will depend on the characteristics of the risk.

subsidies, which are in addition to expense reimbursements provided to private insurers and the costs associated with providing reinsurance, encourage crop farmers to expand production into riskier areas (Orden, 2001; Skees, 2001; Glauber and Collins, 2002). With livestock diseases, some of which are highly contagious, incentives for producers to take on additional risk could exacerbate animal health problems.

Potential Livestock Disease Risk Management Tools

This section presents a number of existing or potential risk management tools that either independently or in combination may be useful in protecting livestock producers against disease-related disasters. These tools include Animal and Plant Health Inspection Service (APHIS) indemnification programmes, private disease insurance, government insurance, wrap-around disaster insurance programmes, business interruption insurance, standing market loss programmes and industry exit assistance.

APHIS Indemnification Programme

APHIS provides leadership in ensuring the health and care of animals and plants by safeguarding US borders against the entry of foreign agricultural pests and diseases. The APHIS indemnification programme attempts to limit the spread of animal diseases and also compensates producers for some losses. Included in the measures to control the spread of disease are depopulation, cleaning and disinfecting. State and federal governments fund indemnities to compensate owners for depopulated animals.

Indemnity payments are typically a percentage of the appraised or market value of the depopulated animals, adjusted for salvage value or other payments received by the owner. Indemnities are generally paid only on the value of the animal, but in some cases, payments are also made to cover additional costs related to depopulation. For example, producers may be indemnified for property or materials that must be destroyed. Producers may be required to clean and disinfect premises, but these costs are rarely compensated. APHIS does not provide compensation for business interruption or herd re-establishment costs.

Private Disease Insurance

Animal disease insurance is offered to a limited market segment of livestock producers. Animal mortality insurance tends to cover multiple perils, but often has several provisions to mitigate moral hazard. This product is generally customized to the producer's needs and therefore is a relatively high transaction cost product. This limits the market to high-value sporting, companion or breeding animals. In many instances these policies specifically exclude coverage for losses due to events indemnified by APHIS.

Any expansion of private disease insurance would likely be targeted to disease outbreaks that are not indemnified by APHIS. Further, coverage would likely be limited to diseases that are not highly contagious and not highly conditioned upon management. Coverage for highly contagious diseases would create potential for huge losses in any given year. Coverage for diseases that are highly conditioned on management would expose the insurer to moral hazard problems. These violations of insurability conditions would greatly increase the transactions costs of offering the insurance coverage. The premium required to cover these costs would likely exceed buyers' willingness to pay for the insurance coverage.

Government Disease Insurance

Government provision of livestock disease insurance can address some violations of insurability conditions. For example, correlated losses are a concern for private insurers because the potential for huge payouts could bankrupt the insurance company. This was a problem with early private-sector efforts to offer multiple-peril crop insurance in the USA, but the federal government cannot go bankrupt like a private company. The government may also be less likely to load premium rates for uncertainty or ambiguity caused by sparse data or transactions costs associated with risk classification or mitigating moral hazard. The government may even subsidize premiums. All of these actions could lead to more economically feasible premium rates.

To avoid duplication, government insurance would cover disease losses not indemnified by APHIS. Further, we assume government insurance would be a standing programme rather than an *ad hoc* programme. Thus, standing authority would exist such that permanent rules and guidelines could be established as in the current crop insurance programmes. This would afford producers the confidence of a standing programme, and USDA the timeframe to implement an equitable and well-designed programme.

However, experience with the federal crop insurance programme indicates government provision does not address all insurability shortcomings. Risk classification remains a difficult challenge in the federal crop insurance programme (Skees and Reed, 1986; Quiggin, Karagiannis and Stanton, 1994; Just, Calvin and Quiggin, 1999) and, given the importance of management in controlling animal disease, may be an even more difficult challenge with livestock disease insurance. Some have suggested limited incentives exist for moral hazard control in the government-facilitated crop insurance programme (Chambers, 1989; Quiggin, Karagiannis and Stanton, 1994; Smith and Goodwin, 1996; Coble *et al.*, 1997).

Each of these problems causes inequities across policyholders in benefits received from the government-facilitated insurance programme. In addition, each may generate incentives for increased risk-taking by insured producers. In the context of contagious livestock diseases, incentives for individual producers to take on more risk can clearly increase the overall risk exposure for the entire sector (including both insured and non-insured producers). Finally, it is important to mention that government-facilitated insurance programmes can be very expensive.

Wrap-Around Insurance for High-Value Animals

APHIS indemnity payments for depopulated livestock are currently based on indemnification schedules developed at the time of the outbreak. However, producers often complain these indemnification schedules do not fully account for differences in the value of depopulated animals. For high-value animals, such as breeding stock or even production animals with highly desired traits, APHIS indemnification may fall far short of the actual value of the depopulated animals.

In principle, owners of high-value animals should be able to purchase insurance (either government or private) that would compensate for at least part of the shortfall in APHIS indemnification. Offering this insurance would be much easier if APHIS indemnification was a standing (rather than *ad hoc*) programme with indemnification schedules that were defined prior to the loss event.

A significant challenge exists in establishing schedules for APHIS indemnification programmes, private or government disease insurance, or wrap-around disaster insurance programmes. If indemnification is too high, diseased animals become more valuable than healthy animals, creating incentives for moral hazard behaviour. If indemnification is too low, producers will be reluctant to disclose potential disease problems. Instead of settling for below-market rates of indemnification on depopulated animals they may try to rush the animals to market while they are still asymptomatic. Thus, indemnification schedules that are either too lucrative or too frugal could exacerbate disease problems throughout the sector.

Business Interruption Insurance Programme

As described earlier, existing APHIS programmes compensate producers for part of the direct losses incurred due to depopulation. However, existing government programmes seldom cover indirect losses. If livestock are in a quarantine zone, producers may not be able to market the animals at the desired weight or during the desired time frame. Disease outbreaks may also reduce consumer demand for livestock products. Depending on the nature of the disease, demand reductions may occur only in the infected region or may affect all producers.

Other business interruption losses include the maintenance cost of non-infected livestock, restrictions on the movement of livestock during the layoff period after cleaning and disinfecting diseased animals, declines in profit margins due to non-acceptance of products and costs of rebuilding livestock herds. The need for business interruption insurance has become a public policy issue in several European countries infected by Foot and Mouth Disease since, in the event of an outbreak, there is a very high likelihood that a producer will be impacted by a quarantine. Insurance companies currently offer business interruption insurance for several industries outside agriculture. It may be possible to modify these existing policies to provide livestock disease business interruption insurance.

Standing Market Loss Programme

We define a standing market loss programme as a government programme that compensates producers for price shocks caused by disease outbreaks. Some animal diseases (particularly those that can be passed to humans) may cause relatively few actual animal losses, but result in significantly lower market prices. A key issue with such losses is that they are extremely correlated because agricultural prices are spatially correlated. As a result livestock prices clearly violate the insurability criterion of non-correlated losses, making the existence of private price insurance highly unlikely. Futures and options markets provide mechanisms for managing price risk on some types of livestock. However, these markets do not exist for all animal agriculture industries. The standing market loss programme envisioned here would be a government programme to mitigate catastrophic livestock price declines attributable to disease events. We presume the federal government has the "deep pockets" to withstand the correlated losses that may occur in the event of a major price shock.

Government-Funded Industry Exit Assistance

The final design we consider is a government-funded industry exit assistance programme. As we envisage it, a programme of this type would facilitate producers leaving animal production because disease has become endemic in the producer's particular location. A programme of this type would be used only in those cases where remediation is cost-prohibitive. The programme would attempt to facilitate the movement of capital and labour into other agricultural or non-agricultural uses. For example, these funds might be used for technical training to assist farm labour in relocating to other employment.

A Proposed Matrix for Integrating Disease Risk Management Tools

We have argued that disease risks will vary in the extent to which they meet insurability conditions. We have also suggested a variety of risk management tools – including, but not limited to insurance products – that could be utilized to help livestock producers manage disease risk. Some of these tools would require a significant role for government while others would not.

In this section we contend no single private or government-facilitated risk management tool will "work best" for all possible livestock diseases. Rather, *the appropriate tool depends critically on the risk characteristics of the disease.* To demonstrate this, we utilize the matrix presented in Table 5.1.

This matrix sorts animal disease risk exposure into two primary categories and ten subcategories. The primary categories distinguish between production losses and market losses. By "production losses" we mean animals that die from the disease and animals destroyed either because they are infected with the disease or may have been exposed to the disease.[3] By "market losses" we mean situations where, as a result of a disease outbreak, producers are unable to market animals or can do so only at reduced market prices.

Under production losses we define five subcategories based on the risk characteristics of the disease and whether indemnification is provided by APHIS. The first subcategory includes those animal diseases completely controllable with proper management. Mastitis in dairy cattle might be an example. The second subcategory includes disease events indemnified by APHIS. The third subcategory includes widespread disease outbreaks not indemnified by APHIS. The fourth subcategory includes localized disease events not indemnified by APHIS. The fifth subcategory includes diseases endemic and persistent to the point where production losses are almost certain.

Five subcategories of market losses are also defined. The first subcategory is market losses resulting from time lags in a producer's ability to repopulate following a required depopulation. A producer whose diseased animals have been depopulated will sometimes not be allowed to repopulate for a specified period of time. For breeding herds, additional

---

[3] We recognize that some livestock diseases do not result in death of the animal or, if the disease is not highly contagious, a need for depopulation. In cattle, for instance, diseases such as brucellosis or tuberculosis may simply reduce weight gain or other performance measures. However, our sense is that, in areas where these diseases are endemic, production losses caused by these diseases are viewed as a cost of doing business rather as a financial risk to be protected against.

Table 5.1. Proposed Matrix for Integrating Disease Risk Management Tools.

Producer Loss Exposure	Risk Characteristics	Insurability Conditions						Potential Risk Management Tool
		Determinable & Measurable Losses	Accidental & Unintentional Losses	Sufficient Information to Conduct Risk Classification	Sufficient Data to Establish Accurate Premium Rates	Losses Sufficiently Uncorrelated to Allow for Pooling	Economically Feasible Premium	
Production Losses	Completely controllable with proper management	Hard to tell if loss was caused by an insured peril or poor management	No	No	Perhaps in some cases, but research required	Yes	No	No justifiable public or private solution
	Outbreaks indemnified by APHIS	Usually	Usually	Perhaps, but may be very expensive	Perhaps in some cases, but research required	Often not	Often not	Wrap-around insurance for high-valued animals
	Widespread outbreaks not indemnified by APHIS	Usually	Usually	Perhaps, but may be very expensive	Perhaps in some cases, but research required	No	Often not	Government insurance
	Localized outbreaks not indemnified by APHIS	Usually	Usually	Perhaps, but may be very expensive	Perhaps in some cases, but research required	Yes	Perhaps	Private disease insurance or government insurance
	Endemic – persistent losses	Usually	No	Yes, high risk for everyone	Yes	No	No	Government exit assistance
Market Losses	Depopulated, lag in repopulating or rebuilding genetics	Yes	Depends on disease	Depends on disease	Difficult	Depends on disease	Depends on disease	Private business interruption insurance
	Missed marketing window due to quarantine	Yes	Yes	Yes	Difficult	Depends on extent of quarantine	Perhaps	Private business interruption insurance
	Localized, short-term lower prices	Yes, if prices are observable	Yes	Yes	Difficult	Yes	Perhaps	Private insurance
	Widespread, short-term lower prices	Yes, if prices are observable	Yes	Yes	Difficult	No	No	Government standing market loss programme
	Long-term loss of market due to endemic disease	Yes, if prices are observable	No	Yes, high risk for everyone	Yes	No	No	Government exit assistance

losses may occur due to lags in acquiring breeding stock and rebuilding the genetic quality of the herd.

The second subcategory deals with missed marketing windows due to quarantine. For some livestock species, animals that grow beyond the ideal size and weight have little market value. Producers who will miss marketing windows due to quarantine restrictions on the movement of animals may destroy the animals rather than continue to incur feed costs.

Disease outbreaks may also cause market losses in the form of reduced demand and lower prices. A third subcategory of market losses occurs when a disease outbreak causes a localized, short-term reduction in demand and thus market price. For example, in 2004 an outbreak of avian influenza on the Delaware, Maryland and Virginia (Delmarva) peninsula caused several countries to temporarily ban imports of chicken from this region. While the disease was limited to a few farms, the price impacts affected all producers in the region.

A fourth subcategory of market losses occurs when a disease outbreak causes a widespread, short-term reduction in demand and market price. The 2003-2004 US experience with Bovine Spongiform Encephalopathy (BSE) in cattle was an example of such an occurrence. The vast majority of economic losses resulting from this BSE incidence were due to a loss of market demand caused by trading partners banning the importation of US beef. The resulting price impact affected all US beef producers without regard to their proximity to the area where BSE was found. A final subcategory deals with long-term losses of market access due to endemic and persistent disease problems or perceptions that production from the region may be unsafe.

Each subcategory is evaluated based on the insurability conditions described earlier. The last column of the matrix presents our suggestion of a risk management tool that may work best for diseases falling in that category. It is important to note that other categorizations and evaluation criteria are certainly possible. We offer this matrix as a starting point for discussions on how to best match disease risks with appropriate risk management tools.

Implementation of the matrix for integrated disease risk management tools to protect producers against risks caused by natural events (including diseases) depends not only on the insurability conditions, but the interaction of the possible designs. Integrating these programmes is challenging. It is easy to envision overlapping designs or gaps in coverage. It appears that significant institutional coordination would be needed. As proposed, products such as wrap-arounds need a well-defined underlying product. To achieve a workable system of insurance as defined in the matrix, both public and private institutions would need to coordinate coverages. Further, in the USA, various public entities such the Risk Management Agency, Farm Service Agency and Animal and Plant Health Inspection Service units of USDA would need to coordinate.

Challenges and Opportunities of Disease Risk Management Tools

In this chapter we have provided an overview of possible approaches to managing the economic implications of livestock disease risk. We note that traditional insurance is based on the assumption the underlying peril meets a number of idealized insurability conditions. We also suggest many livestock diseases deviate significantly from these idealized conditions. We describe a number of potential risk management tools that include but are not limited to insurance. Some of these tools could possibly be offered in private markets. Others would require significant government involvement.

The "take home" message of this chapter is no single private or government-facilitated risk management tool will work best for all possible livestock diseases. Instead, the appropriate risk management tool must be determined based on a careful analysis of the risk characteristics of the disease. The matrix presented in Table 5.1 is an effort to demonstrate this principle and initiate further efforts in delineating the risk characteristics of various livestock diseases. This matrix is, of course, only a starting point. We are confident others will improve on our efforts, but we are convinced that such careful and deliberate analysis of disease risk characteristics will be required to create a portfolio of effective and cost-efficient livestock risk management tools.

References

Arrow, K J. (1964) The role of securities in the optimal allocation of risk-bearing. *Review of Economic Studies* 31, 91-96.

Arrow, K.J. (1996) The theory of risk-bearing: small and great risks. *Journal of Risk and Uncertainty* 12, 103-111.

Chambers, R.G. (1989) Insurability and moral hazard in agricultural insurance markets. *American Journal of Agricultural Economics* 71, 604-16.

Coble, K.H., Knight, T.O., Pope, R.D. and Williams, J.R. (1997) An expected-indemnity approach to the measurement of moral hazard in crop insurance. *American Journal of Agricultural Economics* 79, 216-26.

Cummins, J.D., Lalonde, D. and Phillips, R.D. (2000) The basis risk of catastrophic-loss index securities. Working Paper, Financial Institutions Center, The Wharton School, University of Pennsylvania.

Ellsberg, D. (1961) Risk, ambiguity, and the savage axioms. *Quarterly Journal of Economics* 75, 643-669.

Glauber, J.W. and Collins, K.J. (2002) Crop insurance, disaster assistance, and the role of the federal government in providing catastrophic risk protection. *Agricultural Finance Review* 62, 81-101.

Hardaker, J.B., Huirne, R.B.M., Anderson, J.R. and Lien, G. (2004) *Coping with Risk in Agriculture,* 2nd edition. CAB International, Wallingford, UK.

Just, R.E., Calvin, L. and Quiggin, J. (1999) Adverse selection in crop insurance: actuarial and asymmetric information incentives. *American Journal of Agricultural Economics* 81, 834-49.

Knight, F H. (1921) *Risk, Uncertainty and Profit.* Houghton Mifflin Company, Boston, Massachusetts.

Lewis, C.M. and Davis, P.Q. (1998) Capital market instruments for financing catastrophe risk: new directions? *Journal of Insurance Regulation* 17, 110-133.

Lewis, C.M., and Murdock, K.C. (1999) Alternative means of redistributing catastrophic risk in a national risk-management system. In: Froot, K.A. (ed.) *The Financing of Catastrophe Risk.* The University of Chicago Press, Chicago, Illinois.

Martin, S.W., Barnett, B.J. and Coble, K.H. (2001) Developing and pricing precipitation insurance. *Journal of Agricultural and Resource Economics* 26, 261-274.

Muller, A. and Grandi, M. (2000) Weather derivatives: a risk management tool for weather-sensitive industries. *The Geneva Papers on Risk and Insurance* 25, 273-287.

Orden, D. (2001) Should there be a federal income safety net? Paper presented at the USDA Agricultural Outlook Forum, Washington, D.C.

Quiggin, J., Karagiannis, G. and Stanton, J. (1994) Crop insurance and crop production: an empirical study of moral hazard and adverse selection. In: Hueth, D.L. and W.H. Furtran (eds) *Economics of Agricultural Crop Insurance.* Kluwer Academic Publishers, Norwell, Massachusetts.

Rejda, G E. (2001) *Principles of Risk Management and Insurance.,* 7th edition. Addison Wesley, Boston, Massachusetts.

Richards, T.J., Manfredo, M.R. and Sanders, D.R. (2002) Weather derivatives: managing risk with market-based instruments. Paper presented at the NCR-134 Conference on Applied Commodity Price Analysis, Forecasting, and Market Risk Management, St. Louis, Missouri.

Skees, J.R. (2001) The bad harvest. *Regulation: The CATO Review of Business and Government,* 24, 16-21.

Skees, J.R. and Reed, M.R. (1986) Rate-making and farm-level crop insurance: implications for adverse selection. *American Journal of Agricultural Economics* 68, 653-59.

Smith, V.H. and Goodwin, B.K. (1996) Crop insurance, moral hazard, and agricultural chemical use. *American Journal of Agricultural Economics* 78, 428-38.

Turvey, C.G. (2001) Weather derivatives for specific event risks in agriculture. *Review of Agricultural Economics* 23, 333-351.

Varangis, P., Skees, J. and Barnett, B. (2002) Weather indexes for developing countries. In: Dischel, R. (ed.) *Climate Risk and the Weather Market: Financial Risk Management with Weather Hedges.* Risk Books, London.

von Neumann, J. and Morgenstern, O. (1944) *Theory of Games and Economic Behavior,* 1953 edition. Princeton University Press, Princeton, New Jersey.

Chapter 6

Issues Associated with US Livestock Disease Compensation in the 21st Century

Stephen Ott

Introduction

Controlling the spread of highly contagious livestock diseases or livestock diseases that have human health implications is not the responsibility only of livestock producers, but that of government as well. Government's role in controlling such diseases becomes important when control efforts include the destruction of healthy animals that are at high risk of becoming exposed to or infected with the disease. Individual producers do not have an incentive to destroy animals that do not express any signs of the disease. However, for the public good, the government will order the destruction of these healthy animals to stop or control the spread of livestock diseases. When such destruction takes place the government has an obligation to compensate owners for their losses. The issue then becomes what is fair both to livestock owners and to the taxpayers paying the bill. This chapter will focus on compensation for livestock when the government orders their destruction. Specific topics to be covered are: why compensation is offered, current compensation schemes from the US federal government's perspective, issues that make fair compensation difficult to achieve, and compensation alternatives.[1] This chapter arises from the author's experience in determining fair market value for cattle, deer, elk, koi, poultry and swine.

Why Compensation?

The US federal government has had a long interest in the health of the nation's livestock. In 1884 the Bureau of Animal Industry was established within the US Department of Agriculture (USDA) with the purpose of controlling and eradicating livestock diseases (Wiser, 1987). The first disease of interest was pleuropneumonia, which entered the USA in an imported dairy cow in 1843 (Wiser, 1974). With its authority to quarantine and destroy diseased livestock, the Bureau was able to eradicate the disease by 1892. Other early diseases of interest were hog cholera, Texas fever and tuberculosis. Back then, and today as well, the US federal government has focused on livestock diseases that fall into one of two broad categories:

[1] See Table 6.1 for summary of the issues covered in this chapter.

Table 6.1. Summary of Issues Surrounding Federal Government Compensation of Livestock for Animals that have been Depopulated for Disease Reasons.

Issues	Responses
Why depopulation?	To control or eradicate livestock diseases that are highly contagious or have human health implications.
Why compensation?	To encourage reporting of disease to government officials by livestock producers. Also US Constitution requires federal government to pay when it takes an asset.
How should compensation value be determined?	Fair market value; however, fair market value is not well defined.
What is fair market value?	From a strict economics point-of-view it is the estimated worth of the animal at the time of its taking. To encourage reporting by producers USDA assigns value assuming a disease free status.
Problems associated with appraisals.	Animal health officials lack training in livestock appraisal. Other experts may have financial interest in the appraisal they provide.
Standardized rates as an appraisal method.	A single value would be applied to animals that are grouped together based upon classification criteria.
Standardized rates problem.	If number of categories is few, then it is possible to have a wide range of value within the category. Should a median or mean value be used or something greater?
Getting appraisal values right.	Reduction in number of livestock markets. Commercial poultry production doesn't have markets for live birds. Exotic species can be subject to wide swings in value. Placing a value on pets.
Insurance.	Can act as substitute for government compensation, but the Constitution requires the government to pay for animals destroyed. Expansion of diseases covered by the federal government threatens current industry self-insurance programmes already in place. A more appropriate role would be to cover other losses associated with depopulation and quarantines and where there are maximum limits to individual animal compensation rates. Most likely will require some form of government subsidy as in crop insurance.
Dual-level Indemnity System.	Have low indemnity values in order to satisfy constitutional requirements combined with an optional joint industry/government-funded compensation programme.
Permit/Assurance Bond Scheme.	It is the industry's responsibility to control disease and producers who don't, lose their right to raise and/or market animals or animal products. Permit process could require assurance bonds that the producer would forfeit if disease is found in his animals.

highly contagious diseases such as hog cholera (now known as classical swine fever) or diseases that have human health implications, such as tuberculosis.

While quarantines can help prevent a disease from entering the country or crossing state lines, quarantines by themselves are generally not enough to eradicate a disease. It generally takes destruction of diseased animals and sometimes whole herds to eradicate a disease. With highly contagious diseases, the US federal government will often establish a zone surrounding the infected herd(s) and order the depopulation of all susceptible animals within the zone even if they appear to be healthy. A key component in destruction of these animals is the compensation of owners. The 5th Amendment of the US Constitution states "...nor shall private property be taken for public use, without just compensation". The importance of compensating owners for destroyed animals was recognized in the eradication of pleuropneumonia. To successfully fight this disease, the Bureau of Animal Industry's appropriations were increased. Increased funding enabled it to compensate owners for livestock destroyed and to hire more staff (Wiser, 1987). With cooperation of livestock owners, which was enhanced by compensation, the Bureau successfully eradicated pleuropneumonia within eight years. Thus, compensation was a key component in the eradication of the first livestock disease in the USA.

In addition to the constitutional requirement to pay compensation when animals are either quarantined or destroyed there are economic justifications as well.[2] One such economic justification deals with externalities associated with disease. An externality is where the actions of an individual can either positively or negatively impact others, but the individual can not capture the gain enjoyed by others nor can the others who are negatively impacted collect compensation from the individual (see Shaik *et al.*, this volume). Livestock disease reporting involves an externality. Individually, livestock producers who suspect their herd or flock may have a reportable disease have economic incentives to either ship the animals to market or to wait and hope that their suspicion is false. Either way the whole industry can be impacted negatively. If the animals are diseased and are shipped to market, the disease could be easily spread to other animals. Even if the diseased animals are not shipped to market, the disease can still be spread through feed suppliers and other interfarm traffic. Meanwhile, government and industry are denied precious time to control the outbreak while it is still limited to a few animals.

One way to encourage producers to act positively on their suspicion of disease is to offer compensation for losses associated with the discovery of disease. If livestock owners know that they will be compensated for test-positive diseased animals, then they will more likely contact a veterinarian or animal health official when they suspect disease. The US federal government has taken the view that voluntary reporting of disease is a good thing and needs to be supported. Thus, it has decided to provide compensation to producers whose animals are destroyed in the control of disease.

Current Livestock Compensation

Compensation by the US federal government is governed in part by the Animal Health Protection Act, Subtitle E of the Farm Security and Rural Investment Act of 2002, which

[2] For more information on economic justification for paying indemnity see the chapter by Sumner, Bervejillo and Jarvis (this volume).

states that "...the Secretary shall compensate the owner of any animal, article, facility, or means of conveyance that the Secretary requires to be destroyed under this section" (Public Law 107-171). Furthermore, " ...the compensation shall be based on the fair market value as determined by the Secretary..." and that "Compensation paid any owner under this subsection shall not exceed the difference between the fair market value of the destroyed animal, article, facility, or means of conveyance; and any compensation received by the owner from a State or other source for the destroyed animal, article, facility, or means of conveyance". Thus, the Act establishes two guiding principles for compensation: fair market value and reduction in payment for any other compensation received.

Generally, federal regulations assume fair market value is determined by an appraiser. For example, from Code of Federal Regulations Title 9, Part 50 "Animals for which indemnity is to be paid under this part must be appraised at their fair market value by an appraiser selected by APHIS" (US Code of Federal Regulations, multiple years). The idea is to have someone with expert knowledge of livestock values determine the worth of the animal or herd to be destroyed and provide detailed justification of those figures. In addition, the appraiser would be someone without any financial interest in the results of the appraisal, e.g. they cannot benefit from the values they determine. However, in practice, finding individuals that have both expert knowledge and financial independence can be difficult. This is especially true in exotic species where subtle differences can greatly influence fair market value. Individuals with the required expert knowledge often have some type of financial interest in their appraisal. Some experts either own or operate an auction market and may have had, or hope to have, business dealings with the owner. With exotic species, experts are often owners themselves and any valuation they provide has the potential of being applied to their own animals.

Compensation Issues

Determining Fair Market Value

At the heart of compensation is fair market value. However, neither the Animal Health Protection Act nor other federal regulations dealing with livestock disease for which compensation is to be paid define fair market value. All that is stated is that fair market value shall be determined by appraisal. With fair market value being determined by an appraiser with few or no guidelines, the appraisal process can become very subjective. Subjective appraisals have the potential for causing ill feelings towards USDA by livestock owners if they feel their animals have been appraised too low and the reasoning for the valuation is capricious. Alternatively, animals could be appraised too high which can result in over-payment and a waste of taxpayer dollars.

Economically, fair market value is known with certainty only when there is a transaction between a knowledgeable and willing seller and buyer. At all other times, fair market value can only be estimated and thus subjected to the assumptions made by the individual estimating the value. Any estimate of fair market value is subject to quality, spatial and temporal assumptions. For example, is the value of a specific animal based on its own specific characteristics, including location and day of appraisal, or can the animal be grouped with other animals of the same type with similar though not exactly the same characteristics, across a wide geographic area, and over a period of time? How this question is answered has implications for how the government organizes its efforts. If each animal must be appraised for its specific characteristics, including spatial and temporal characteristics, then the appraisal process will be complex and labour intensive as each animal is individually appraised. If, on the other hand, animals that are similar can be grouped together, then appraisal could be done by standardized or uniform rates that would greatly simplify the appraisal process.

In addition, for animals either diseased or exposed to or suspected with the disease there is a timing issue. Are the animals to be valued in their current diseased state or assumed to have a disease-free status? US federal regulations are not clear, but disease-free status is often assumed. However, from a strict economic point of view, fair market value should take into consideration the animal's disease status.

An alternative approach to fair market value is "compensation value". Compensation value would be a value that encourages livestock producers and owners to report their suspicion of disease to their veterinarian or animal health officials. To this end, compensation value for diseased animals could be above true fair market value. The idea behind compensation value is the recognition that offering compensation encourages the reporting of disease. With true fair market value taking into consideration the animal's disease status, compensation value in many cases will be above true fair market value. In reality, what the USDA often refers to as fair market value is really a compensation value that is equal to fair market value assuming disease-free status. With this definition of fair market value, claims by livestock owners with diseased animals that they didn't receive full value for their animals should be minimized.

Standardized Appraisal Values

An alternative to appraising animals on-site by someone with knowledge of value is the use of standardized appraisal values. Standardized appraisal values are not based upon the expert opinion of the appraiser, but on measurable characteristics of the animal; characteristics that would classify animals into categories. Animals in each category would then be valued the same.

Characteristics could include such items as age, sex, breed, productivity level (e.g. milk production per cow), coloration and region of the country. Characteristics would be customized to each species. An expert or group of experts would then determine value for each category or classification. The process could be further simplified so that the values for various categories are a function of some reported price series so that as market prices change, so does the standardized appraisal value.

The purposes of standardized appraisal values are several. First, they facilitate efficient and uniform compensation payments whenever USDA orders animals to be destroyed. With standardized appraisal values, the appraisal task is simplified to proper classification of animals into categories. Once an animal is classified, its value is the value for that classification, i.e. a standardized appraisal value. Such efficiency could prove critical in an outbreak of a foreign animal disease where rapid response would be vital to the successful elimination of the disease. Second, standardized indemnity values can reduce the potential of large excesses in compensation payments. Finally, standardized indemnity values can reduce USDA Veterinary Services' operational costs associated with compensation. Instead of having to pay for private professional appraisers, appraisals can now be done by Veterinary Services' field personnel.

A key to successful implementation of standardized appraisal values is assuring that they are viewed as equitable compensation values. One way to improve their acceptance is to solicit input from those with marketing knowledge of the species in question. Such individuals would include university livestock experts as well as representatives from livestock industry associations. Working in partnership with such individuals should improve the quality and accuracy of the standardized appraisal values by ensuring that the important characteristics are included and valued appropriately. Having outside experts provide assistance in the development of standardized appraisal values should make it easier to obtain the second key to successful implementation, support of industry associations. Industry associations can promote and thus increase the acceptance of standardized indemnity values by communicating with their membership that the standardized appraisal values are fair for a wide cross-section of the membership. For example, industry representatives were consulted in the design of the Accelerated Pseudorabies Eradication Program and poultry disease outbreaks in Virginia and California. All three standardized appraisal value plans were well received.

Another factor influencing acceptance of standardized values as a fair representative of market value is a proper understanding of fair market value. Producers sometimes confuse fair market value with cost of replacing with a newer asset. For example, a dairy farmer who has a third-lactation cow destroyed for tuberculosis will most likely seek to replace the animal with a springer heifer.[3] Consequently, dairy farmers may feel that when the government orders their third-lactation cow destroyed they should be paid enough to purchase a springer heifer. This is similar to expecting an insurance company to pay enough to purchase a brand new model of your three-year-old car after it has been totalled in an accident. The insurance will pay "blue book" for a three-year-old car minus any deductibles. Likewise, livestock owners of depreciable breeding stock should expect only "blue book" value for their older breeding animals. Most likely this will take some education effort on the part of USDA.

Educating producers about appraisal in general can be another key to successful implementation of standardized appraisal values. As mentioned earlier, fair market value is known only when a sale actually takes place. Consequently, there can be valid differences of opinion of the worth of any particular animal. How producers respond to these differences in opinion affects the desirability of using standardized appraisal values, especially in a large disease outbreak situation where individual animal or herd characteristics may have little effect on the value estimated. Producers may react negatively to any value that is less then their own estimation of value by appealing to have their animals appraised by an appraiser. If a large number of producers choose to appeal against the appraisal values offered, then much of the efficiency gained by using standardized appraisal values is lost. Consequently, there is incentive for the US federal government to educate producers that a) appraisals are an estimate of value and the amount offered should be accepted unless it is grossly unfair and b) that the appraisal offered by government is a firm offer and not some starting point for further negotiation of value.

Availability of Sufficient Information for Estimating Value

Lack of market prices complicates the determination of appropriate indemnity values. As livestock production becomes more industrialized with marketing contracts between producer and processors, open markets that handle large numbers of animals are dwindling

---

[3] Springer heifers are heifers that are late in their first pregnancy. They will soon give birth and start producing milk. They have all of their potential milk and calf production in front of them, and are generally worth more than cows that have already gone through a couple of lactations.

in number. Markets with many sellers and buyers are efficient in price discovery and not easily manipulated by a single buyer or seller. Such markets are said to be liquid. When the number of buyers is few or number of animals offered for sale is few there is some doubt as to whether the sale price observed is a fair representation of true value. Consequently, the disappearance of liquid markets makes it more difficult to determine fair market value. This is especially true in the poultry industry where the first price observed is often a wholesale carcass price and even that may be based on a relatively small percent of birds harvested.

When market prices are limited, the income appraisal approach may be used to determine fair market value. The income appraisal approach uses revenue and cost-of-production data and this approach was used in the 2002 low-pathogenic avian influenza (LPAI) outbreak in Virginia in chicken broilers and turkeys. An example of how the income appraisal approach works:

For meat birds,
processed bird wholesale value
 – plant processing cost
 = live meat bird value
 – total meat bird cost-of-production
 = profit margin
 × profit margin allocated to meat bird operation
 = profit of meat bird production
 + total meat bird cost-of-production
 = compensation value of meat bird at slaughter age.

For breeder birds,
profit margin
 × profit margin allocated to breeder bird operation
 × number of meat birds produced per breeder bird
 = gross value of breeder bird
 – breeder bird production costs
 + breeder bird salvage value
 = compensation value of breeder bird.

Fair market value for younger meat birds is a simple linear decline to initial placement value, where initial meat bird placement value equals fair market value at slaughter age minus production costs. As the breeder lays eggs (produces meat birds), its fair market value declines until it reaches its salvage value. Sexually immature breeder birds are valued as a linear relationship between their initial placement cost and their value at beginning of lay.

The income appraisal approach has a couple of limitations. First, value of breeding animals can be very elastic relative to changes in output prices. Small percentage changes in output price can result in a relatively larger percentage change in profit margins and thus value of the breeding animal. Second, if output prices are low enough, profit margins can go negative which would result in a negative value for breeding animals. Consequently, a minimum value needs to be established, such as one based upon cost of production.

It is also difficult to determine fair market value for exotic animals or minor species. A particular problem is the susceptibility to wide market price swings, in part due to their few numbers and the faddish nature of their popularity. When an exotic species becomes popular, the demand for animals becomes very great and prices rise. New owners have the expectation of selling all offspring as breeding animals at high prices. Expected high prices for offspring are then used to justify even higher prices paid for breeding pairs. As long as additional people enter the business, prices remain high. Once additional people stop entering the business, the production of breeding animals soon exceeds demand. Now the offspring must be valued for the products they produce: meat, fibre, antler, etc., which is much less than their worth as breeding animals. Consequently, animal value drops and if there is not a developed market for the final products, the drop in prices can be very great. The presence of a disease which the government is trying to eradicate exacerbates the boom-bust characteristic of exotic animals. Depopulation of several animals can influence the price if total numbers are relatively few to begin with. On the other hand, if part of the eradication process includes movement controls, then prices can be negatively impacted as potential buyers are eliminated from the market place as out-of-state buyers can not move the animals onto their operations.

Probably the hardest group of animals to determine fair market value for is pets. During the 2002-03 Exotic Newcastle Disease (END) in California pet birds were destroyed and compensation was paid. A common approach for valuing pets is to base value on the replacement cost which equals prices charged by businesses selling pets to the general public, i.e. a retail store price. But economically speaking, fair market value is based upon what the owner could sell his pet for. Most likely such a price would be less than pet store prices, i.e. the owner would receive a wholesale price. Therefore, pet indemnity value should be based on prices pet stores pay for their inventory. Another factor to consider is depreciation. Just as a dairy cow depreciates in value with each completed lactation, so will the value of a pet decline as it ages.

An alternative approach for determining the value of pets is a value based on cost of annual ownership or maintenance. The estimated annual cost of keeping a pet is multiplied by expected number of years of remaining life to obtain an indemnity value. This approach is based upon the assumption that annual value of the pet to the owner is at a minimum equal to its annual upkeep; otherwise the owner would not have the pet. This method could be useful for those pets obtained from animal shelters or obtained free from other pet owners.

Of course, the above discussion on valuing pets is based upon a utilitarian approach and no value is placed upon the emotional attachment or benefit the owner may receive from the pet. Determining value for emotional attachment would be rather difficult and not consistent with the fair market value approach of what an owner could receive for his pet. However, taking a strictly utilitarian approach in determining value can reduce public acceptance of depopulation of their pets. Consequently, governments, with possible assistance from industry associations, will have to engage in public education campaigns as to the need to depopulate pets and that value offered has to be reasonable and cannot take into account the emotional attachment humans feel towards their pets.

Expanding Compensation Eligibility

In situations where whole herds or flocks are depopulated there are generally other losses besides the animals depopulated. There are costs associated with euthanasia and disposal. Feed may be destroyed. Buildings and equipment have to be cleaned, and if cleaning is not possible or cost effective then sometimes these items are destroyed, as well. Of course, any asset ordered to be destroyed by government results in an indemnity paid to the owners. The US federal government will often assist owners in the euthanasia and disposal of their animals by doing the work or paying at least partial costs of having it done privately. At times the US federal government has paid some of the costs associated with cleaning and disinfection. Reasons for government involvement with, or paying for, euthanasia, disposal, cleaning and disinfection are the desire to see that the animals are killed in a humane fashion, that disposal is done in a way that minimizes spread of disease and environmental concerns, and that cleaning and disinfection is done properly.

There is a desire by industry to expand items eligible for compensation beyond these direct costs. A popular item that could increase the amount paid is indemnity for lost income or downtime, the period from depopulation until full production has been reestablished. The USDA has received comments that compensation of animals doesn't begin to cover all the losses associated with depopulation. Lost income associated with downtime or business interruption is sometimes cited as a loss just as great as the value of the animals depopulated. However, much of the downtime can be a result of management decisions which should not represent a claim against the government. For example, if a dairy is depopulated the owner receives enough compensation to replace his herd with like animals. But instead of repopulating with the same age distribution of cows already in milk production, the owner decides to repopulate with bred heifers. Consequently, the period of no or low milk production has been expanded from the two to three weeks of cleaning and disinfection to several months as he waits for the bred heifers to calve and begin milk production.

Another form of lost income deals with the secondary or indirect effects of controlling a disease outbreak. In a disease outbreak situation there can be a quarantine which would prohibit movement of livestock into or out of an area. Often livestock are born on one operation and raised on another. Livestock movement restrictions can upset the orderly flow of livestock from one operation to another. Alternative buyers may have to be found, often at a price discount, or an operation may be without any livestock as animals are prevented from entering the quarantine zone. In either case, there is economic loss for which livestock owners would like to receive compensation. Allied industries that serve the affected livestock industry can experience reduced income as the demand for their services is diminished when animals are depopulated. While such losses are indeed economic consequences of a disease outbreak, they are hard to quantify, which would make any compensation of such losses difficult to administer. In addition, livestock owners outside of the quarantine zone may benefit as demand for their animals may increase when animals within the quarantine zone are depopulated. Thus, sorting out indirect winners and losers of a disease outbreak would be rather difficult, and if done improperly, result in more inequities than not paying any of these indirect losses. Thus, the US federal government generally does not get involved in compensation arising from indirect losses.

Another reason for not compensating indirect losses is to provide incentives for owners to practice good biosecurity in order to minimize the chance of disease. If the federal government paid all indirect costs associated with a disease outbreak, then the incentive to incur the costs associated with good disease management or biosecurity is greatly reduced. Thus, some portion of any downtime or other direct losses should be viewed as reasons to practice good biosecurity.

One form of lost income that might warrant indemnity payment is lost income by contract growers. The structure of livestock agriculture is changing as owners of livestock contract with others to produce the livestock. Contract growers are very common in the poultry industry and becoming more common in the swine industry. In the beef industry, custom feeding of cattle where the cattle are not owned by the feedlot is also common. Within the dairy industry, contract heifer-raising is becoming more common. Unfortunately, current US federal regulations do not recognize the role contract growers have in animal agriculture, and as such, they are not eligible for compensation. Yet, such growers are the ones with daily contact with the animals and would be the first to notice any signs of disease.

Determining appropriate compensation for contract growers can be a complex process. Contract growers are paid by owners for providing two or three main services: labour, housing and sometimes feed. The simplest approach would be to pay contract growers for the income they would have received from the owner. A problem of equity arises when some animals within an industry are raised by contract growers while other animals are not. For the same type of animal, total compensation received would be greater for those animals raised by a contract grower as total compensation paid equals value of the animal as well as lost contract grower income. One solution to the equity problem is to compensate owners who do not use contract growers for their housing and labour costs as well. This would treat all operations equally, but at the cost of increased taxpayer outlays. Alternatively, only owners would be paid for the lost value of their animals, but require owners that use contract growers to honour their contracts and pay their growers as if the animals had not been depopulated. Such an alternative would most likely involve some type of payment scheme that would help ensure that the contract growers were indeed paid according to their contracts.

Compensation Alternatives

As previously stated, an economic justification for the government to compensate livestock owners when their animals are depopulated is to encourage producers to report potential disease problems. The potential consequences of delayed reporting can be substantial. Thus, paying producers is one way to minimize the externality of not reporting. Another way to minimize the externality is to have livestock owners internalize the externality. Several methods for internalizing the societal costs associated with not reporting disease include insurance, industry-generated compensation funds and a permit/assurance bonding scheme.

Insurance

Insurance is a possible alternative to compensation from the government for losses associated with livestock diseases. With livestock disease insurance, the livestock owner receives compensation from the insurance company based upon the insurance policy's contract. A major advantage of livestock insurance is that compensation is no longer the government's responsibility, but becomes a business decision of the livestock owner. The owner, within limits, could set the value of his animals, deductible level, and percent of loss coverage and pay the resulting premium. The insurance company would have a financial interest in the owner practicing good biosecurity and general herd health. Reduced premiums for good biosecurity practices would encourage livestock owners to improve their biosecurity practices. Thus, livestock insurance products that provide coverage in the event of major disease outbreaks are in the public interest and should be encouraged.

However, livestock insurance faces many obstacles that must be overcome before a majority of food animal producers insure their animals. A large obstacle is the basic premise of insurance and how insurance companies generate profits. The basic idea of insurance is to protect against very large losses. For a livestock owner, this may mean the death of several animals in his herd. Livestock owners transfer the risk of high losses to an insurance company. The insurance company accepts the risk and the potential payout of compensation for a fee known as a premium. The insurance company generates profit when the expected value of premiums received is greater than the expected value of payouts for losses. Conversely, livestock owners collectively will pay more in premiums than what they receive in payouts for losses. Insurance then becomes another expense of doing business. Increased costs are not very attractive to livestock owners where profit margins may be small.

Another large hurdle related to insurance as a substitute for federal compensation is the "takings" issue. In control of a disease outbreak, animals are often destroyed to stop the spread of disease. Even healthy animals may be destroyed if they are at high risk of being exposed to or infected with the disease. The destruction of these animals represents a taking by the government and thus the government must provide compensation to the owners for the value of the animals taken. With the US federal government bound by the US Constitution to pay owners for the taking of their animals, incentives to have livestock insurance are greatly reduced. Compounding the problem further is that current US law states that compensation paid to owners shall be reduced by compensation received from other sources[4] (Public Law 107-171). With the federal government making payment for any animal taken due to disease, the incentive for owners to purchase livestock disease insurance or to participate in industry self-insurance funds is reduced.

Uncertainty faced by insurance companies is another hurdle they must overcome before livestock insurance for diseases becomes popular. Uncertainty makes it hard for any business to plan. Though it appears that insurance companies deal with uncertainty, they actually deal in risk. Risk differs from uncertainty as risk can be measured in probabilities. For example, insurance companies can estimate the risk of future house fires or auto accidents based upon past occurrences of such events. Foreign animal diseases, which haven't been in the USA for several decades, are much harder to quantify – both the potential of their occurrence and their severity. Thus, it is difficult to figure out the proper amount of premium to charge livestock owners.

Contributing to the uncertainty faced by insurance companies is the role of the government and the actions government may or may not take. Inconsistency in dealing with diseases discourages demand for livestock insurance. For example, in combination with the state of Virginia, which ordered the depopulation of poultry in 2002 due to low-pathogenic avian influenza (LPAI), the US federal government made indemnity payments that covered some of the losses associated with destroyed birds. In later outbreaks of LPAI, the federal government did not make any indemnity payments.[5] With the federal government positively responding to a major LPAI outbreak, owners might conclude that the government will respond in a similar way again to a major LPAI outbreak and thus the purchase of insurance for LPAI, if such were available, would not be perceived as a good investment.

Alternative disease control methods employed by the government, or government indemnity values that vary greatly among similar animals, make it hard for both livestock

---

[4] Currently insurance is viewed as another source of compensation. For insurance not to be considered as another source of compensation, a change to the Animal Health Protection Act will most likely be required.

[5] The Virginia outbreak was different from other LPAI outbreaks in the large number of flocks that were positive with the disease and that this outbreak had the real potential to mutate into high-pathogenic avian influenza. The other LPAI outbreaks were confined to few operations which didn't warrant a federal response as the states were able to handle the events.

owners and insurance companies to plan. Such uncertainty increases the gap between the premium level the insurance companies feel they must charge and the premium level producers are willing to pay. Standardized indemnity values that are transparent in how they are determined can at least help reduce the uncertainty surrounding compensation paid by the government.

Livestock insurance, if it can be created, has value for livestock owners. First, insurance can provide extra compensation for high-valued animals. US federal compensation sometimes has per-animal limits, such as $3000 for dairy cows with tuberculosis (US Code of Federal Regulations, Title 9, Chapter 1, Part 50). Insurance could provide compensation above the maximum value that the government can offer.

More importantly, insurance can be useful in providing protection from losses associated with lost production (downtime), increased costs or reduced revenues due to lost markets: losses not normally compensated by the US federal government. After a premise has been depopulated it must be cleaned and disinfected before it can be repopulated. Cleaning and disinfection costs are not always paid for by the government. Also, disinfection may involve a period of downtime before repopulation occurs. Downtime, especially if it lasts several weeks, greatly delays when the operation will start producing revenue again. Making matters worse, repopulation may be delayed due to a general quarantine that prohibits the import of livestock. Consequently, an operation that was depopulated early in an outbreak might need to wait several months before being allowed to repopulate. Such a long downtime could result in losses greater than the value of the animals destroyed.

Another potential loss that livestock owners face is lost markets due to quarantine controls. Because of quarantine controls, livestock owners may not be allowed to ship their animals to their normal market outside of the quarantine zone, even though the disease is not on their operation. Instead, they must find markets within the quarantine zone, and if there is an oversupply of animals within the quarantine zone, market prices may be discounted. Quarantine controls also can affect livestock owners outside of the zone area as they may not be allowed to ship any animals into the zone. Once again, the owner would have to find alternative markets.

The US federal government can encourage the use of insurance by livestock producers in a few important ways. First, they should allow insurance payments to supplement any traditional indemnity payments received. If insurance is viewed as another form of compensation, then the indemnity to be paid by the US federal government is reduced by the insurance payment received. Instead, insurance should be viewed as a business contract between two firms and the transfer of monies between such entities a matter of contractual agreement. Second, the federal government can promote livestock insurance by subsidizing premiums paid by producers and paying some of the operating costs of the insurance companies. It has taken such subsidies for insurance to gain widespread acceptance from crop producers (Dismukes and Vandeveer, 2001).

Industry-Generated Compensation Funds

Livestock producers can assess themselves levies to generate funds for compensation. In some parts of the USA, where there are major concentrations of poultry companies, the companies have a self-insurance programme where they assess premiums themselves to provide indemnity funds. These funds allow the industry to offer indemnity to the birds' owner and thus depopulate the flock in less time than government can. This rapid response helps to minimize the spread of the disease and speeds up its eradication. Often such funds will cover backyard flocks whose owners have not contributed to the industry indemnity fund.

Since industry-generated compensation funds are a form of self insurance, the same issues that affect insurance apply to industry-generated compensation funds. However, it is possible to modify the concept of industry-generated funds to overcome some of the problems associated with insurance. This involves a two-tiered indemnity structure.[6]

The federal government would have relatively low fixed indemnity rates that would be available to all producers. Producers could then opt into a higher indemnity payment schedule by participating in an industry/government-sponsored indemnity programme. Industry would collect monies from participating producers with the federal government providing matching monies. The fund would have to be actuarially sound or, at a minimum, be able to cover a given percentage of the animal population. An industry/government commission would set charges and indemnity values. In some ways, it is similar to insurance, but instead of private insurance companies there would be an industry/government commission.

An advantage of the partially industry-funded two-tiered compensation scheme is that the indemnity compensation rates are set by the industry/government commission. Since the fund used to pay indemnity must be actuarially sound, producers will directly face the trade-off between levies collected and indemnity levels. Another advantage is that industry could select how many different diseases to cover in such a system. They may choose more than what the US federal government is currently involved with. Of course, industry would face the trade-off between levies collected and number of diseases covered.

Permit/Assurance Bonding Scheme

A combination of permits and assurance bonding is another way to internalize the externalities associated with livestock diseases. With the permit/assurance bonding scheme, it is the owners' responsibility to keep their animals free of disease, especially for diseases that have human health implications. Under this scheme, the federal government could require permits to either own or market livestock. Failure to keep their animals free of disease could result in fines and/or loss of such permits. As part of the permit process, livestock owners could be required to post assurance bonds that are redeemable when the animal is sold.[7] Assurance bonds could be designed to cover an operation for a given period of time. Buyers would have an incentive to make sure that animals being purchased are indeed disease-free as they would want to be able to market the animals later on and redeem their assurance bonds. Such an incentive might result in buyers demanding proof by the seller that the animals are test-free of diseases. The permit/assurance bonding scheme would save the federal government millions in compensation payments and generate demand for services of private veterinarians engaged in certifying disease-free status.

---

[6] Several European countries have various combinations of government- and industry-funded compensation schemes. For more detail, see the chapter by Van Asseldonk *et al.* (this volume).

[7] For more detail on the use of assurance bonds see Thomas and Hanson (2004) and Thomas and Randall (2000).

Summary

Paying owners compensation for destroying their animals in order to control disease has been an important tool in the eradication of livestock diseases in the USA. Not only does the US Constitution require compensation whenever the government takes an animal away from its owner, economically it makes sense to do so as well. Individually, owners of diseased animals have little incentive to report diseases, which is detrimental to the industry as early detection is important in the control of transmittable diseases. Compensation for animals destroyed becomes the financial incentive for reporting livestock diseases. Determining the right compensation level can be problematic. If too low, owners may decide disease reporting is not worth it. If too high, the industry may become too complacent in disease prevention since the government pays for most or all of the financial losses associated with the disease. Insurance, industry-generated compensation funds and permit/assurance bonds are alternatives to government-financed compensation. These alternative schemes internalize to the individual the potential industry loss from slow or non-reporting of disease.

References

Dismukes, R. and Vandeveer, M. (2001) US crop insurance: premiums, subsidies, and participation. US Department of Agriculture, Economic Research Service, *Agricultural Outlook,* December.

Public Law 107-171. Farm Security and Rural Investment Act of 2002, Title X, Subtitle E, Animal Health Protection Act.

Thomas, M.H. and Hanson, T.R. (2004) Evaluating a new decision protocol with assurance bonding for releasing potentially invasive exotics: the case of the black carp (*Mylopharyngodon piceus*) in Mississippi. *Proceedings of Aquaculture 2004*. World Aquaculture Society, Honolulu, Hawaii, March.

Thomas, M.H. and Randall, A. (2000) Intentional introductions of nonindigenous species: a principal-agent model and protocol for revocable decisions. *Ecological Economics* 34, 333-345.

US Code of Federal Regulations. (multiple years) Title 9. Animal and Animal Health Products. Chapter 1 – Animal and Plant Health Inspection Service, Department of Agriculture. Part 50 Animals destroyed because of tuberculosis. Part 51 Animals destroyed because of brucellosis. Part 52 Swine destroyed because of pseudorabies. Part 53 Foot-and-mouth disease, pleuropneumonia, rinderpest, and certain other communicable diseases of livestock and poultry. Part 54 Control of scrapie., Part 55 Control of chronic wasting disease. [Accessed 2004.] Available from: http://www.access.gpo.gov/nara/cfr/cfr-table-search.html#page1.

Wiser, V. (1974) Protecting American agriculture: inspection and quarantine of imported plants and animals. USDA, Economic Research Service, Agricultural Economic Report Number 266.

Wiser, V. (1987) Part I: Healthy Livestock – Wholesome Meat. In: Wiser, V., Mark, L. and Purchase, H.P. (eds). 100 Years of Animal Health, 1884-1984. Associates of the National Agricultural Library, Inc., Beltsville, Maryland.

Chapter 7

Conceptual Issues in Livestock Insurance

Calum Turvey

Introduction

Recent epidemics of animal diseases coupled with the incidents of terrorism have raised issues concerning the risks facing livestock producers and how those risks can be managed through insurance products. In the autumn of 2001, the Risk Management Agency (RMA) of the US Department of Agriculture (USDA) approved livestock gross margin (LGM) and livestock gross revenue (LGR) insurance policies for swine. In addition to revenue-based insurance products there is also considerable interest in livestock insurance products against on-farm diseases and catastrophic losses from the market effects of particular pathogens such as Foot and Mouth Disease (FMD).[1]

A key policy question is concerned with the appropriate role of agricultural insurance in the USA (and elsewhere) to reduce losses from animal diseases and market prices. Is a joint policy possible, or should production and market-related risks be insured separately? Should insurance policies cover catastrophic risks due to natural or bioterrorist outcomes and can catastrophic risks in the livestock market be reinsured?

A case in point is the December 23, 2003 discovery of the first documented US case of Bovine Spongiform Encephalopathy (BSE) in a cow from a Washington State dairy herd. BSE, also known as mad cow disease, is a degenerative neurological disease that affects cattle. Human consumption of meat from an animal infected with BSE is believed to result in a similar condition in humans known as variant Creutzfeldt-Jakob disease (vCJD), a rare and fatal degenerative brain disease with no known treatment or cure. The US beef industry represents about $40.76 billion in gross output (in 2000) and this output supports an additional $188.4 billion of economic activity (NCBA 2003). In total, US producers raise over 27 billion pounds of beef each year of which approximately 10%, valued at over $3.5 billion in 2003, is exported primarily to Japan, Korea, Mexico and Canada. One of the more significant impacts of BSE was the immediate evaporation of these exports markets, which caused dramatic decreases and volatility in prices. In addition, there were fears that some consumers would

[1] Several definitions of catastrophic risks have emerged. Schlesinger (1999) refers to catastrophic risks as extreme events found and rarely occurring in the extreme tails of a probability distribution. Likewise, Duncan and Myers (2000) define catastrophe as an infrequent event that has undesirable outcomes for a sizeable subset of the insured population. To be insurable, Kunreuther (2002) points out that the risks have to be identifiable in probability space, and if it occurs the extent of loss must be calculable.

stop eating beef altogether. With consumers spending about $60.3 billion on beef each year, the Food Policy Institute found via a nation-wide poll of consumer perceptions in January 2004 that 4% of US consumers would reduce or stop eating beef. The potential impact of such a change is $2.412 billion, and perhaps as much as $11.14 billion in indirect costs on an annualized basis (Schilling, Hallman and Turvey, 2004).

The purpose of this chapter is to establish some basic principles related to livestock insurance and to illustrate how these principles can be applied in practice using many techniques available today. First, a general model is used to illustrate the complexity of the risks at the farm level, and several possibilities for insuring all risks are discussed in a qualitative way. Second, a general model is developed to illustrate the principles of frequency, intensity and duration for developing livestock disease insurance. Third, a more specific class of net revenue insurance models are presented and empirically evaluated. These models assume certainty in production and feed use, but allow for variability in livestock and feed prices. Monte Carlo approaches to calculating the value of several conventional and path-dependent livestock net revenue insurance possibilities are illustrated assuming the existence of a futures market. Fourth, insuring catastrophic market risks arising from the introduction of a disease that would cause market livestock prices to evaporate is modeled as a jump process with a disease arriving at unknown times, but with known frequency. Calculation of a Poisson-induced indemnity as an insurance product could be considered in addition to conventional livestock or revenue insurance, or the revenue insurance should be adjusted to include the probability of catastrophic market risk.

Theoretical Considerations for Livestock Insurance

The purpose of this section is to explore in a very general way the randomness affecting a livestock operation. It is first assumed that the farm-level operations involve the production of cattle and the use of feed maize as an input. The net revenues per head are given by:

(1) $$R = \theta\, p - \omega\, f$$

where p is the price of livestock and f is the price of feed. The symbols θ and ω refer to production coefficients on output and feed. For example, if θ represents 1 pound of growth then ω represents the amount of feed required for 1 pound of growth. If (1) is viewed as an expectation then the total derivative of equation (1) illustrates the total possible change in net revenues from all sources of risk, as shown in equation (2).

(2) $$dR = \theta\, dp + p d\theta - \omega\, df - f\, d\omega$$

For the purpose of this chapter it is assumed that $d\theta=0$, $d\omega=0$ while dp and df are random variables with expected values of zero, with standard deviations σ_p and σ_f respectively, and covariance $\sigma_{p,f}$. The variance of net revenue is then:

(3) $$VAR(dR) = \theta^2\, \sigma_p^2 + \omega^2\, \sigma_f^2 - 2\, p\, f\, \sigma_{p,f}$$

and the joint distribution of net revenue is:

$$(4) \qquad E(R) = \iint [\theta p - \omega f] \, g(p,f) \, dp \, df$$

where g(p,f) represents the joint probability distribution between cattle prices and feed.[2] However, equations (1) through (4) represent the variability in net revenue per animal if the animal survives. If the animal dies from disease the expected payoff changes and expected revenue is given by:

$$(5) \qquad E(R) = \begin{cases} \iint [\theta p - \omega f] \, g(p,f) \, dp \, df & \text{if deathloss} = 0 \\ 0 & \text{if deathloss} = 1. \end{cases}$$

The question remaining is whether it makes sense to design livestock insurance products that attempt to capture the complex risks of equations (4) and (5). In other words, should a single insurance contract be constructed to protect farmers against both net revenue loss originating from systematic market risk and nonsystematic farm-level diseases, or should separate insurance contracts be written for each source of risk?[3] The degree of correlation between market risks and on-farm risks is also problematic. If the market risks are statistically independent from on-farm disease risks, then attempting to provide a net revenue insurance policy that attempts to encapsulate all sources of net revenue risk would be futile. Rather, it would be sensible, as suggested by Hart, Babcock and Hayes (2001), to focus net revenue products on the correlated livestock and feed prices as one product, and animal diseases as another product. On the other hand, in many instances disease occurrence can not only affect death loss on an individual farm, but can also have negative impacts on market prices. The impact of disease on market prices will depend largely on whether the disease is harmful to consumers in domestic and foreign markets (e.g. BSE), or will cause trade sanctions from foreign markets (e.g. FMD).

There are then, three candidate insurance products for livestock producers. The first, livestock production insurance, would protect farmers from loss and business interruption due to illness or death, as well as provide recovery of veterinary costs due to on-farm diseases. The second, net revenue insurance, would protect farmers against losses from the market place. The third, catastrophe insurance, would protect farmers against extreme price losses due to the emergence of a disease that correlates with rapid decreases in market prices. The first two policies arise from the statistical independence between market prices and farm disease, while the third arises from the remote probabilities of a catastrophic epidemic of a disease that will be negatively correlated with the market price of livestock. The principles of these three insurance contracts are discussed in the following sections.

---

[2] There are of course legal avenues when it comes to the purchase of poor feed. Civil liability may substitute for insurance.

[3] Endogenous, non-systematic risks are controllable at the farm level, and unlike the exogenous, systematic risks, can be influenced by the individual producer. Moral hazard is an issue. Would the mere existence of a revenue insurance contract that covers lost productivity be sufficient to cause farmers to act less diligently in mitigation?

Principles of Livestock Production Insurance

Separating livestock production insurance resulting from disease should be founded on three basic principles: frequency, duration and intensity. Frequency refers to the likelihood that, in any given year, a disease will occur in the herd. Some diseases occur more frequently than others, and therefore all things considered equal, a more frequent liability will cost more. Duration refers to the length of time (e.g. the number of days) that a herd is infected. The longer a pathogen remains in a herd, infecting more animals, the greater the loss will be. The third principle is intensity. Intensity refers to the degree by which the herd is infected as a function of duration. Not all pathogens have the same intensity. A mild pathogen might infect a herd slowly or result in only moderate losses over a period of time, whereas a more aggressive pathogen with high intensity will result in greater losses sooner. The more susceptible a herd is to a pathogen the greater will be its intensity. The World Organization for Animal Health (OIE) classifies diseases according to contagion and intensity.

The frequency and intensity represent randomness. For example, if a pathogen appears only once in five years it will have a 20% prior probability of appearing at any time without warning. Likewise, the duration is a random variable. The duration can be one day or two weeks, again depending on random factors, and other factors such as population medicine and inoculation. The structure of a candidate loss function is presented in equation (6)

$$V(f,\lambda,\beta) = 1000\ f(t) \int \lambda^{(-\beta)}\ g(\lambda)\ d\lambda. \qquad (6)$$

In (6) the valuation is based upon an indemnified value ($1,000) although any unit of measurement can be used. The function f(t) is the probability of occurrence and represents the frequency principle. The parameter λ represents the duration, and its probability distribution function is $g(\lambda)$. In general $g(\lambda)$ will be a negative exponential or gamma type distribution. The power function $\lambda^{(-\beta)}$ captures the intensity. The higher the value of β the greater is the intensity associated with the duration, λ. For example if $\beta=0$ there is no loss associated with the pathogen. If $\beta=0.5$ the intensity is moderate, but if $\beta=2$ the intensity is high. Essentially, the higher the intensity the faster the $1,000 value will be driven to zero.

To illustrate how such a loss function works, assume that f(t)=0.30 so that the pathogen arrives on average three out of ten years. When it arrives it has a mean duration in the herd of 14 days with a standard deviation of 14 days. Assume that $g(\lambda)$ is a gamma distribution with a negative exponential shape so that a short duration has a higher probability of occurring than a long duration. Subtracting the outcome in equation (6) from $1,000 provides an estimate of expected losses. The indemnity function is therefore used to generate the cost of insurance per $1,000 of revenue. Using Palisade Corporation's @RISK software, the cost to the livestock producer per $1,000 of revenue is $180 for $\beta=0.5$, $235 for $\beta=1$, and $264 for $\beta=2$. The maximum indemnity in all three cases approaches $1,000 asymptotically. Since the frequency variable is a prior probability, the cost of insurance is directly and linearly related to frequency. For example, by dividing the above results by 3, the resulting premiums would represent a frequency of occurrence of 1 in 10 years rather than 3 in 10 years.

Of course the forgoing represents, in a very simple way, the essential elements of pricing livestock production insurance. The premium values will differ if a different intensity function is used, if the duration period is changed, or if the frequency changes. However, the results do illustrate several salient points. First, the more frequently a disease occurs, the higher the cost of insurance. The longer the duration of a disease in the herd, the greater will be the rate of infection and hence the premiums, and lastly, the susceptibility of the herd to the disease will also lead to increased premiums. In order to achieve actuarial soundness it is fundamental that measures of frequency, intensity and duration be known with reasonable scientific certainty. Failure to accurately represent these key variables could lead to premium over- or underestimates relative to the true risks.

There is also an important policy issue about livestock insurance. It is quite clear that sound population medicine, herd health management and best management practices will affect all three factors. Frequency will be lower, duration will be shorter and intensity will be smaller. All three of these factors indicate that mitigation through inoculation or antibiotics will reduce production risks and hence insurance costs. In light of this, consumer perceptions of food safety risk that inhibit, or even laws that prohibit, inoculation can lead to a greater incidence of disease and higher costs of insurance to livestock producers. For example, even though a vaccine for FMD is available, it is not used on North American herds because the presence of the antibodies from the vaccine cannot, given current technology, be distinguished from a live virus. Japan and other trading nations will therefore not accept vaccinated animals. Notwithstanding the importance of trade, and the high risk of losing foreign markets versus the relatively low risk of an FMD outbreak, a policy of prohibiting vaccination increases the likelihood of an FMD outbreak and, all other things being equal, higher insurance costs.

Principles of Net Revenue Insurance

Including only the price risks for livestock and feed, we can construct a proximate measure of net revenues that can be defined over input and output prices. Assuming only one source each of input and output price risks, the relevant probability distribution (as presented in equations 4 and 7) is given by

(7) $$E(R) = \iint [\theta p - \omega f]\, g(p,f)\, dp\, df$$

where g(p,f) represents the joint probability distribution function for input and output prices and the expectation is measured assuming fixed coefficients for θ and ω. To solve this problem assume that p and f follow correlated geometric Brownian motions

(8) $$df = \alpha_f f\, dt + \sigma_f f\, dw_f$$

(9) $$dp = \alpha_p p\, dt + \sigma_p p\, dw_p$$

where α_f and α_p are the drift rates and σ_f and σ_p the volatilities of feed and output prices. The terms dw_f and dw_p are Wiener processes and the covariance between feed and livestock prices is

(10) $$\mathrm{cov}(f, p) = \rho\, \sigma_f\, \sigma_p.$$

Using Ito's Lemma on equation (1) and equations (8), (9), and (10) the partial differential equation for the change in net revenues is

(11) $$dR = (\theta\, \alpha_p\, p - \omega\, \alpha_f\, f)\, dt + \theta\, \sigma_p\, p\, dw_p - \omega\, \sigma_f\, f\, dw_f.$$

With expected value at time T (e.g. the date T years from the date that the insurance contract is written) defined by the drift

$$(12) \qquad E(dR) = (\theta\, \alpha_p\, p - \omega\, \alpha_f\, f)\, T$$

and variance

$$(13) \qquad VAR(R) = (\theta^2 f^2 \sigma_f^2 + \omega^2 p^2 \sigma_p^2 - 2\, \theta\, \omega\, f\, p\, \rho\, \sigma_f\, \sigma_p)\, T.$$

Since equations (8) and (9) are jointly lognormally distributed by definition, equation (11) is jointly normally distributed with a mean change described by (12) and variance given by (13). In (13), variance is influenced by feed conversion ratios and the covariance between livestock and feed prices. The variance measures in (13) are measures of levels of prices, but from (8) and (9), they represent the variance in the percentage change in prices. The feed conversion ratio is a parameter, fixed by the terms of the contract, so there is no possibility of moral hazard in feeding regimes. In terms of covariance, a positive correlation between feed and livestock prices will result in a reduction in overall variance, while a negative correlation will result in an increased variance.

Because (11) is normally distributed, and not a geometric Brownian motion, it is not possible to develop an insurance product based on a Black or Black-Scholes framework (Black, 1976; Black and Scholes, 1973). Nonetheless, equation (11) represents time-dependent or path-dependent behaviour in that price movements follow a random walk between the times t=0 and T at expiration. Furthermore, since (8) and (9) follow Geometric random walks then they can jointly provide the path or evolution of net revenues over time. This feature can be exploited in a number of ways. First, using the normal distribution on time T payoffs we can develop a simple net revenue insurance product. However, with knowledge of the underlying stochastic structure it is possible to consider a number of useful 'exotic' or path-dependant derivative products that combine both output and input price risks. These exotic options include (but are not limited to) Asian, look-back and barrier options.[4]

Asian Options

A path-dependent option is defined as one whose payoff at expiration, T, is contingent on the path or evolution of prices prior to expiration. An Asian option is an option whose payoff is dependent on the average price or revenue over some period T – t, where t can take on any value from $0 \leq t < T$. For example, define

$$(14) \qquad J = \left[\frac{1}{t_2 - t_1}\right] \int_{t_1}^{t_2} R(t)\, dt$$

[4] Further discussion of exotic and Asian options can be found in Hull (1997), Wilmott (1998) or Wilmott, Howison and Dewynne (1995). The models presented in this section relied on the notations and models found in the latter two books. For specific applications of revenue insurance to agriculture see Turvey (1992), Stokes (2000), Stokes *et al.* (1997), Tirupattur *et al.* (1997) and Yin and Turvey (2003). A similar analysis has been done by Hart, Babcock and Hayes (2001).

where t_1 represents the beginning of an averaging period, t_2 is the end of an averaging period, and R(t) is the realization of equation (4) at time=t. An option based on E{MAX[0, X – J]} with J defined by (14) is referred to as an Asian option. It is path dependent because the payoff depends on the evolution of R(t) over the time horizon t_1 to t_2.[5] Since J represents the average realization of R(t) the Asian option states that if the average realization of R(t) (i.e. J) is less than the strike price X (or revenue insurance coverage level) at expiration, then the insurance pays the difference between the strike and the average. An Asian option will generally be lower in price than the basic option since the reference value J, taken as the average across time, will generally eliminate extreme highs or lows making it more likely to be in the money, but less likely to have a large payoff. Nonetheless, for livestock revenues that are, on average, lower than the strike over a given period of time, this type of option can provide considerable protection.

An alternative form of an Asian option is to set the strike value as the average realization. This is referred to as an average strike option. Here the payoff function is given by MAX[J – R(T), 0]. Suppose that net revenues have on average been higher than R(T), which implies that net revenues are falling as expiration approaches, then the average strike option will pay off. If net revenues are rising, and are above the average net revenues for the time period considered, then R(T)>J and the option expires worthless. These kinds of options are invaluable to the producer who does not want to do worse than the average realization in the given year. Note, however, that this is truly path dependent and time dependent since the strike value changes as market conditions change. In contrast, the conventional Asian option fixes the strike or target revenue, and pays off only if average realizations are below the strike.

Look-back Options

A second type of path-dependent option is called a look-back option. This option has a payoff at expiration that is equal to the difference between the maximum value recorded over the time horizon and the value at expiration. From t=0 to T, let J=MAX(R(t)) be the maximum valued occurrence. Then the payoff at expiration is MAX[J – R(T),0]. Essentially, if over the life of the contract the value of equation (14) valued at the prices f_t and p_t ever exceeds the expiration value based on f_T and p_T, a payoff is made. The look-back option differs from the Asian option in one significant respect. The payoff is on the extremes rather than on the average. Therefore, the cost of a look-back option will be significantly higher than that of an Asian option, but will provide a higher expected payoff in a regime in which R(t) is falling as expiration approaches.

Barrier Options

A third type of path dependant option is referred to as barrier options. A 'knock-in' barrier option is initially worthless at t=0 and then becomes active only when a particular economic condition arises. For example, if a cattle rancher places some reservation value on R(t), say R*, then if ever R(t) falls below R*, a put option is triggered with a strike price X. This is referred to as a 'down-and-in' barrier put option and becomes more valuable as the option

[5] If the payoff is dependent only on the realization at T, that is R(T) above, it is not path dependent.

moves into the money (as R(t) falls). Alternatively a knock-out option can be written such that a put option with strike price X, originally alive at $t=0$ is knocked out, or made worthless if at any t, $R(t)>R^*$. This is referred to as an 'up-and-out' barrier put option and it becomes less valuable as the option moves out of the money (as R(t) rises above X).[6]

Such options are valuable because they limit the exposure to unnecessary time value. The probability set is comprised of two basic elements. The first is the likelihood that the option will become active or inactive (the probability that it will cross the barrier at R*), and the second is the probability that when active it will expire in the money. While active, the barrier options value is the same as that of a conventional option but has a value of zero when inactive. As long as the probability of crossing the barrier at R* is positive, the barrier options will always have a value less than that of conventional options.

The path-dependent options discussed above can be expanded in definition to examine a subset of options of use to livestock net revenue insurance. For example, the variable J in the Asian option (equation 14) can be averaged over the time interval from $t=0$ to $t=T$ or any other suitable interval. In many stabilization and crop insurance programmes the price attached to yields is often set equal to the average commodity price in the primary harvest month, or over a two-to-three month period. Likewise, a livestock producer may want to accept slightly higher risk by setting J in equation (14) as the average over a shorter period of time, say t_1 to T, with $t_1>0$. Obviously if $t_1=T$, then the value of the option will be identical to a plain vanilla option, which will be of higher risk.

Barrier options can provide particularly interesting flexibility for the net revenue model discussed since the barrier can not only be established relative to net revenues R* , but p* or f* as well. In other words, if a cattle producer knows that the greatest uncertainty is in feed costs, then he may wish to establish a barrier option which will activate or deactivate if, and only if, the price of maize crosses a barrier at f*. Likewise, if cattle price is more important, then the barrier can be tied to the price of cattle at p*. More complex structures can also be envisioned: for example, a barrier option that is activated if $(p<p^* \text{ OR } f>f^*)$, or more complicated yet, $(p<p^* \text{ AND } f>f^*)$.

Monte Carlo Approaches to Options Pricing

This chapter examines a number of net revenue options. While many of the options such as plain vanilla and path dependent have available solutions, these solutions are sometimes complex and cumbersome. Alternatively, Monte Carlo approaches can be used.

Data and Estimation

The Monte Carlo simulations assume the existence of futures contracts for cattle and maize and the insurance contract is based on the performance of the futures markets rather than the cash market. That is, the approach used does not necessarily eliminate basis risk. Summarized in Table 7.1, the data represent 950 matched daily observations from 1996 through February 7, 2001 on the nearby futures price. The futures contracts include live cattle and grain

---

[6] We are concerned here with put options only. For call options the equivalent barriers are 'down-and-out' for the knock-out call, and 'up-and-in' for the knock-in call.

Table 7.1. Sample Statistics.

Statistic	Maize Price [a]	Live Cattle Price [b]
Mean	272.33	65.61
Median	258.13	65.68
Mean/Median	1.06	1.00
Mode	219.00	66.98
Std. Dev.	76.99	3.24
Min.	178.50	54.80
Max.	548.00	73.64
Range	369.50	18.84
Range/Median	1.43	0.29
Skewness	1.54	–0.17
Coeff. of Variation	3.54	0.05
Log Change Mean growth	–0.0006	0.0001
Log Change Volatility	0.0236	0.0133
Annualized Growth	–0.15	0.02
Annualized volatility	0.37	0.21
	Correlation Matrix	
	Maize Price	Live Cattle Price
Maize Price	1	–0.56353
Live Cattle Price	–0.56353	1

[a] $/bu
[b] $/cwt

maize traded on the Chicago Mercantile Exchange (CME) and the Chicago Board of Trade (CBOT).

The sample means and range are given in Table 7.1. In the last two columns the annualized geometric growth rate and volatility based on a 250-day trading year are presented. The results show that maize faced a general decline of about 15%/year while live cattle had an increase of about 2%/year. The price of maize ranged from $5.48/bu to $3.69/bu while the ranges for live cattle were $73.64/cwt to $54.80/cwt.

On average, volatility exceeded 20% per year. The most volatile commodity was maize at about 37%, while the livestock contracts had annualized volatility of about 21%. Table 7.1 also provides the correlations between the commodities. The correlation between daily changes in maize and live cattle prices was –0.56.

The correlations are important to what follows. Recall that the variance of the net revenue is negative in correlation, meaning that an actual negative correlation increases variance. This result implies that, quite generally, a percentage increase in the price of cattle corresponds with a percentage decrease in the price of maize. Since a decrease in the price of maize corresponds with a reduction in cost, it also contributes to an increase in net revenues. That is, a negative correlation between a revenue item and a cost item will ultimately increase overall variability. Finally, the modelling and pricing approach used requires that cattle and maize prices follow a geometric Brownian motion.

Monte Carlo Simulations

This section describes the initial conditions for the Monte Carlo simulations and the modelling approach used. The prices for cattle and maize were \$0.70/pound and \$2.50/bushel respectively. These prices are within the neighbourhood of current prices as well as the prices used to calculate historical volatilities. The historical volatilities were 0.20 and 0.30 for cattle and maize. Because futures contracts are used as the underlying risk instrument, it is assumed that the underlying risks can be spanned and therefore a risk-neutral valuation is used and the risk-neutral growth rate was set to 5%.[7]

For purposes of these simulations a 120-day horizon was used. Assuming an average daily gain of 4.58 pounds, a stocker animal can be fed from 500 pounds to 1050 pounds, for a 550 pounds gain. Assuming further a feed conversion rate of 4.5 pounds of feed per 1 pound of gain implies that 2,475 pounds of maize is required. Converting pounds to bushels is accomplished by dividing 2,475 pounds by 39.6 pounds/bushel. This suggests that 62.5 bushels of maize are required to achieve the desired weight. The initial conditions are thus established. The initial revenue expectation is 550 pounds × \$0.70/pound = \$385 and the initial cost expectation is \$156.25 for a net revenue expectation of \$228.75.[8]

The simulations were made operational using the @RISK software. At expiration (T=120 days) the revenue measure was calculated from equation (15)

$$\text{(15)} \qquad \text{Revenue}_T = Q\left(p_T - \left(\frac{4.5}{39.6}\right) f_T\right)$$

where the prices of maize and cattle evolved dynamically with a correlation coefficient of –0.57.

Options prices and simulations were calculated for:

1. Uninsured net revenue (the base case)
2. A put on net revenue, with strike price equal to t=0 expectation of \$228.75

---

[7] The assumption of risk-neutral valuations follows from the proposition in Cox and Ross (1976), and Cox, Ingersoll and Ross (1985). If the underlying risks can be traded then a hedging regime can be constructed to eliminate risk. Under such a condition the natural growth rates in the price series are replaced by the risk-free rate. If instead the prices were on non-traded feeds or livestock, the problem becomes somewhat more complicated since the risk-neutral growth rate would be set to the actual growth rate (or drift rate) less the market price of risk. See Yin and Turvey's (2003) comment on Stokes, Nayda and English (1997).

[8] In this model only the net gain is considered. This naively assumes that the purchase price of the calf is sunk. However, another form of the model would be to set revenue expectations at total weight (1,050 pounds) so that the net revenue would be initialized at 1,050 × 0.70 – 156.20 = \$578.8. Using gross weight rather than net weight will increase the cost of insurance since overall variability will increase.

Table 7.2. Results for Net Revenue Insurance Simulations: Plain Vanilla Products ($US).

Name	Base Case	Revenue Put	Live Cattle Put plus Maize Call	Live Cattle Put	Maize Call
Mean	234.30	234.30	234.30	234.30	234.30
Std. Dev.	80.81	49.82	48.79	62.09	66.68
Skewness	0.08	1.67	1.60	0.72	0.54
Kurtosis	3.16	5.74	5.71	3.93	3.38
Minimum	–67.92	199.39	195.51	31.53	60.60
Maximum	561.44	532.08	528.20	543.19	546.45
5th Percentile	103.04	199.39	195.51	147.73	137.09
95th Percentile	367.36	338.00	334.30	349.12	352.54
Insurance Cost	0.00	29.36	33.24	18.25	14.99

3. A put on the cattle price with strike = $0.70/pound and a call on the maize price with a strike of $2.50/bushel
4. A put on the cattle price with strike of $.70/pound and no call on maize price
5. A call on the maize price with strike of $2.50/bushel with no put on the cattle price
6. An Asian put option on average net revenues with a strike of $228.75
7. A put option on the average strike, where the strike price becomes a random variable
8. A look-back option with put payout based on a strike equal to the maximum net revenue observed over the 120 days
9. A down-and-in barrier option with barrier set at 0.90 × 228.75, and
10. An up-and-out barrier option with barrier set at 1.10 × 228.75.

The Monte Carlo approach used 10,000 iterations. Each iteration comprised the generation of 120 days of prices and net revenues, and the calculation of the net revenues and options payout for that particular iteration. The value of the option was taken as the average payout across all 10,000 iterations. The procedure involved two steps. The simulations were first run to capture the values of the various option premiums. In the second step, the model was run again, using the same seed value as the first, to capture the net effects of the insurance. The net effects were estimated as net revenue plus option payout less the cost of the option.

Results From Net Revenue Insurance

The results of the analyses are presented in Tables 7.2 and 7.3. Using conventional options to hedge, Table 7.2 shows that the unhedged position has the highest overall standard deviation, as expected. The skewness of approximately 0, and kurtosis of approximately 3, confirm the normality of the net revenue distribution. In terms of variance the greatest amount of risk reduction is with the put option on cattle plus the call option on maize, an expected result given the negative correlation between the two prices. However, in terms of downside risk, the row indicating 5th Percentile reveals that insuring net revenue directly will have a slightly better result. A 5% chance of revenues falling below $199.39 dominates a 5% chance of revenues falling below $195.51. Likewise, since the insurance costs of net revenues are lower

Table 7.3. Simulation Results for Exotic Net Revenue Insurance Products ($US).

Name	Net Revenue	Net Revenue Put	Asian Put	Avg. Strike Put	Down-and-In	Look-back Put on max.
Mean	234.30	234.30	234.30	234.30	234.30	234.30
Std. Dev.	80.81	49.82	64.62	64.33	49.94	51.37
Skewness	0.08	1.67	0.59	0.60	1.66	1.10
Kurtosis	3.16	5.74	3.61	3.62	5.71	4.32
Minimum	–67.92	199.39	52.28	36.31	177.06	171.13
Maximum	561.44	532.08	544.62	544.18	532.22	551.80
5th Perc.	103.04	199.39	141.13	140.85	199.53	171.25
95th Perc.	367.36	338.00	350.54	350.18	338.14	332.71
Option Price	0.00	29.36	16.82	17.26	29.22	9.64

than the insurance costs of independent puts and calls, the upside potential is also dominant. The last two columns examine the conventional use of one option or the other. Variance is lower for the put and call scenarios than the base case, but is higher than insuring both, so the benefit of insuring net revenues is evident. Likewise, the downside risk assessment at the 5th percentile level indicates that downside risk is higher under these two scenarios than the net revenue scenarios, but these strategies still dominate the no-insurance case. The upside is higher for these strategies since the cost of the insurance is lower. In terms of skewness, the net revenue insurance policy dominates, followed by the put-plus-call strategy, and then the individual option strategies. The cumulative distribution functions for these scenarios are presented in Fig. 7.1. The net revenue insurance distribution is truncated at the strike and there is an imperfect truncation for the put-plus-call strategy. The individual options strategies are characterized by continuous distribution functions, but they are not truncated. Rather, the distributions reveal a shift of probabilities from the downside towards the central core. The distribution of the Asian put is very similar to that of the put option on the average strike price, thus the lines overlay.

The four exotic options are presented in Table 7.3. The down-and-in option most closely resembled that of the net revenue put. Recall that the down-and-in option only becomes activated if revenues hit a barrier or threshold. For these simulations this barrier was set at 90% of the strike, so the result simply states that in the majority of cases net revenues fell below the barrier. The option price of $29.22 is only slightly lower than the $29.36 value of the net revenue insurance and reflects a very low probability that the barrier set would not be breached.

The Asian option, with a value of $16.82, reduces risk by approximately 50% relative to net revenue insurance. However, it does protect the downside by shifting probabilities from the lower partial moments to the mid-partial moments as can be seen by the increased skewness. The minimum revenue under the Asian option was $52.28 compared to –$67.92 for the uninsured case and $199.39 for the net revenue insurance case. With net revenue insurance there is a 95% chance of exceeding $199.39 but with the Asian option there is a 95% chance of exceeding $141.13. The results for the average strike option are very similar to that of the Asian, but it is worth reiterating the differences. With the Asian option a payout is made if the average net revenue across 120 days falls below the strike price, which is fixed.

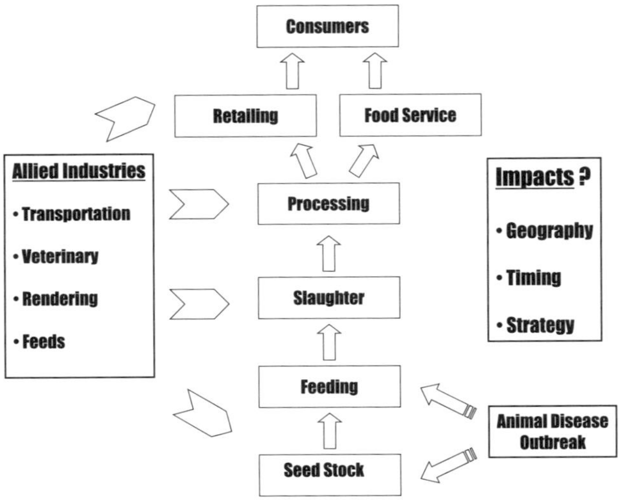

Fig. 7.1. Cumulative Distribution Functions for Net Revenue Insurance.

In contrast, the average strike put recalculates the strike for each iteration, setting the strike equal to the 120-day average. If the average strike, representing average revenues, exceeds the net revenue at expiration, a payout is made. The average strike option ensures that the producer at least receives the average of revenues, whereas the Asian option insures that the producer does no worse than the average. The probability space of the payoff functions for these options will differ under identical states of nature, but the aggregated outcomes across all states of nature are similar because in both cases the payout is based on the average, and the distribution of revenue itself is normal.

The last exotic is the look-back option. This option looks back over the 120 days and picks the maximum net revenue observed. If this net revenue exceeds the net revenue at expiration then a payout is made. In terms of downside risk protection this option is more skewed than the Asian options. Its minimum was $171 and is more positively skewed than the average options. However, its cost, at $9.64, is quite low relative to the other options types. The cumulative distribution functions are presented in Fig. 7.2. The distribution of the down-and-in barrier option is very similar to that of the net revenue insurance, thus the lines overlay.

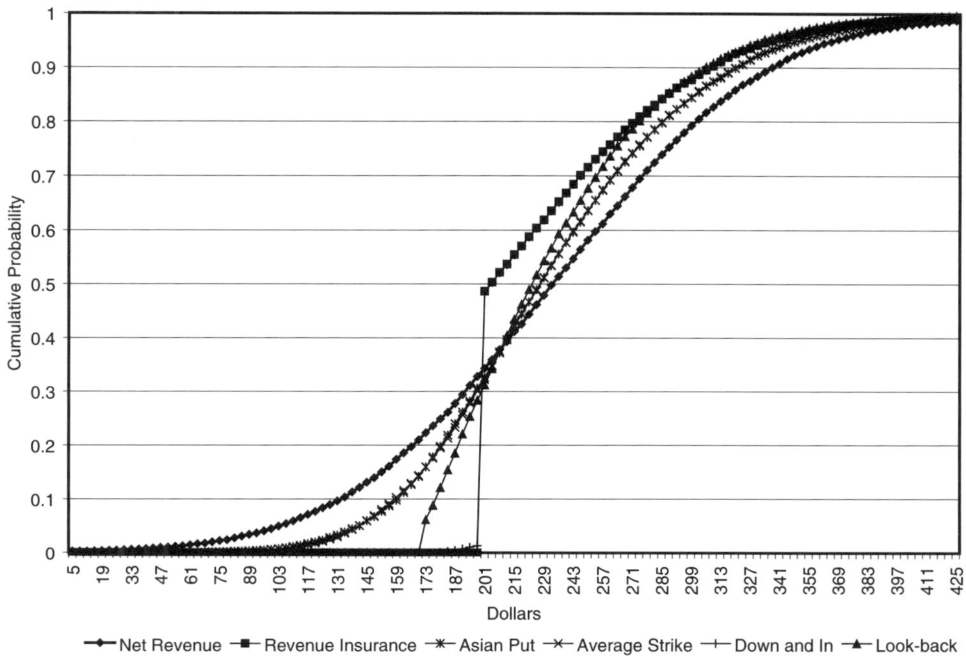

Fig. 7.2. Cumulative Distribution Functions for Net Revenue Insurance.

Principles of Catastrophic Insurance

In this section, a Poisson probability model is presented that can easily be used to calculate the losses from a rapid decline in market prices due to the emergence of an infectious disease. The finding of BSE in the USA in 2003 caused an immediate and precipitous decline in market prices as consumer concerns about food safety caused demand to fall. A finding of FMD would have a similar effect as export demand falls and domestic supply increases. A common approach to measuring jump processes in prices is to define the stochastic differential equation as:

(16) $$\frac{dp}{p} = \alpha_p\, dt + \sigma_p\, dw_p - dq$$

where $$dq = \begin{cases} 0 & \text{with probability} = 1 - \lambda dt \\ \theta & \text{with probability} = \lambda dt. \end{cases}$$

A jump arises when there is a discontinuous change in one of the variables (e.g. price) (Wilmott, Howison and Dewynne, 1995). Accordingly, equation (16) states that the occurrence of the event with probability λdt results in a loss of θp. If the event does not happen (with probability $1 - \lambda dt$) then the price path follows that of the original Brownian motion. We can then write:

$$(17) \qquad \frac{dp}{p} = (\alpha_p - \lambda\,\theta)\,dt + \sigma_p\,dw_p$$

with

$$(18) \qquad E(dp) = (\alpha_p - \lambda\,\theta)\,dt + \sigma_p\,dw_p$$

and

$$(19) \qquad VAR(dp) = (p^2\,\sigma_p^2 - p^2\,\theta^2\,\lambda)\,dt.$$

Equation (18) identifies the drift of the price process. Under normal economic conditions the mean change in prices is given by α_p. In the event of a disease outbreak, the drift is adjusted downward by the jump factor $\lambda\theta$. Equation (19) gives the variance term, now comprised of two separate, but uncorrelated, components. The first term is the instantaneous variance of the normal price process, whereas the second term is the additional increase in variance due to the possibility of a shock to prices.

Under a general jump process multiple events can occur, but in terms of livestock prices a single jump will be sufficient to cause widespread price reductions. Since such jumps are not considered in any of the revenue insurance possibilities discussed in the previous section, this section outlines a simple approach to considering the impacts of an event.

The simplest approach would be equivalent to a knock-out option. In this context a knock-out option is one option that substitutes, or knocks out, another option when a specific event happens. For convenience, suppose that the event is the occurrence of FMD or BSE on US soil. Furthermore, suppose that in the event of an occurrence it is expected that prices will fall by $(1 - \theta\%)$. If the current price is p_0, then should the event happen, the payout is $p_0 - (1 - \theta\%)p_0$ or θp_0. Let $F(Qp,t)$ represent the actuarial value of a price-insured revenue insurance option available to farmers and θQp_0 be the value of a payout if BSE or FMD occurs. An example of $F(Qp,t)$ is the price insurance product in the fourth column of Table 7.2. The value of the knockout option is then $G(Qp,\lambda,t) = MAX[F(Qp,t), \theta Qp_0]$ or

$$(20) \qquad G(Qp,\lambda,t) = \begin{cases} F(Qp,t) & \text{with probability} = 1 - \lambda dt \\ \theta Qp_0 & \text{with probability} = \lambda dt \end{cases}$$

and

$$(21) \qquad G(Qp,\lambda,t) = (1 - \lambda dt)F(Qp,t) + (\lambda dt)\theta Qp_0$$

As an example, assume that the probability of a disease outbreak is 5%/year and when that event happens prices are expected to fall by 75%. Then a \$70/cwt price falls to \$17.50/cwt for a payout of \$52.50. The probability of this occurring is 5%/year or 1.67% over a 120-day period, so the expected cost is \$0.875/cwt. For a 550-pounds gain as assumed in the previous section the marginal cost per animal is \$4.813.

In the bottom row, fifth column of Table 7.2, it is shown that the cost and expected payout of a price-based insurance product is F(Qp,t)=$18.25 for 5.5 cwt of gain. Under the knockout option policy this occurs with a 98.33% probability, while the disease event with a payout of $52.5/cwt × 5.5cwt=288.75, occurs with a 1.67% probability. The value of the knockout option is the probability-weighted sum of the two payouts, i.e. G(Qp,λ,t) = (0.9833 × 18.25) + (0.0167 × 288.75) = $22.77 or $4.14/cwt.

Based on these assumptions, the incremental increase in the cost of insurance is about 24.8%. But the assumptions are explicit and unproven. The assumption that FMD, for example, will appear 5 out of every 100 years is higher than the actual probabilities based on recent history. However, the probability is likely elevated with the rise of incidence in the UK, EU and elsewhere, as well as the rising concern about agroterrorism. Likewise, the assumption of a drop of 75% is unproven. For example, the 2003 discovery of a single case of mad cow disease caused an immediate reduction in live cattle futures prices of over 20%, but once it was established that the cow was an isolated incident and originated in Alberta, Canada, livestock prices increased rapidly. Nonetheless, with an annualized volatility of livestock prices of about 21% (Table 7.1) or about 12.14% ($0.21 \times (1/3)^{0.5}$) for the 120-day period under discussion, a 75% drop in price implies a drop of about 6.18 standard deviations (Z = 0.75/0.1214=6.18), an occurrence that would simply not happen under normal market conditions. Nonetheless, the belief that a contagious disease outbreak such as FMD or BSE will cause a precipitous decline in beef prices requires, as a matter of probability, supplementary consideration of such an event occurring.

Conclusions

This chapter has examined the problem of providing revenue insurance for the livestock industry. To provide revenue insurance requires insuring a minimum of four separate sources of risk: productivity, selling prices, feed quality and input prices. The characteristics of risk between the four categories differ significantly. Productivity is subject to pathogenic and ecological risk. Disease outbreaks, herd health and population medicine are all factors of importance. The characteristic of disease risk differs from price risk, since disease outbreaks arrive periodically at unknown times and with intensity and duration that are also random. Some of these factors are controllable through extraordinary herd and veterinary management practices. Productivity losses are to a certain degree reversible, although reversibility does come at a cost. Feed quality risk is probably the least important since it is easily reversible, although again with some cost. Productivity losses due to feed quality are more probably settled through legal channels than insurance mechanisms.

In terms of pricing livestock insurance, this chapter argued that three essential principles should be considered. First, the frequency of a disease outbreak measures the likelihood that in any given year an outbreak will occur. Second, given an outbreak, the duration of the outbreak is critical. The duration measures the number of days that the herd is infected. The longer the duration the greater will be the damage and hence the premiums. Finally, the third principle is intensity. Intensity measures how susceptible the herd is to the disease. Low susceptibility will result in only moderate losses, but high susceptibility or intensity will result in large losses. A representation of the loss function and an example of premium setting using a gamma distribution in exponential form for generating randomness in duration, and a power function form for intensity, were provided to illustrate the basic concepts.

This chapter presented in more detail an approach to hedging net revenues when output price and feed costs are random. Taking the position that proximate net revenue can be insured using available data on livestock and feed prices, a general net revenue insurance product was developed. The model requires the assumption of Brownian motion in cattle and feed (maize) costs, and through this assumption it was shown that net revenues are approximately normally distributed. Although net revenue can be insured using simple notions of conventional options pricing, the empirical component to the chapter examined an array of possible products using Monte Carlo simulations. The products chosen included a put option on net revenue, a put option on cattle prices, a call option on input costs, an Asian option, an option on an average strike price, a look-back option and a barrier option. The point of presenting the variants to insuring net revenues was to illustrate that there are many alternative structures available to insuring proximate net revenues, each with its own unique cost and benefit in terms of downside risk reduction.

Based on the conditions imposed it was shown that net revenue insurance provided the greatest benefit to risk reduction. A revenue insurance based upon a put on cattle prices and a call on input cost was also shown to be effective as were the look-back and barrier options. Options, on the average, offered low-cost revenue protection but with slightly higher downside risk. However, in practice the notion of protecting average net revenues over a period of time may be attractive to many farmers. A model comprised of using only a put or only a call provided the least downside risk protection. In the USA, the Risk Management Agency has recently piloted revenue insurance for hogs. The details and principles of this chapter support those efforts and illustrate how flexible, and somewhat exotic, revenue options can reduce cattle producers' revenue risks.

Consideration was also given to methods for pricing catastrophic insurance in the event of a disease outbreak such as FMD or BSE. The model followed the conventional approach of incorporating a jump process into the standard Brownian motion price process. Based on an assumption that either one of these diseases could occur 5 out of every 100 years, and that when such an occurrence arrived the price of livestock fell by 75%, it was shown that the catastrophe premium increased the premium of a simple price-insured revenue insurance product by about 24%. In other words, even though the likelihood of an outbreak is low, the magnitude is sufficiently high to be of economic significance. However, the analyses were based on two unknown data points, namely the likelihood of occurrence and the magnitude of loss. Clearly the cost of catastrophe insurance will increase or decrease if either one of these variables increase or decrease, and a recommendation for further research is to empirically examine and enumerate the costs of catastrophe using historical precedence and perhaps more sophisticated insurance models. For example, space considerations restricted us from examining the effect of catastrophic jumps for most of the models we examined.

The chapter presented a series of models that are mathematically feasible and consistent with certain insurance objectives. Nonetheless the models do present some qualitative shortcomings. First, when calculating expected losses from an animal disease, the frequency, intensity and duration must be known with reasonable certainty in order to ensure that premiums are actuarially sound. Research needs to be undertaken to quantify these variables for animal diseases. Second, the empirical model was based only upon the net revenue gain from feeder to finish, but in reality some producers may not like the risk exposure that this presents and might prefer insuring all of the productivity and the feed costs. This is a minor adjustment to the empirical model, and is easily captured in the mathematical model. Perhaps more critical is the naivety in which feed costs were expressed. In the theoretical and empirical model, it was assumed that the feed price risks were based on average daily gains and a single crop. In reality feed rations are more complex and may include prices on crops such as hay and forage that are not traded. The theoretical model can handle this added complexity, but the pricing formula becomes more complex. In addition it was assumed that the relevant price series was a futures contract, but some farmers may prefer insurance on the local cash price. This too creates theoretical and empirical problems since commodities bought or sold in the cash market are non-traded in the sense of risk-neutral or arbitrage-free insurance pricing. Models of proximate net revenue insurance that include the cash market risk would have to also include the market price of risk, rather than the risk-free rate, in the growth rate equations for pricing. Third, the idea of introducing catastrophic risk was based on the design of a jump process that is specified as a knockout option. This will work for current revenue insurance programmes, but it should be added that catastrophe insurance could also be offered independently from other insurance products. Still, what is the likely impact? When mad cow disease was discovered in 2003, prices fell by only 20% in the USA whereas in other countries they evaporated. Research is needed to identify the possible price reductions and also the permanency of those reductions. We have experimented with computable general equilibrium models to get an idea of price effects (Huff *et al.,* 2003) due to FMD and research along this line would be helpful. Still, in the absence of observable data, care must be taken to ensure that supply-and-demand elasticities as well as import-export linkages are sound.

Finally, the models and representations in this chapter should be considered within the context of a new area of study that will evolve over time. Despite the prevalence of animal diseases within flocks and herds, the idea of revenue insurance has not received a lot of academic consideration. This chapter's intent was to provide a set of principles from which livestock insurance can be designed. As indicated in the text, disease insurance based on the properties of frequency, intensity and duration is easier to state as a set of mathematical principles than to implement in practice. Research must be undertaken to document these variables for each of the diseases that are endemic in the USA and effort must be taken to identify probabilities for diseases that are not endemic. There is a role for APHIS in this task, but there is also a need for academic scrutiny, farmer buy-in and consideration by the Risk Management Agency. Consideration must also be given to whether insurance should be publicly or privately provided. Unlike market (price) risk, which is borne by the farmers and within the control of farmers, animal diseases, especially those with large epidemic possibilities, could impose disastrous or catastrophic externalities on the farm sector, with considerable and significant welfare losses for consumers. While one might argue that consumers should bear any of the market losses from animal diseases, an equally sound argument can be made for public intervention to provide stability.

References

Black, F. (1976) The pricing of commodity contracts. *Journal of Financial Economics* 3, 167-179.

Black, F. and Scholes, M. (1973) The pricing of options and corporate liabilities. *Journal of Political Economy* 81, 637-659.

Cox, J.C. and Ross, S.A. (1976) The valuation of options for alternative stochastic processes. *Journal of Financial Economics* 3, 145-166.

Cox, J.C., Ingersoll., J.E. and Ross, S.A. (1985) An intertemporal general equilibrium model of asset prices. *Econometrica* 53, 363-384.

Duncan, J. and Myers, R.J. (2000) Crop insurance under catastrophic risk. *American Journal of Agricultural Economics* 82, 842-855.

Hart, C.E., Babcock, B.A. and Hayes, D.J. (2001) Livestock revenue insurance. *The Journal of Futures Markets* 21, 553-580.

Huff, K., Meilke, K. and Turvey, C. (2003) Issues in modeling bio-terrorism in the agrifood sector. Paper presented at WCC-72 Western Coordinating Committee on Agribusiness, Las Vegas, Nevada.

Hull, J. (1997) *Options, futures and other derivatives.* Prentice Hall, Inc., Toronto.

Kunreuther, H. (2002) The role of insurance in managing extreme events: implications for terrorism coverage. *Business Economics,* April, 6-16.

NCBA. (2003) Economic fact sheet – US cattle and beef industry. National Cattlemen's Beef Association, Denver, Colorado. [Accessed 2003.] Available from http://www.ncba.org.

Schilling, B.J, Hallman, W.K. and Turvey, C.G. (2004) Sanity prevailing in the mad cow case. Selected paper presented at the Joint Northeast Agricultural and Resource Economics Association and Canadian Agricultural Economics Society Annual Meeting, Halifax, Nova Scotia, Canada.

Schlesinger, H. (1999) Decomposing catastrophic risk. *Insurance: Mathematics and Economics* 24, 95-101.

Stokes, J.R (2000) A derivatives security approach to setting crop revenue coverage insurance premiums. *Journal of Agricultural and Resource Economics* 25, 159-176.

Stokes R., Nayda, W.I. and English, B.C. (1997) Pricing of revenue assurance. *American Journal of Agricultural Economics* 79, 439-451.

Tirupattur, V, Hauser, R. and Boyle, P.P (1997) Theory and measurement of exotic options in US agricultural support programs. *American Journal of Agricultural Economics* 79, 1127-1139.

Turvey, C.G. (1992) Contingent claim pricing models implied by agricultural stabilization and insurance policies. *Canadian Journal of Agricultural Economics* 40, 183-98.

Wilmott, P, (1998) *Derivatives: The Theory and Practice of Financial Engineering.* John Wiley & Sons, New York.

Wilmott, P, Howison, S. and Dewynne, J. (1995) *The Mathematics of Financial Derivatives.* Cambridge University Press, Cambridge, UK.

Yin, S. and Turvey, C.G. (2003) The pricing of revenue assurance: comment. *American Journal of Agricultural Economics* 85, 1062-1065.

Chapter 8

Data Requirements for Domestic Livestock Insurance

John W. Green, James L. Driscoll and Megan L. Bruch

Introduction

The potential market for livestock insurance in the USA is large. The size of the domestic herd defines this market. This market would require private policies at competitive prices or joint industry/government programmes modelled after other countries and described in the other chapters. No matter which insurance approach is used, livestock mortality data are essential in determining potential payouts on insurance policies and, thus, the premiums necessary for insurance carriers to make reasonable profits. The adequacy of currently available death loss data is assessed for the purpose of government indemnification and/or private sector provision of policies. The adequacy of the data relates to three potential risk management strategies and to data requirements for reasonable insurance underwriting and rating.

New developments in data collection and animal tracking are being pursued in the USA by producer and government coalitions. New information and better monitoring changes the risk associated with producer practices and government programmes and, thus, affects the risk insured by government via indemnity programmes and/or by the private sector via insurance policies. The principal government programme currently being developed is the US Animal Identification Plan (National Institute for Animal Agriculture, 2003).

The Potential Market for Livestock Insurance in the USA

The size of the livestock herd can be defined using production and slaughter data collected by the US Department of Agriculture's National Agricultural Statistics Service (USDA-NASS). Production data reflect the annual inventory of livestock while slaughter data summarize the number of live animals moving into food products and other commodities.

Production Sector Information

Production data for livestock and poultry include inventories, death losses, number of operations, livestock movements and sales activity at auction markets. These data are collected by the periodic Census of Agriculture and through annual inventory sampling conducted by NASS and include balance sheet estimates where the ending inventory of animals in each state on January 1 is equal to the net of births, in-shipments, number marketed, farm slaughter, deaths and ending inventory from the previous year (NASS, 2002). These balance sheets are available for cattle and calves, hogs and pigs, and sheep and lambs.

Cattle (Beef and Dairy)

The USA had about 97 million head of cattle and calves on 1 January 2002. Texas had approximately 14% of total cattle and calves and ranked first among the states in all categories except milk cows. Ten states (Texas, Kansas, Nebraska, Oklahoma, California, Missouri, South Dakota, Iowa, Wisconsin and Colorado) had 57% of all cattle and calves. Steers weighing 227 or more kilograms totalled 15.8 million with approximately 41% located in Texas, Kansas and Nebraska. The top ten states (Texas, Kansas, Nebraska, Oklahoma, Colorado, Iowa, South Dakota, California, Missouri and Minnesota) had 74% of all steers. Milk cows that had calved accounted for 9.1 million head with 32% located in California and Wisconsin. In 2001, the USA had 814,400 beef cow operations, 97,560 dairy operations, and 92,000 feedlots. About 2,110 of the feedlots had a capacity of at least 1,000 head.

Swine

The hog population is concentrated in the upper Midwest and in North Carolina. As of 1 December 2002, the USA had 59 million hogs and pigs with 53 million classified as market hogs. Iowa, North Carolina, Minnesota and Illinois were the top four states in inventory with approximately 59% of total hogs and pigs, 51% of breeding swine and 60% of market swine. In 2001, the USA had 81,330 swine operations.

Sheep

Sheep are raised primarily in the Midwest and western regions of the USA. The USA had 6.7 million head of sheep and lambs on 1 January 2002. Texas ranked first with approximately 17% of total sheep and lamb inventory and 18% of total breeding sheep. For total market sheep and lambs, Texas ranks second with approximately 14% of inventory. California had the largest inventory of market sheep and lambs with approximately 24% and the second greatest inventory of total sheep and lambs with approximately 12%. In 2001, the USA had 65,120 sheep operations.

Goats

Goat production is concentrated in three states: Texas, Arizona and New Mexico. These states had a total of 312,000 goats in 2001 with 250,000 in Texas. Texas also had 1,250,000 angora goats. The total number of goat operations has not been estimated.

Poultry

The USA had 441 million chickens, excluding commercial broilers, on 1 December 2001. Most of those birds, 338 million, were classified as layers. Iowa ranked first with approximately nine percent of chickens and ten percent of layers. Ohio had nine percent of each category.

Georgia, Pennsylvania and Indiana rounded out the top five states for chicken inventory. For layers, California, Pennsylvania and Indiana also ranked in the top five states.

In 2002, estimated turkey production was 272.4 million. North Carolina and Minnesota produced the most turkeys with 17% and 15%. Arkansas was third with 11% of production.

The total number of broilers produced between 1 December 2000 and 30 November 2001, in states producing more than 500,000 broilers, was 8.4 billion head. Of this total, approximately 15% were produced in Georgia, 14% in Arkansas and 12% in Alabama.

Livestock Slaughter in the USA

Most of the total inventory of commercial livestock in the USA is processed at slaughter plants into commercial meat products and by-products. A significant portion is maintained as a breeding population to preserve the inventory of market animals. A much smaller number are maintained as pets, for research, to improve genetics and for miscellaneous other purposes.

The total number of federally inspected slaughter plants for cattle, calves, hogs, sheep and goats was 2,270 in 2001.

Cattle Slaughter

In 2001, federally inspected cattle slaughter totalled 35 million head. Nebraska ranked first in cattle slaughter with approximately 22% of the total. Kansas and Texas followed closely with about 21% and 18%, and Colorado was fourth with seven percent of cattle slaughter.

Federally inspected slaughter of steers and heifers in the USA was 28 million head in 2001. Kansas and Nebraska each accounted for 23% of steer slaughter. Kansas had 28% of heifer slaughter while Nebraska had 25%. Texas ranked third in steer and heifer slaughter with 20% of each. Colorado was fourth in slaughter of each class with eight percent of steers and 11% of heifers.

US dairy cow slaughter totalled 2.6 million head in 2001. Wisconsin ranked first with approximately 28%. California and Pennsylvania followed in the rankings with 19% and 11%, respectively. Other cows slaughtered, i.e. cows of non-dairy breeds, totalled 3.1 million head. Nebraska accounted for 23% of other cow slaughter. Texas ranked second with 20%, and Georgia and Minnesota followed with 11% and nine percent.

Hog Slaughter

Federally inspected hog slaughter totalled 96 million head in 2001 and two million hogs were slaughtered in non-federally inspected slaughterhouses. Iowa ranked first in federally inspected slaughter with approximately 28% of the total. North Carolina followed with about ten percent, and Illinois and Minnesota each had nine percent.

Federally inspected slaughter of market hogs in the USA totalled 93 million head for 2001. Iowa ranked first in federally inspected slaughter with 29%. North Carolina ranked second with ten percent, and Illinois and Minnesota followed with about nine percent each. Sow slaughter in the USA reached three million head in 2001. Tennessee ranked first with approximately 22% of sow slaughter. Illinois, Iowa and Wisconsin followed in the rankings with 19%, 14% and 13%.

Sheep Slaughter

Sheep slaughter for 2001 totalled 3.1 million head that were federally inspected and 100,000 head non-federally inspected. Colorado ranked first in federally inspected slaughter with approximately 35%. Iowa and Texas followed with about 15% and 14%.

Federally inspected slaughter of lambs and yearlings in the USA totalled 2.9 million head for 2001. As in total sheep slaughter, Colorado ranked first with approximately 37%. Iowa and Texas were next with 16% and 14% of lamb and yearling slaughter. Mature sheep slaughter in the USA was 100,000 head in 2001. Michigan ranked first with approximately 18% of all mature sheep slaughter.

Poultry Slaughter

In 2001, 8.6 billion were slaughtered, of which 8.4 billion were young chickens. Turkey slaughter totalled 269 million birds with 267 million being young turkeys. Duck slaughter was 26 million birds. Georgia and Arkansas ranked first and second, respectively, in young chicken slaughter with approximately 1.2 billion birds each. Alabama was third with just under one billion. Minnesota ranked first with 40 million young turkeys slaughtered and Virginia slaughtered 37 million birds. North Carolina was a close third with 35 million young turkeys slaughtered.

Diseased animals are sometimes discovered at high-volume market animal slaughter plants. More frequently, they are discovered at lower volume slaughter plants that specialize in cull animals, former breeding animals or animals that, for some other reason, are sent individually or in small groups to a cull plant. Other animals die at birth or become diseased or die on the farm and are disposed of through rendering services.

Data Sources for Livestock Mortality

Livestock mortality data are an essential element in determining potential payouts from insurance policies. The NASS publishes death and predator loss data for the USA, by species (NASS, 1996; 2000; 2001).

Cattle (Beef and Dairy)

Death losses in the USA for 2001 were estimated at 1.7 million cattle and 2.5 million calves. The states with the greatest cattle and calf inventories reported the highest total death losses. Average state death loss for all states, as a percent of the 1 January 2002 cattle herd, was 1.9%, with state ranges from one to six percent. The same average for calves was 2.8% with a range of one to 4.5%. Alaska had the highest cattle death loss with 6.1%. Delaware ranked second with 2.7%. For calves, Michigan ranked first with a 4.5% death loss followed by Alaska and Wisconsin with 4.3% and 4.2%.

The infrequently published cattle predator loss reports provide additional death loss data estimated by state (NASS, 1996; 2001). The 1995 report provides losses divided into two classes, cattle and calves, and seven causes: predators, digestive problems, respiratory problems, calving problems, weather, poison and theft. Predation killed 21,200 cattle and 96,200 calves. Digestive problems killed 197,000 cattle and 666,200 calves. Respiratory problems killed 418,950 cattle and 784,400 calves. Weather resulted in the death of 134,000 cattle and 283,000 calves. Poisoning destroyed 32,000 cattle and 16,800 calves while 8,000 cattle and 11,700 calves were lost to theft. In 2001, this report only includes predator loss data, reporting 21,000 cattle and 126,000 calves.

In the 2002 APHIS National Animal Health Monitoring System (NAHMS) dairy report, the abortion rate for participating operations averaged four percent of term pregnancies (calves born alive plus abortions) (APHIS, 2002). Live calves were born to almost 90% of cows and heifers. Death loss for heifers born alive, but that died before weaning, was 8.7%, primarily due to digestive problems. Respiratory problems also caused significant deaths. Between weaning and calving, another two percent of heifers died, with over half due to respiratory problems. Unknown causes, calving problems, mastitis and lameness/injury accounted for a 4.8% death loss in cows.

Swine

Death losses for swine in 2001 were estimated at 7.3 million by NASS (NASS, 2002). Iowa and North Carolina had approximately 46% of all deaths. Death losses by state as a percentage of the 1 December 2001 inventory averaged 10.6% with a range of two to 37%. Texas experienced the highest death loss with 37%. Mississippi and California each had a 27% death loss.

According to the NAHMS Swine 2000 Survey results, the average litter had 10.9 pigs; ten were born alive and 8.9 were weaned (APHIS, 2000). Between December 1999 and May 2000, more than half of pre-weaning deaths resulted from being laid on by the sow. Starvation caused approximately 17% of the deaths and scours (enteritis) accounted for another nine percent. The average sow and gilt death loss was 2.5% to 3.7%, depending on herd size, with larger herds having a higher mortality rate.

Death loss in the nursery averaged 2.6% according to the NAHMS study. The most deaths, 28.9%, were caused by respiratory disease. Starvation and scours represented 13.3% and 12.6%, respectively, of total deaths.

During the period between 1 December 1999 and 31 May 2000, the NAHMS study found an average death loss of 2.9% of pigs in grower-finisher operations. Operations with more animals had a higher percentage loss. Respiratory problems caused the highest death loss at 39.1%.

Sheep

Death losses, reported by NASS for 2001, were estimated at 270,200 head of sheep and 471,300 head of lambs (NASS, 2001). Texas was estimated to have 15% of total death losses for sheep and also for lambs. South Dakota, Colorado and Iowa had the next greatest losses of lambs. Montana, South Dakota, California and Utah followed Texas in sheep death losses.

Death losses for sheep and lambs as a percent of the 1 January 2002 total inventory were also calculated for each state by NASS. For the states, the average losses for sheep were 4.7% with a range of 1.5% to eight percent. Indiana, New York and Nevada had the highest losses. The average losses for lambs averaged nine percent with a range of 1.5% to 16%. West Virginia had the greatest death loss percent, followed by Nevada with a 15% death loss.

The NASS reports provide some additional death loss estimates by state (NASS, 2000). In 1999 and 1994, the reports provide predator losses separated by sheep and lambs. The USA lost 77,000 sheep and 196,000 lambs to predators in 1999.

According to the NAHMS 2001 Sheep report, death loss of adult sheep in 2000 in the range flock system was five percent of the 1 January 2000 inventory (APHIS, 2001). Predators, lambing problems and old age caused over half of these losses. Lamb death loss for 2000 in range flocks was about 11% of lambs born alive. Predators caused the highest percentage of lamb deaths, approximately 60%.

Death loss for adult sheep in farm flocks was five percent of the 1 January 2000 inventory. Over half of these losses were caused by old age, predators and lambing problems. Lamb death loss for 2000 was about nine percent of lambs born alive in farm flocks. Predators also caused the most lamb deaths in farm flocks at 25.5%. Respiratory problems accounted for 16.5% of deaths and digestive problems caused another 15.3%.

Feedlot death loss of lambs and sheep in 2000 was only 2.2% and 4.4%. Digestive problems caused 33.5% of these sheep deaths while respiratory problems caused 54.4%.

Goats

The NASS predator loss report provides additional death loss data (NASS, 2000). In 1999, the report provides predator loss data for goats by type of predator. Texas, New Mexico and Arizona lost 61,000 goats to predators in 1999. The greatest losses were attributed to coyotes, bobcats and dogs.

Poultry

The USA experienced an 11.6% loss of young turkeys in 2002 (NASS, 2003). The western region, with 15.1%, had a significantly higher percentage death loss than the national average while the south central region had the lowest loss rate with 10.8%.

Adequacy of Mortality Data

The adequacy of mortality data depends on the purpose for which they are being used. Aggregated data may be adequate for evaluation of domestic state and federal programmes, but less useful in structuring insurance policies for groups of producers.

Assessing the quantity and quality of data to determine premium rates for individual producers first requires definition of an insurance model. Determining how disease risk to the US livestock herd and to the food supply will be covered must be debated irrespective of the three strategies (listed below) considered in the APHIS study for the Risk Management Agency (RMA) (APHIS, 2003).

1. Indemnity payment by a government agency: published indemnity values for condemned animals and products, destruction and disposal costs, cleaning and disinfection, and welfare care and feeding of animals awaiting destruction.
2. Alternative value insurance: coverage for the value of animals greater than the published indemnity values.

3. Consequential loss insurance: coverage for other losses not paid by indemnity or alternative value insurance, including loss of markets and business interruption.

Those strategies focus on how the insurance plan(s) are financed, i.e. the sources of funds, whether 1) wholly governmental (federal, state, local or some combination thereof), 2) wholly producer financed, including commercial industry operating costs and profits, or 3) some combination of governmental funds and producer payments.

The choice of financing from among these alternatives determines the quantity and quality of data required by government and industry. A plan for indemnity coverage financed wholly by government funds, with ultimate financial liability for covered losses falling upon the US Treasury, can be developed using the aggregate data currently available. However, any amount of producer risk management using commercial insurance in a risk-bearing environment requires more detailed data to provide assurances to commercial underwriters that the policies will yield a profit consistent with market expectations and that catastrophic losses can be broadly dissipated, i.e. that viable reinsurance options are available.

Thus, the role of government in the reinsurance market determines the minimum level of data, both in quantity and quality. Lesser precision is needed if government is ultimately responsible for catastrophic losses that exceed the threshold the commercial market is willing to bear. Finally, a completely commercial environment requires detailed and accurate data to properly assess and price potential risk, both to an individual producer and for catastrophic losses on a wider scale.

The "budget exposure" to the government to compensate producers for losses caused by specific diseases can be approximated with available aggregate data. Budget exposure refers to the potential average payments by government that would be necessitated by enactment of specific legislation intended to assure availability of funds, as opposed to current legislative authorities that require declaration of a disease emergency before specific funds can be utilized. A broad, comprehensive government programme does not target risk to individuals or species; therefore, more detailed death loss data are not required.

However, currently available data do not meet the detail and quality requirements for development of a commercial insurance market. The data do not provide reliable estimates of the risk exposure in individual producer operations, with the possible exception of data describing the number of individual depopulations carried out because of exotic and endemic diseases. It may be possible to develop useful business interruption policies for outbreaks where depopulation is necessary. A joint financial arrangement between the federal government and commercial insurance carriers might result in useful policies based on available data. Such an arrangement could enable the government to meet public and animal health objectives by providing incentives for producers to report diseased animals before the disease can spread. This approach recognizes that positive incentives, as well as legal requirements with penalties, are necessary to ensure the full cooperation of producers in containing the spread of animal diseases.

The incidence of some diseases, for example, mastitis in dairy cattle, represents a normal and expected cost of operation and is well within the limits of a reasonable policy deductible. Such risks are better self-financed than insured. There is no need to collect data to develop insurance policies covering normal and expected operating costs. Development of insurance policies to cover endemic exposures requires specific data collected within the framework of an appropriate probability sample. For some situations, a simulation model developed using expert knowledge may be possible.

There are several programmes underway to develop more and better data. Better data will enable greater participation by the private sector in livestock insurance. Government agencies can also use better data to assess the efficiency of state and federal safeguarding programmes, including protection against acts of terrorism.

Animal Identification and Bioterrorism Data Developments

Safeguarding the US animal herd from disease is the responsibility of both producers and government, especially the federal and state governments working together. The residual risk resulting from how well producers and government carry out their responsibilities must be assessed by both government and the insurance industry. A more comprehensive domestic livestock risk management programme will not develop without better data. Figure 8.1 depicts this situation graphically.

Insurance policies must be priced to provide a profit to the insurance industry, yet be competitively priced so that producers can afford to purchase the policy. If producers cannot justify the premium in terms of the risk protection offered by the policy or if they believe that financial assistance from the government will be forthcoming in a major catastrophe, they

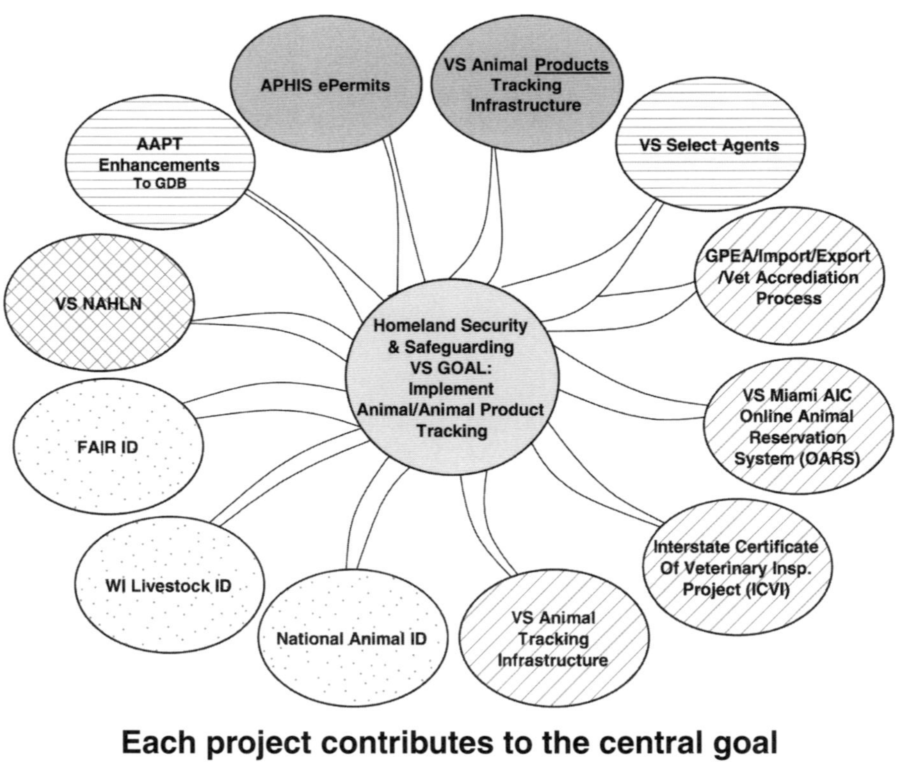

Fig. 8.1. Residual Risk in Livestock Insurance.

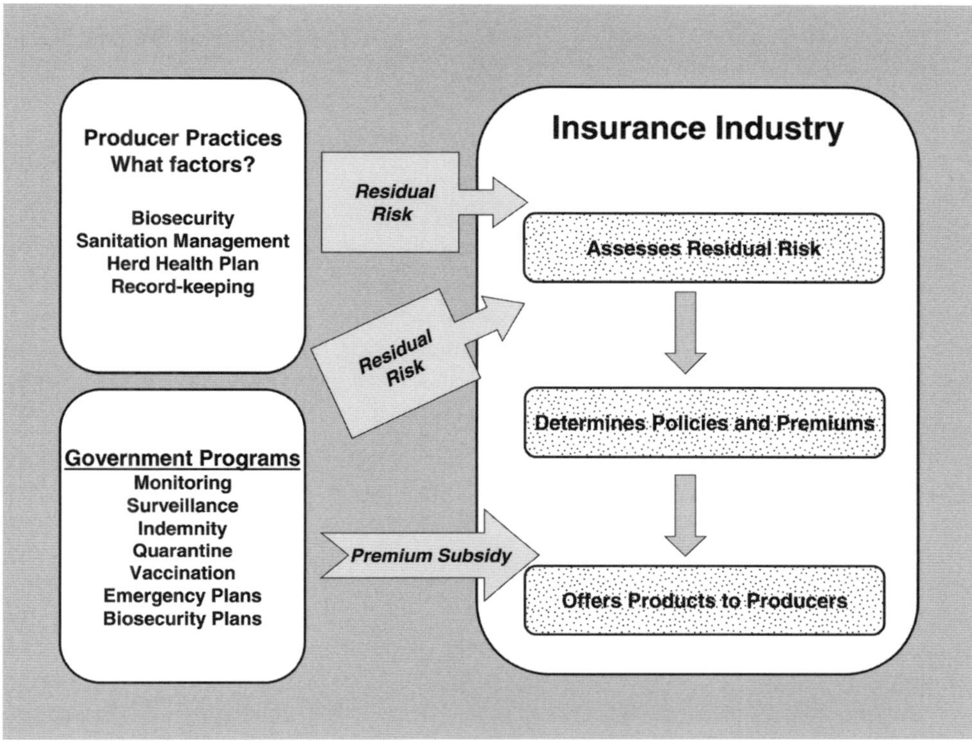

Fig. 8.2. The Animal and Animal Products Tracking System.

will not purchase the protection. In this situation, the insurance industry may ask government to subsidize the cost of the premium. This is the procedure used with crop insurance in the USA, where the USDA subsidizes approximately 60% of the premium cost.[1]

Insurance company representatives evaluate individual producer practices when determining whether a policy will be offered to the requesting producer. The insurance company representative may require additional management protections to reduce disease risk on individual premises. Government programmes, however, are less easily monitored so that the insurance industry cannot easily determine the residual risk resulting from government programmes.

APHIS, through its Veterinary Services, cooperates with the Department of Homeland Security to develop information systems measuring the effectiveness of government programmes in safeguarding the US animal herd from foreign animal diseases, the spread of endemic animal diseases, and the risk of terrorist acts on the US food supply. Jointly funded projects develop these information systems. Figure 8.2 presents the Animal and Animal Products Tracking System while the following discussion summarizes several of the key

[1] Profit is not built into crop insurance rates. If government subsidizes a livestock insurance product, it is not likely that profit loading in the premium will be allowed.

component projects. The colours of the ovals in Fig. 8.1 are not relevant to this discussion but they do group projects according to the source of funds.

US Animal Identification Program

The USA, following the lead of Canada, the EU and Australia, is initiating an animal identification and traceback system.[2] All premises on which animals reside or through which they are marketed are being located. Simultaneously, individual animals or groups of animals are being identified. Databases and software are being developed to provide the capability for 48-hour traceback in the event that a specific animal disease is discovered in the domestic herd or elsewhere in the food chain. Periodic summaries of the data will allow for greater efficiency in the allocation of resources to prevent the entry of livestock diseases into the USA and increase the reliability of monitoring and surveillance programmes.

Data Acquisition and Use

The many forms used by APHIS to monitor disease eradication programmes are being made available on the World Wide Web and the information from the forms is being directed into databases. This allows for quicker and more accurate monitoring of programme success. Interstate animal movements can be summarized from the Interstate Certificates of Veterinary Inspection. Animals arriving in the USA subject to quarantine can be housed and monitored. Permits for the import of animals and animal products can be monitored and tracked to their destination should a disease emergency arise. Exports of US animals and animal products can be tracked to the port of embarkation should our trading partners require assistance in assessing the risk of US products to their food supply or should regionalization be required to maintain export markets.

Monitoring and Surveillance of the Domestic Animal Herd

The Veterinary Services use the Generic Data Base (GDB) to track the activities of state and federal veterinarians as they visit farms, ranches, market centres, slaughter plants and other premises where livestock are gathered. The diagnostic laboratory network tests blood and other fluid samples taken from animals, with the results entered into the National Animal Health Laboratory Network database and coordinated with the GDB. Accredited private veterinarians are tracked using the Veterinary Accreditation System. This interactive web-based system is also used for announcements/warnings affecting the health of the domestic herd and for communication with private veterinarians, the most important link in safeguarding the health of the domestic food animal population.

---

[2] The USDA has been asked by producers to cooperatively develop an animal identification system that will allow the movements of a diseased animal, no matter where it is discovered, to be traced back to the premises of birth. In addition, all other animals that it might have infected must also be located. This entire tracing process must be completed within 48 hours after a positive disease test for the animal is confirmed.

Biologic Terrorism

The Biologic Terrorism Emergency Reporting System uses the above systems, and others, including geographic mapping, to prevent infection of food animals and contamination of the domestic food supply. All of these projects represent a major shift in documentation of the efforts of APHIS in safeguarding the health of the domestic livestock population and derivative food supply. The federal government is working to lower the risk of disease in the livestock population, thus lowering the cost of livestock insurance policies and the amount of subsidization they might require.

Data Requirements for Reasonable Insurance Underwriting and Rating

Insurance underwriting refers to a process of classifying risk into homogenous subgroups of a population. An example is the classification system used for automobiles, wherein the age of the driver, the driving record, type and model of the automobile, its use, locale where garaged, and other characteristics define the amount of premium charged. That premium represents the insurance company's estimate of the total amounts that will be paid to persons in that subgroup for covered losses, the costs to the insurance company associated with that subgroup, and profit.

Greater population heterogeneity requires more and better data for acceptable risk assessment and profitable underwriting. This process creates a need for more detailed and better data because inadequate classification of risk segments in the client base creates the potential for adverse selection. This is a situation in which the pool of insured clients is skewed toward those most likely to incur a loss and require indemnification. The condition arises because the premium cost represents an estimate of the average amount of payment that might have to be made by the insurance company. A heterogeneous population has a greater spread (greater variability) of risk characteristics among clients. Therefore, an average premium cost that is reasonable for the average client will be too high for some clients and too low for others. Thus, it can easily be postulated that the clients purchasing insurance will not be a random sampling of all possible clients. In fact, it can be shown that clients at greater risk will purchase insurance while clients subject to lesser risk will not make the purchase, thus upsetting the assumptions of average payouts used to price the insurance product.

Data needs for underwriting also are a function of the indemnification model. Consider, for example, the Group Risk Plan of crop insurance presently offered by the Federal Crop Insurance Corporation (FCIC). This model determines the indemnity (payment to the insured person) on the basis of an aggregate yield (a county average) for a crop year relative to an expected average yield. This model requires no individual data for underwriting since the actions of a single individual have no substantive impact on the aggregate.[3] The Group Risk Plan is best suited to the insurance needs of producers whose annual yields are highly correlated with the county average. That is, yields must move in the same direction and change in about the same magnitude. Low correlation results in random outcomes in which a producer might be indemnified in a high-yield year (for that producer), but not indemnified in a year when that producer's yields are low.

---

[3] This model requires certain minimum planted acreage in a county to maximize the probability that multiple producers are producing the crop.

Alternatively, an indemnity model that compensates individual growers for reductions in their annual yield requires more information about each individual to properly assess that person's potential for losses. The relative frequency of low yields and the average amount of loss differ among individual producers, depending, in part, on their management acumen.

Another method used to evaluate the risk profile is to examine the extent to which losses are related to management. To illustrate, assume that all producers in an area are equally affected by some adverse weather event. In this situation, there is no need to distinguish among individual producers. However, some producers typically manage better than others, even in adverse situations. This outcome can be related to physical attributes, such as location, but the managerial ability of the individual often plays a major role in the outcome. Again, there is a need to obtain data that adequately discriminate among individuals in order to properly assess relative risk. In the crop insurance programme, the average yield of an individual, relative to the county average yield, provides a measure of relative risk. Producers with lower average yields will have losses more frequently and the premium rate structure will reflect this difference. In this situation, the underwriting is completed automatically by comparing the producer's average yield to the county average yield.

Developing premium rates for insurance products first requires data to estimate the probability that a loss event will occur. Frequency is an estimate of the percentage of the insured population that will incur a loss in a given policy period. Severity, the second variable, is an estimate of the amount or percentage of insured value that will be paid to the average person reporting a loss. Returning to our earlier example of automobile insurance, we might find that the average frequency of loss is 50 per 1000 vehicles per year (5%). With respect to those 50 accidents, the average loss payment might be \$10,000. Thus, the premium amount would be 5% × \$10,000, or \$500 per year. For commercial insurance, this premium must be increased (loaded, in insurance terms) to cover the operating costs of the company and a profit margin. This simple example does not cover the entire range of issues considered in setting an appropriate premium rate, but it does illustrate the basic points. Data adequate to estimate these two parameters are an absolute prerequisite.

In many situations, the two parameters are not estimated separately. The product of average frequency and average severity (the pure premium rate) also can be estimated as a 'loss cost', which is the amount of indemnity payments divided by the insured liability. Using this procedure, a 'trigger' is compared to a realized amount, such as annual survival rate (total herd population minus death loss). The trigger represents a threshold that must be penetrated before any indemnity would be paid, i.e. it is comparable to a deductible. For example, assume the deductible is 5%, and the annual death loss incurred by the producer in a particular year is 11%. The threshold for survival rate is 95% (100% minus the 5% deductible), and the survival rate for that year is 89% (100% minus the 11% death loss). The amount of loss payable, expressed as a loss cost, is 95% minus 89% divided by 95%, or 6.32%. Assume we have 20 years of these observations. The loss cost in some years would be zero percent (death loss was less than 5%) and would be greater than zero in other years. An average of all 20 annual loss costs, including the zero values, provides an estimate of the average loss cost. This average loss cost is an estimate of the required premium rate.

Developing adequate premium rates for individually triggered agricultural insurance policies is hampered by the paucity of adequate records for individual operations. Few producers maintain detailed records for prolonged periods of time; in fact, many may have only a few years of production experience. But agricultural risks differ materially among years. In some years, conditions are favorable and there are few or no loss events. In other years, the losses may be pervasive. Consequently, defining the average amount of loss on an individual basis is a daunting task. One method to overcome this problem is to utilize a long-term series of observations at some aggregate level (such as a county), supplemented with available observations from individual producers for the most recent years. Using appropriate statistical techniques, the higher variance of the individual observations can be 'boot-strapped' to the long-term data series to estimate the range of probable outcomes that would have been observed if the data had been available.

Conclusions

The market for livestock insurance in the USA is large enough that the private sector is developing instruments to protect producers against risk. There are millions of animals of the major commercial meat and meat product species. The majority of animals are produced on thousands of individual farms and sent directly to slaughter. Breeding animals are housed to maintain the commercial slaughter herd.

Many variables will influence the development of a comprehensive risk management programme for commercial livestock producers in the USA. A major determinant will be the evolution of the federal government indemnity payment programme. Private sector programmes will either supplement the federal indemnity programme or replace it.

Assessing the quantity and quality of data to determine premium rates for individual producers first requires definition of a risk management strategy. The role of government in paying indemnity and in the reinsurance market determines the minimum level of data, both in quantity and quality. Less precision is needed if government pays indemnity and is ultimately responsible for catastrophic losses that exceed the threshold the commercial market is willing to bear. Currently available data do not meet the detail and quality requirements for a commercial insurance market. The data do not provide reliable estimates of the risk exposure in individual producer operations, with the possible exception of data describing the number of individual depopulations carried out because of exotic and endemic diseases.

Safeguarding the US animal herd from disease is the responsibility of both producers and government, especially the federal and state governments, working together. The residual risks resulting from how adequately producers and government carry out their responsibilities will be assessed by the insurance industry and the federal government as policies are developed for livestock producers. These policies must be priced to provide a profit to the insurance industry, yet be competitively priced so that producers can afford to purchase the policy. The insurance industry may ask government to subsidize the cost of the premium.

Insurance company representatives evaluate individual producer practices when determining whether a policy will be offered. The insurance company may require additional management protections to reduce disease risk on individual operations. Government programmes, however, are less easily monitored so that the insurance industry cannot easily determine the residual risk resulting from government programmes.

Insurance underwriting and premium rates depend on the insurance strategy, the heterogeneity of the group or subgroup, and the variability in year-to-year outcomes. Developing a premium rate requires data to estimate the frequency and severity of loss. Data relating to the risk characteristics of individuals in the population to be insured are also needed, as well as data describing the effectiveness of government safeguarding programmes.

Information requirements will depend on the structure of the comprehensive programme, but will certainly be greater than the data currently available. Safeguarding the domestic animal herd and food supply is also expanding beyond considerations of accidental introduction of a foreign animal disease.

References

APHIS (2000) *Swine 2000 Part III: Swine Health & Environmental Management in the United States*. US Department of Agriculture, Animal and Plant Health Inspection Service, Centers for Epidemiology and Animal Health, Fort Collins, Colorado.

APHIS (2001) *Sheep 2001: Part IV: 2001 Sheep Feedlot Health and Management*. US Department of Agriculture, Animal and Plant Health Inspection Service, Centers for Epidemiology and Animal Health, Fort Collins, Colorado.

APHIS (2002) *Dairy Cattle: Part III: Dairy Cattle Health and Health Management Practices in the United States*. US Department of Agriculture, Animal and Plant Health Inspection Service, Centers for Epidemiology and Animal Health, Fort Collins, Colorado.

APHIS (2003) *Disease Risk Management Products for Livestock and Poultry Producers*. US Department of Agriculture, Animal and Plant Health Inspection Serice, Centers for Epidemiology and Animal Health. Research Report submitted to USDA-Risk Management Agency under FDIC-APHIS Interagency Agreement Number (2002) 02IA0831-015.

NASS (1996) *Cattle Predator Loss*. US Department of Agriculture, National Agricultural Statistics Service, Washington, D.C.

NASS (2000) *Sheep and Goats Predator Loss*. US Department of Agriculture, National Agricultural Statistics Service, Washington, D.C.

NASS (2001) *Cattle Predator Loss*. US Department of Agriculture, National Agricultural Statistics Service, Washington, D.C.

NASS (2002) *Meat Animals Production, Disposition, and Income: 2001 Summary*. US Department of Agriculture, National Agricultural Statistics Service, Washington, D.C.

NASS (2003) *Turkeys Raised*. US Department of Agriculture, National Agricultural Statistics Service, Washington, D.C.

National Institute for Animal Agriculture (2003) *United States Animal Identification Plan*. National Identification Development Team, Version 4.1, December 23.

Chapter 9

Public and Private Schemes Indemnifying Epidemic Livestock Losses in the European Union: a Review

Marcel A.P.M. van Asseldonk, Miranda P.M. Meuwissen, Ruud B.M. Huirne and Eckhardt Wilkens

Introduction

Epidemics in livestock, such as Foot and Mouth Disease (FMD) and Classical Swine Fever (CSF), may inevitably affect many farms at the same time, causing severe economic losses in the EU and other parts of the world.[1] Around the world, many countries are officially free of FMD (without vaccination), other countries vaccinate against the disease (preventive vaccination), and in some countries FMD is endemic. In the EU, preventive vaccination has been prohibited since 1991. After 1991, FMD epidemics occurred in Greece and Italy (1993, 1994 and 1996) and the UK, Ireland, France and the Netherlands (2001-2002). CSF is also a highly infectious virus posing a threat to livestock production. At the beginning of the 21st century, CSF is still endemic in many parts of the globe. Successful eradication has been achieved in many countries, including those of North America, Australia and parts of northern Europe, and in the absence of vaccination, has contributed to a totally susceptible swine population (Edwards *et al.*, 2000). Preventive vaccination was stopped in all EU member states in the early 1990s (Westergaard, 1991). In the wild boar population CSF is still endemic in Germany and Italy (Fritzmeier *et al.*, 2000). In the 1990s large CSF outbreaks (more than 40 farms infected) occurred in the Netherlands (1997), Germany (1993, 1994, 1995, 1997, 1999, 2000), Belgium (1994) and Italy (1995, 1996, 1997) (Laevens, 1998; Handistatus II of the OIE, 2001).

Member states are obliged to apply the control measures laid down in EU directives if an outbreak arises of so-called 'List-A diseases' (Office International des Epizooties, 1998). The basis for these measures originates from EU Council Directives 85/511/EEC and 80/217/EEC. Measures include 1) stamping-out of infected herds; 2) pre-emptive slaughter of contact herds; and 3) the immediate establishment of surveillance zones around such herds. In these zones, animal movements are restricted and to a large extent prohibited. Depopulated farms may repopulate 21 days (FMD) or 30 days (CSF) after the cleaning and disinfecting of the

[1] This review is based on the 15 European Union member states before the enlargement that took place with 10 new countries joining in 2004.

farm (7-10 days after diagnosis), or, after the lifting of restriction zones (lifted only after clinical and serological tests). The latter generally takes much longer than 21-30 days. As an example, during the 1997/98 epidemic of CSF in the Netherlands, many pig farms were in restriction zones for more than six months.

After obtaining EU approval, member states may take additional control measures. If restriction zones with a movement standstill lead to severe animal welfare problems on the farms (possible with 25-kg piglets on farrowing farms, and on farms with 110-kg hogs and veal calves), so-called welfare slaughter is generally applied. Also, a more comprehensive pre-emptive slaughter scheme can be applied to control the disease more effectively. Furthermore, all susceptible animals within a large area around the infected herds might be vaccinated (emergency vaccination, "ring vaccination"). For example, in the 2001 FMD epidemic in the Netherlands, the Dutch government decided on a temporarily complete movement standstill in the whole of the country, including horses and poultry, and the transport of feed and animal products, such as manure and milk. Also, herds within a 1-km radius of contact herds were pre-emptively culled. Furthermore, all susceptible animals within a large area around the infected herds were vaccinated and destroyed afterwards. A "welfare measure" taken during the 1997-1998 CSF epidemic in the Netherlands included a breeding prohibition.

Obviously, these livestock epidemics can have large economic consequences not only for farmers but also for other various parties of the production chain in terms of direct and consequential losses. *Direct losses* comprise the value of the animals destroyed under depopulation and welfare control measures and the costs of organizational aspects, such as the monitoring of farms in restriction zones. *Consequential losses* that arise at farm level can comprise one or more of the following categories:

Business interruption: business interruption occurs because farm buildings become (partly) empty due to stamping-out and welfare slaughter or breeding prohibition, and stay empty until restriction zones are lifted.

Losses related to established restriction zones: farms in restriction zones face (potentially long) periods in which animals (e.g. finishing pigs and veal calves) and manure cannot be transported from the farm. These periods are characterized by animal welfare problems, extra feeding costs, and emergency measures for housing of pigs and storage of manure. Such losses will vary widely across farms and are therefore complicated to measure. Milk from dairy farms in restriction zones might be collected taking into consideration strict hygienic measures.

Additional repopulation costs: these losses include extra costs of animal health problems.

Losses from emergency vaccination: given a situation, in which vaccinated animals are destroyed, losses might arise from the above categories (business interruption, repopulation costs). However, for reasons of social acceptability, the rendering of vaccinated animals is under debate. With future epidemics, meat and milk from vaccinated animals may be destined for the local market, which likely leads to extra costs and/or lower prices. Something similar may be applied to animals under welfare slaughter programmes.

Price effects: livestock epidemics can have a rather severe impact on prices, especially meat prices. The impact depends on aspects such as the size of the epidemic (duration, size of restricted area), reactions of other countries (closure of borders, increased production), and whether vaccination is applied (which generally leads to long periods of export limitations).

Table 9.1. EU financing schemes covering direct and (part of) consequential losses resulting from livestock epidemics.

	Contribution sector direct losses[1]	System consequential losses[2]		
		Private	Public–Private	Public
Austria	+	–	–	–
Belgium	+	–	–	–
Denmark	–	–	–	+
Finland	–	–	–	+
France	–	–	–	+
Germany	+	+	–	–
Greece	+	–	+	–
Ireland	–	–	–	–
Italy	–	+/–	–	–
Luxembourg	–	–	–	–
Netherlands	+	+	–	–
Portugal	–	–	–	–
Spain	–	–	+	–
Sweden	–	+	–	+
United Kingdom	+/–	+	–	–

[1] –: no levy, +: (partly) levy or compulsory insurance scheme.
[2] –: not available, +: (partly) coverage and in the case of private insurance scheme more than 1% participation (ratio between insured livestock and registered livestock).

The direct losses are (partly) compensated by governments (national and European). Consequential losses are almost always completely borne by the farmers themselves if not insured privately. In some member states, the consequential loss exposure is transferred by means of private insurance schemes. This chapter reviews how EU member states finance the direct losses and consequential losses, focusing on state support (national and European) and risk transfer by means of private insurance schemes. First, a review of the compensation scheme per EU member state for direct losses is provided. Second, existing insurance schemes per EU member state are elaborated upon. Finally, the need for and feasibility of developing "a European system" is discussed.

Compensation of Direct Losses

The veterinary budget of the EU refunds, in most cases, 50% of the costs of compulsory and pre-emptive slaughter, 70% of the costs of welfare slaughter and 50% of the costs of organization (see Council Directive 90/424/EEC; Ministry of Agriculture, Nature Management and Fisheries). The financing of the non-EU compensated part of the direct losses differs between the EU member states.

A number of member states finance the direct losses entirely from the national budget (see Table 9.1). Member states that have no statutory or voluntary levies to establish an emergency fund are Denmark, Finland, France, Ireland, Italy, Luxembourg, Portugal, Spain,

Sweden and the UK. In general, the national government pays only for the value of the animals that are compulsorily culled. The commercial impact of movement restrictions or other controls is not eligible for compensation. In Denmark a further 20% of the animal value is paid, funded by the farmers by means of a levy, to cover the loss of income from the herd if a whole herd is culled. In some member states and for specific diseases only non-diseased animals are compensated. For example, as far as epidemic poultry diseases are concerned, such as Avian influenza and Newcastle disease, only birds culled which are non-diseased are compensated in the UK (at 100% of their market value). A further exclusion is the fact that sometimes it is not clear *a priori* whether the losses are compensated from the national budget. For example, the UK government compensates for Aujeszky's disease in pigs, in line with other epidemic livestock diseases, 100% of the animal's market value. The only difference is that when Aujeszky's disease was in the country, a levy was paid on all pigs at slaughter to cover the compensation costs. This levy is no longer collected but the legislation is in place to collect it again should the disease re-enter the country.

Other member states have set up some form of a statutory system to co-finance the non-EU compensated part of the direct losses (Austria, Belgium, Germany and the Netherlands). These public-private financing schemes have a compulsory fund structure in which all farmers pay a levy. Risk financing by means of a levy system is based on pooling over time within the sector. Payments to the fund can be organized through up-front payments (deposits) or through assessment payments after an epidemic, or both. However, most member states have opted for an assessment approach. These latter systems have no annually fixed levies. The government will finance the compensation payments in advance. The input of the government will however be repaid over the following years. Therefore, after an epidemic, the levy is set according to the amount that the government paid in advance for the sector. Note that the levy can, and in most cases will, also vary between species.

In the case of co-financing to complement the public payment, the amount that is financed by the sector can be proportional, non-proportional or both. If risks are shared between the sector and the national government by means of a proportional contract (i.e. *pro rata* contract) the levy is specified as a fraction of the coverage. With non-proportional contracts, the national government indemnifies only claims in excess of a particular threshold (see Fig. 9.1).

Member states that have a proportional levy system to establish an emergency fund are Austria, Belgium and Germany. In Germany the fund is established by the Bundeslander (the

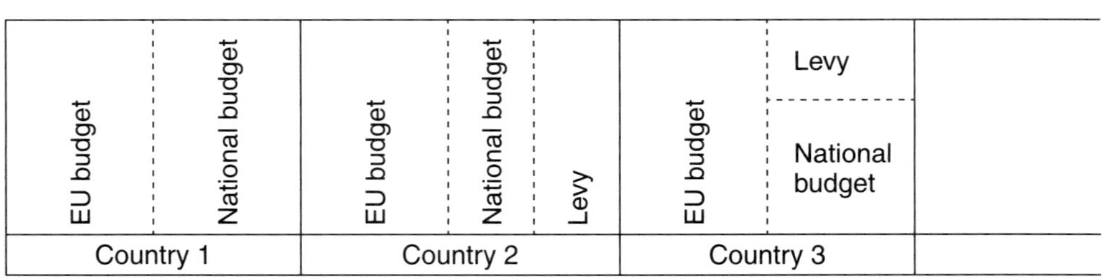

Fig. 9.1. Example of existing financing schemes covering direct losses of livestock epidemics in the EU.

same holds for Austria). The Bundeslander draws up the detailed rules of the application. The programme is run by an administrative council that decides the level of the levy. The administrative council is made up of farmers and ministry representatives. The compensation payments are made from the available funds and the Ministry of Agriculture will pay for the costs if the fund runs out of money. The levy is only used to co-finance the EU veterinary measures following a disease outbreak. It therefore only pays for culled animals under EU veterinary measures. No compensation is paid to farmers in surveillance zones. The Lander and the levy fund each pay half of the remaining non-EU compensated part (see also Fig. 9.1). In Belgium the levy can vary depending on the level set by the government and is differentiated on the basis of species and farm size. With respect to pig production, the premium is differentiated on the basis of farm structure (e.g. open versus closed pig production). Note that differentiation between species and farm structure is determined post-epidemic and based on occurred losses per species as well as on subjective criteria.

The private bank guarantee system in the Netherlands (i.e. levy system) is in the form of a non-proportional contract. Within this system the government can withdraw capital without prior approval of the farming sector from a private bank to co-finance actual losses. Any capital provided by the bank is paid back with interest by the primary sector through assessment payments over a certain time horizon. For example, Dutch pig producers will have to cover potential losses for up to 227 million euro in the five-year period 2000-2005 (the same amount and time horizon holds for the cattle sector).

The system in place in Greece is not a levy system but the sector co-finances the non-EU compensated part of the direct losses. The Greek government operates a compulsory agricultural insurance scheme by the Greek Agricultural Insurance Organization (ELGA). ELGA has the objective of organizing and implementing programmes of proactive protection and ensuring the production and assets of agricultural enterprises. More specifically, insurance with ELGA includes compulsory insurance against damage that is caused to animal assets of farmers. Persons who own breeding stock, poultry or domestically produced products and by-products of animal origin shall be subject to insurance. ELGA is funded by an "income from special insurance contributions" (of which the fee is 0.5% of the value of the sold livestock production) and this constitutes the major financial source.

In summary, currently the values of the animals that are compulsorily culled are compensated by means of a public or public-private financing vehicle. While some member states finance the direct losses from the national budget, other member states have set up some form of statutory assessment system. The amount that is payable by the farmer depends mainly on whether or not there were major epidemics in previous years.

Compensation of Consequential Losses

In this section, the current EU financing schemes covering consequential livestock diseases are reviewed. The results are obtained from the literature and a survey among members of agricultural representatives of Comité Européan des Assurances (CEA).

Some governments (e.g. Denmark, Finland, France and Sweden) provide financial assistance for consequential losses without requiring payments from the sector (see Table 9.1). Such government programmes can either be formalized by a formal public compensation scheme or by ad hoc disaster payments. In the case of a formal public programme, the risks covered are specified *a priori,* while disaster relief programmes usually operate only after a widespread disaster on an *ad hoc* basis (Skees and Barnett, 1999). Affected producers are uncertain whether or not and to what extent they will be indemnified by means of the disaster relief programme.

In Denmark a formal public programme is in place whereby the government pays a further 20% to cover the loss of income from the herd. In Sweden a farmer receives a formalized compensation from the government that is calculated as the difference between the actual profit and the expected profit if the farm was still engaged in production. Compensation is 50% in the case of Salmonella and Paratuberculosis. For diseases such as FMD, CSF and BSE the compensation for consequential losses is 100%.

The Finnish and French governments can provide financial assistance for consequential losses via *ad hoc* disaster payments. The current standard policy in Finland is to compensate business interruption losses when authorities have ordered culling of the herd. Since there is no money reserved for it, the policy is subject to available finance arrangements. When epidemics occur in France these losses are paid by public authorities according to the financial possibilities and importance of the epidemic.

A number of member states have a form of public-private partnerships whereby both governments and the private sector contribute to cover consequential losses. Within such a partnership the government commonly functions either as an insurer or as a reinsurer for the subsidised consequential loss policy. Another option is that the government subsidizes the premiums directly. In cases when the government is the primary insurer, the public insurance policies are often retailed by the private sector. The private companies receive a commission but retain no loss risk (Skees and Barnett, 1999). In the case of a public-private partnership with governmental reinsurance, the private insurer both retails and services the insurance policy, while retaining a part of the loss risk (Meuwissen *et al.*, 2003). However, the policies are reinsured not solely through the reinsurance market but also by the government, either as a quota share or stop-loss provision. Quota share provisions specify what percentage of premiums and loss exposure the private company will retain, with the remainder being passed on to the reinsurer. Stop-loss provisions specify the maximum amount of loss that the company will have to cover before the reinsurer covers the additional losses (Skees and Barnett, 1999). Public-private partnerships, in the sense that governments subsidize a consequential loss policy, are scarce in EU member states. An example is the Spanish "AGROSEGURO" system where farmers can insure against disease outbreaks, though only for cattle (limited to only breeding animals) as well as sheep and goats. The insurance covers the difference between the actual level of aid farmers receive when an animal is culled and the market value. These policies are subsidized by the government up to a maximum of 41% in the case of cattle and 32% in the case of sheep and goats.

In some member states, the absence of governmental assistance for consequential losses has led to the creation of private insurance schemes for some types of livestock production. There are two broad approaches to insurance provision – namely mutual and limited insurance companies (Harrington and Niehaus, 1999; Rejda, 1998) – both of which have been employed in business interruption insurance schemes. Both types of insurance companies are allowed to create policies according to which premiums can be (partly) collected after claims. Since mutuals are owned by the insured farmers, there is likely to be a broader support for differentiation of premiums and deductibles since colleague farmers impose these measures. A further implication of farmers owning the mutual is that they are very critical of what type of farmers are accepted as members of the insurance pool. Existing mutuals therefore apply strict underwriting criteria, for example by requiring certain certificates such as GAP (Good

Agricultural Practices). Note that, in theory, limited insurance companies can also apply these enhancements. Farmers would probably be more willing to participate in a pool created by farmers themselves, although it may well be the case that they could be more prone to invest in a solvent, large and well-known insurance company.

Few private insurance schemes exist on the European market to cover the risk of consequential losses from livestock epidemics (e.g. Germany, Italy, Sweden, the Netherlands and the UK). Those that do exist are either extensions of general livestock insurance policies or specific policies of limited insurers and mutual insurers. Many standard livestock insurance policies in Europe indemnify farmers for animal losses as a result of a number of perils, but some have been extended, sometimes as an option, to cover at least a part of consequential losses from epidemics. Most general livestock insurance schemes cover death and emergency slaughter due to illness. Also, the risk of accident, theft, contamination of products, and fire and storm are amenable in most policies for pooling. A more comprehensive policy that also covers more farm-related diseases (e.g. mastitis and claw disorders) is rare, but does exist.

The additional consequential loss coverage can have the following contract specification: 1) based on a proportion of the insured sum of the value of the livestock, or 2) based on the period with business interruption or 3) based on the actual losses (a typical business interruption coverage). In most cases the farmer chooses, within a certain range, the value of the livestock and the daily gross margin. The indemnification of the third approach is based on the difference between the actual gross margin after the loss event and the insured gross margin. The insured need to have accurate accounting records for a number of subsequent years. In general, farms that are confronted with losses as a result of decreased market value of their products, but are not infected with an epidemic disease or are not in a movement standstill zone, are not eligible for compensation. In addition, economic losses as a result of only movement standstills do not trigger indemnity payments given a coverage based on a proportion of the insured sum or period with business interruption.

As with all private insurance policies, the policies exclude direct losses that are met by the public sector. Additional constraints include a probationary period, a maximum coverage period, a multi-year policy term, a maximum insured amount, a maximum indemnification amount and a deductible. The premium is in some policies differentiated on the basis of risk parameters (e.g. open or closed farms and between regions).

Examples of insurance schemes that are not arranged as typical business interruption coverage are those in the Netherlands and the UK. The additional coverage in the Netherlands (only available for cattle) is either a proportion of the insured sum of the culled animals (ranging from 10% up to 30%), or is based on the period with business interruption (in some policies a limited period). In general, the indemnification is based on the number of cows that die or are culled. Covered perils are Brucellosis, BSE, Contagious Bovine Pleuropneumonia, FMD, Rinderpest, Tuberculosis and Vesicular Stomatitis. The participation level is less than 10%. The insurance scheme in the UK covers FMD, CSF and SVD and pays a selected percentage (usually 25%) of the government compensation. For Aujeszky's there is also insurance to contribute towards consequential losses that work in the same way as for FMD, CSF and SVD. Insurance is also available to complement the compensation for animals culled due to bovine tuberculosis, brucellosis and BSE in cattle. The compensation and insurance schemes are more complicated for these diseases. The NFU (National Farmers Union) Mutual Insurance Society does not provide any schemes covering consequential losses for poultry diseases. Approximately 10% of farmers insure against FMD, 10-15% of dairy farmers insure

Table 9.2. Example FMD coverage consequential loss insurance in Germany (R+V Ertragsschadenversicherung).

Computation of claim	Before claim	Period of claim	Claim
Delivered milk (kg/year)	292,500	165,000	127,500
A: Milk revenue loss (euro/year)			47,175
B: Reduced variable costs (euro/year)			11,475
C: Expenses livestock replacement (euro/year)	13,300	93,500	80,200
D: Compensation payments relief fund (euro/year)		83,000	83,000
E: Additional expenses (euro/year)[1]		8,000	8,000
Total loss (euro/year) (A – B + C – D + E)			40,900
F: Franchise (euro/year)			5,880
Indemnification (euro/year) (Total loss – F)			35,020

[1] Expenses for 1) disinfection and cleaning; 2) special disposal of animal feed and milk; 3) additional veterinary costs, medicines and laboratory diagnostics.

against tuberculosis, and approximately 10% of dairy farmers insure against brucellosis. For the other insurable diseases, less than 5% of farmers insure against them.

In contrast to the previous schemes, the Italian insurance scheme covers not only business interruption losses as a result of depopulation, but also losses in movement standstill zones will trigger indemnity payments. The additional consequential loss coverage is only available for dairy cows and sheep. The level of participation is very limited (<5%). Covered perils are Brucellosis, FMD, Tuberculosis and Leucosis.

In Germany the private "Ertragsschadenversicherung" indemnifies farmers against the full range of consequential losses as one of the coverage options, including those resulting from movement standstills. The indemnification is based on the difference between the actual gross margin after the loss event and the insured gross margin, taking into account a deductible, and a maximum covered period of 12 months. Losses as a result of movement standstills are indemnified at a 50% level. The insured need to have accurate accounting records for three years. The premium is a percentage of the insured sum, which is based on the value of the livestock and the gross margin per year. The farmer chooses, within a certain range, the value of the livestock. The standard gross margin differs between the performance levels. Participation level of farmers with dairy cows is approximately 50%, cattle 30%, sows 42% and hogs 23%. Insurance for sheep is not available. Insurance for poultry will not be continued (participation in 2003 was approximately 5%).

A coverage example of the "R+V Ertragsschadenversicherung" is depicted in Table 9.2. In the case of the EVT-S policy, the following perils are covered: 1) notifiable animal diseases such as FMD, BSE, tuberculosis, salmonellosis, leucosis; and 2) accidents such as rupture of the slatted floor, and breakdown of the ventilation system. In the case of the more comprehensive EVT-N coverage, the following perils are also covered: 3) other contagious diseases such as mastitis, BVD/MD, diseases of claws, pneumonias; 4) contamination by substances such as PCB; and 5) theft. Indemnified eligibility is based on changes in production and expenses in the case of damage, including: 1) compulsory culling of the livestock combined with a lockout period of three months; 2) reduced milk production under a milk quota because of empty buildings and replacement with pregnant animals; 3) loss of raised animals after rearing; 4) increased expenses for health management (veterinary, medicines, laboratory diagnostics); 5) expenses for disinfection; 6) special expenses for disposal of contaminated feed and milk; and 7) replacement costs for animals are considered by the relief fund for epizootics.

The example is based on a liability coverage of the loss of income for a dairy farm of 45 dairy cows, an average milk production of 6,500 kg/cow/year (delivered annual milk volume is 292,500 kg) and a gross milk price of 0.37 euro/kg. The annual net premium amounts to 365 and 470 euro for the EVT-S and EVT-N policies. This assumes a decreased milk production of 127,500 kg as a result of an FMD infection at the farm, resulting in a difference in milk revenue of 47,175 euro. In combination with other losses, the overall loss is assumed to be 40,900 euro, resulting in an indemnification of 35,020 euro.

Discussion of Current and Prospective Financing Schemes

Prospective Financing Schemes

Livestock epidemics, such as epidemics of CSF and FMD, can result in substantial losses for governments, farmers and all the other participants of the livestock production chain involved. Governments (national and European) generally bear the largest part of the direct losses, such as the value of destroyed animals and organizational costs. A number of member states finance the non-EU compensated part of the direct losses entirely from the national budget. Other member states have set up some form of statutory system to co-finance the direct losses. These public-private financing schemes have a compulsory fund structure in which all farmers pay a levy. In the case of co-financing being used to complement the public share, the amount that is financed by the sector is either proportional or non-proportional.

Only a limited number of member states offer free public disaster assistance or compensate above the value of the animals that are compulsorily culled to cover part of the consequential losses. Public-private partnerships, in the sense that national governments subsidise a consequential loss policy, are scarce. In some other EU member states, the absence of governmental assistance has led to the creation of private insurance schemes for consequential losses for some types of livestock production. The current applied consequential loss coverage can be based on the actual losses incurred, on an estimation of the loss based on the period with business interruption or on a fixed amount. Indemnification of the first approach is based on the difference between the actual gross margin after the loss event and the insured gross margin. The insured need to have accurate accounting records for a number of subsequent years. In general, farms that are confronted with losses as a result of decreased market value of their products but are not infected with an epidemic disease or are not in a movement standstill zone are not eligible for compensation.

Based on the survey, it can be concluded that producers do not commonly take up private policies that are specifically designed to cover consequential losses (a positive exception is the consequential loss insurance in Germany). Causative factors can be found related to both the demand and supply sides. With respect to the demand side many producers evaluate animal health as a less important business risk. Other problems are cognitive failure to assess probability and extent of low-probability-high-consequence risks and the fact that schemes are new and farmers may need time to adopt (the familiar adoption lag). With respect to the supply side, problems are in the field of research and development costs, and difficulties in dealing with asymmetric information problems and systemic risks (maybe insurers have been very conservative in setting premium rates until they accumulate a reasonable claim history).

Prospective Financing Schemes

Due to various developments (such as the enlargement of the EU and budgetary constraints), the current risk-financing system for livestock epidemics is being reconsidered. Emerging schemes should as much as possible fulfill the following requirements: 1) no disturbance of markets; 2) compatible with WTO (World Trade Organization) agreements; 3) run by the private market without official EU participation; and 4) applicable to the entire EU.

Regardless of the requirement of no market distortion and no official EU participation, all EU member states finance (partly) direct losses and some partly finance consequential losses from the national budget. WTO agreements increasingly restrict the amount of subsidies that are allowed. To be classified under the green box (i.e. the allowed forms of support) the support measures have to fulfill certain conditions. The conditions relate to the absence of price support, little or minimal trade-distorting effects and no effects on production. Public payments for relief programmes (made available either directly or by way of financial participation via levy schemes) have to fulfill basically the same conditions. There are, however, a number of differences. Payments can only be triggered by a production loss resulting from a disaster, which is specifically recognized by the government. Payments can be based on losses of livestock but also on depressed income, and can compensate up to 100% of the total costs. *Ad hoc* disaster relief programmes need to be evaluated on a case-by-case basis as to whether they are in line with the WTO agreements. Public financial participation in insurance is classified as green box compatible if: 1) insurance relates to income shortfalls based on a reference period. The payments may not relate to the type or volume of production or the prices related to such production or to the factors of production employed; and 2) income loss is more than 30% and the payment amounts compensate for less than 70% of the farmers' income loss (European Commission, 2001).

The current (partly) public levy programmes seem legitimately in the green box of the WTO-framework. Governments could facilitate the development and adoption of an insurance scheme for consequential losses. If reinsurance problems are hampering the development of insurance products or result in relatively high premiums, a public-private partnership for the reinsurance of the risk could be considered (possibly in combination with a "Pool Re Europe" design or other alternative risk transfer solution). To prevent inefficiencies, such public-private partnerships need to be properly designed to meet strict criteria.

Given the specific risk under consideration, a mandatory system to finance direct losses is essential in order to facilitate alertness and rapid disclosure in the case of an outbreak of an epidemic. In contrast, a consequential loss compensation scheme might be voluntary (Van Asseldonk and Meuwissen, 2004). Harmonization of EU financing schemes is desired because it will at least level the playing field between farmers within the EU member states. However, such an EU policy that would encompass harmonization of schemes is difficult to accomplish and not foreseen in the near future.

References

Edwards, S., Fukusho, A., Lefèvre, P.-C., Lipowski, A., Pejsak, Z., Roehe, P. and Westergaard, J. (2000) Classical swine fever: the global situation. *Veterinary Microbiology* 73, 103-119.

Fritzmeier, J., Teuffert, J., Greiser-Wilke, I., Staubach, Ch., Schlüter, H. and Moennig, V. (2000) Epidemiology of classical swine fever in Germany in the 1990s. *Veterinary Microbiology* 77, 29-41.

Handistatus II of the OIE. (2001) Office International des Epizooties. [Accessed 2003.] Available from http://www.oie.int/hs2/report.asp.

Harrington, S.E. and Niehaus, G.R. (1999) *Risk Management and Insurance.* Irwin McGraw-Hill, New York.

Laevens, H. (1998) Epizootiology of classical swine fever: experimental infections simulating field conditions, and risk factors for virus transmission in the neighbourhood of an infected herd. PhD dissertation, University of Ghent, Belgium.

Meuwissen, M.P.M., van Asseldonk, M.A.P.M. and Huirne, R.B.M. (2003) Alternative risk financing instruments for livestock epidemics. *Agricultural Systems* 75, 305-322.

Office International des Epizooties. (1998) Diseases Notifiable to the OIE. [Accessed 2003.] Available from http://www.oie.int/eng/maladies/en_classification.htm.

Rejda, G.E. (1998) *Principles of Risk Management and Insurance.* Addison Wesley, New York.

Skees, J.R. and Barnett, B.J. (1999) Conceptual and practical considerations for sharing catastrophic/ systemic risks. *Review of Agricultural Economics* 21, 424-441.

Van Asseldonk, M.A.P.M. and Meuwissen, M.P.M. (2004) Financing strategies to manage epidemic animal disease costs in the UK. *Farm Management* 11, 686-697.

Westergaard, J. (1991) The effect of the EEC internal market on trade of animals and animal products, disease situation and control programmes. In: Eriksson (ed.) *The Importance of Animal Disease for Trade, Food and Public Health in an Integrated Europe.* 8-29. Stockholm, Report Number 56.

Chapter 10

Designing Epidemic Livestock Insurance

Miranda P.M. Meuwissen, Marcel A.P.M. van Asseldonk, Jerry R. Skees and Ruud B.M. Huirne

Introduction

In designing insurance schemes for epidemic diseases such as Foot and Mouth Disease (FMD) and hog cholera, there are several typical aspects of the schemes that need to be addressed. Probably the most important aspect is the large influence of *human behaviour* on the course of an epidemic. Applying proper sanitary rules and obeying enforced animal movement restrictions are crucial. Second, losses of livestock epidemics are generally *systemic*: even if the number of outbreaks remains limited, the type of control measures taken results in a large number of farms affected. These control measures include the enforcement of surveillance zones, the pre-emptive culling of farms and, possibly, emergency vaccination programs. A complicating factor is that governments are to an increasing extent withdrawing from providing free disaster coverage for such catastrophic risks. Third, there is a lack of adequate historical *data* on which premium rates can be based. Epidemics do not occur frequently and the type of control measures taken is changing over time. Finally, agribusiness itself is changing. Farmers' interest in insurance is increasing due to the decreasing number and increasing size of farms, and due to imposed quality and safety assurance programs.

This chapter deals with 1) designing an epidemic livestock insurance in such a way that farmers get the right incentives to behave in the interest of the collective; 2) designing a risk-financing model that deals with the systemic character of the risk and that considers a diminishing financial role for governments over time; and 3) rating the insurance by means of expert information and Monte Carlo simulation modelling to address data problems. The chapter has a European perspective and estimated premium rates apply to FMD and hog cholera in the Netherlands. The Netherlands is one of the most densely populated livestock areas in Europe both for cattle and pigs (Michel and Windhorst, 2003). Livestock production chains have extensive animal traceability systems (Meuwissen *et al.,* 2003c), and sanitary measures throughout the chain are at a rather high level.

Incentives for Risk Management

The occurrence and extent of livestock epidemics largely depends on the risk management of farmers. Table 10.1 shows the behavioural aspects of farmers that are crucial in minimizing

Table 10.1. Important behavioural aspects of farmers for minimizing the risk of epidemics and a list of insurance contractual and organizational incentives for motivating such behaviour.

Behavioural aspects	Insurance contractual incentives	Insurance organizational incentives
Risk prevention	Risk classification with strong price discrimination Deductibles, but only for consequential losses Advance payments in combination with additional assessments Additional assessments also apply to farmers who quit farming after an epidemic	Annual retention for insurance pool Local organization Sector organization Mutual insurance company No possibility to purchase insurance if epidemics occur in close proximity
Rapid disclosure	No compensation of sick and dead animals Full compensation of healthy animals	Obligatory insurance for direct losses
Compliance with movement standstill	Insurance coverage for losses from movement standstills and emergency vaccination	Link between risk classification and governmental penalty system
No deliberate infection	Compensation at the lowest value of cost of production and market price	

the total losses of an epidemic, and insurance contractual and organizational incentives that can motivate such behaviour.

Risk Prevention

Risk prevention of livestock epidemics aims at preventing an epidemic from occurring, and, when there is an epidemic, minimizing its extent. Risk classification with strong price discrimination is likely to encourage such risk-preventive behaviour. For this, risk classification should be based on risk factors that are manageable by the farmer himself, such as the number of animal contacts, the type of farm, the presence of hobby animals and the quality of sanitary measures taken. Also, in order to minimize related transaction costs, classification criteria should be transparent and easy to monitor, which reduces the applicability of the quality of sanitary measures. An example of a risk-financing system for livestock epidemics in which farms are classified according to their epidemic disease risk can be found in the public-private animal health fund of Belgium (see Van Asseldonk *et al.,* this volume). In this system farms are classified according to their number of animal contacts.

A further way of stimulating risk prevention is by a limited use of deductibles. Although standard practice in most lines of insurance, deductibles should be carefully applied in epidemic livestock insurance schemes: incentive problems may arise if large deductibles apply to direct losses from culled animals (see the section below on "rapid disclosure"). On the other hand, for consequential losses, such as business interruption after depopulation and business interruption resulting from movement restrictions, deductibles are a useful instrument for moral hazard reduction and risk prevention.

Also the timing of premiums influences farmers' risk awareness and related incentives for risk prevention. Paying the whole amount in advance is likely to have a larger preventive effect than paying the full rate through additional assessments: advance premiums continuously make farmers aware of the risks they face. On the other hand, advance payments may lead to moral hazard since premiums have already been paid anyway, while additional assessments are a continuous threat in case things go wrong. A combination of both advance and additional payments in one scheme is likely to produce the largest preventive effect, especially if farmers who quit their business after an epidemic still have to fulfill their share of additional assessments.

Incentives for risk prevention are further influenced by the way the insurance is organized, for instance with respect to the retention (or "deductible") of the insurance pool. The retention should preferably have some annual basis. An annual retention stimulates farmers each year to prevent the risks – in contrast to a multi-annual retention in which there are no incentives for risk prevention once the retention threshold is exceeded. Multi-annual retention levels are currently in place in the Netherlands for the animal health funds per sector. For instance, the retention for cattle farmers is 225 million euro for the total period of 1999 to 2004 (Meuwissen, Van Asseldonk and Huirne, 2003a).

Incentives for risk prevention can in addition be affected by the extent and ownership of the insurance pool: local pools, pools that are organized per sector and mutual pools, which are likely to have a positive effect on the incentives for risk prevention. A direct relationship among pool members reduces problems of asymmetric information – and related costs of monitoring and verification. Also, there is likely to be broader support for risk classification and a critical underwriting policy since colleague farmers – instead of anonymous insurance companies – impose these measures. In addition, at the time of an epidemic, such pools make proper loss assessment easier because of social control and familiarity of colleague farmers with production circumstances.

A further insurance organization incentive is to stop selling insurance coverage if epidemics occur in close proximity – and to also communicate this clearly in "peacetime". Still selling insurance when epidemics occur in nearby countries would lead to moral hazard and adverse selection (Meuwissen, Van Asseldonk, and Huirne, 2003b).

Rapid Disclosure

In order to encourage immediate disclosure, animals that are sick or already dead at the time of culling should not be compensated. Healthy animals, however, should be fully compensated in order to prevent farmers from selling them before disclosing the outbreak. Also, insurance for direct losses should be obligatory. Only in such a way can *all* farmers be given incentives for alertness and rapid disclosure in case of an epidemic.

Compliance with Movement Standstill

Too often, established movement standstills are not immediately acted upon: farmers still try to get their animals out of the restricted area. Although maybe logical from an individual farmer's perspective, it is a major risk for spreading the disease to other areas. For this reason there should be a link between insurance premiums and indemnities and some national penalty system that acts upon breaking the rules. What may also help is to provide insurance coverage for losses related to a movement standstill. This becomes especially relevant with the application of emergency vaccination programs in restricted areas, since this will likely lead to a substantial decrease in the value of the animals and their products.

No Deliberate Infection

With extended periods of movement standstills and related supply and delivery problems, it may be more attractive for a farmer to have an infected herd, receive full compensation and have the stables empty for a while. To prevent incentives for deliberate infection, culled animals and destroyed animal products should be compensated at the lowest value of cost of production and actual market price.

Dealing with the Systemic Risk

Livestock epidemics generally involve many farms at the same time, i.e. epidemic disease risks are systemic. Insurance companies, especially small and local pools, have problems dealing with such risks. Also, adequate reinsurance capacity is not usually available when the scale of the systemic risk is large. The capital market is not well acquainted with epidemic disease risks. In such circumstances, governments are often financially involved in providing disaster coverage. However, national governments are starting to withdraw from such involvement and are no longer automatically providing free disaster coverage for catastrophic agricultural risks. Also, the EU is likely to become more critical about its co-financing policy, as the EU is enlarged and the current "50% of all direct losses policy" seems no longer affordable.

Figure 10.1 proposes an alternative risk-financing model for livestock epidemics in the EU. The model starts out with a private part as well as a public part. Public EU and national budgets respond to the current lack of reinsurance capacity in the private market.

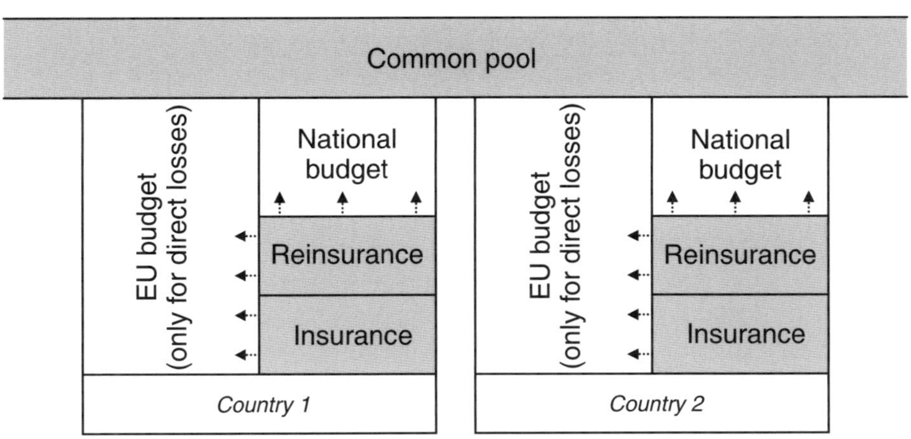

Fig. 10.1. Proposed risk-financing model for livestock epidemics in the European Union. Shaded areas reflect private markets, non-shaded areas are public budgets. Arrows indicate a reduced role for public budgets over time.

The role of public budgets is, however, limited by a common pool organized by the private sector and is envisioned, over time, to be (largely) taken over by private markets as well. The organizational aspects of the (re)insurance segment from Fig. 10.1 have already been discussed in the previous section. The insurance could be organized through a common sector pooling mechanism that operates with advance payments and additional assessments and that offers obligatory coverage for direct losses and voluntary coverage for consequential losses.

The goal of the common pool is to take up the "open end" currently borne by governments. The rationale behind such a multi-country pool is an EU-wide recognition of co-financing policies by the European Commission and a common interest in the prevention of livestock epidemics. The pool furthermore facilitates sector initiatives in funding (part of the) catastrophic coverage for epidemic disease risks. For the pooling idea to work, a few principles apply. First, the ceding insurance companies should establish a common pool. Second, the pool, as a partnership of the insurers, determines the premiums to be paid and assumes the settlement of claims. Third, insurers transmit (part of) the premiums to the pool and the pool indemnifies (part of) the loss via the insurer to the policyholder. Fourth, the pool manages the reinsurance risks. And, fifth, after the year-end, profits and losses are distributed to the involved insurers.

For private markets to be able to further take over the public part once additional capacity is organized in the market, risk-sharing agreements between the private and public sectors must be set up appropriately from the beginning. This "appropriateness" relates to 1) the type of threshold used for the involvement of the public budgets, i.e. a loss-based threshold or a threshold based on some other value; and 2) whether coverage from the public budget is provided at actuarially fair rates or at zero costs.

As for the type of threshold used, loss-based thresholds can be problematic since governments have generally less information than (re)insurers about the size of losses. This may lead to rent-seeking behaviour of (re)insurance companies (Skees and Barnett, 1999). Ambiguity about the exact involvement of public budgets complicates the opportunities for private markets to take over (part of) this commitment. Establishing thresholds on an independent and transparent index would help to overcome these problems. An index could, for instance, be based on the animal mortality rate or the number of vaccinated animals.

With respect to costs at which governments provide their coverage, the opportunities for private markets to take over (part of) the capacity are enhanced if governments charge for the provided capacity at rates that will cover their costs, though "actuarially fair rates" from a government perspective likely differ from those in the private sector. There are, however, also some arguments for governments *not* to charge actuarially fair rates. One of the arguments is the moral hazard in government behaviour: many catastrophes can be either prevented or magnified by government policies (or lack thereof). Having governments financially responsible for some losses might be an incentive for them to put into place appropriate hazard management measures. A further argument is that if governments bear part of the catastrophic risk, political pressure to provide disaster relief on an *ad hoc* basis will likely be less, which in turn benefits the demand for the insurance. For these reasons and because governments are only starting to withdraw from providing free coverage, there will be more support for alternative risk-financing models if governments provide their coverage not for free but at some lower cost. Again, in order to encourage private market opportunities, one should aim at full transparency.

Risk Premiums

This section establishes risk premiums for direct losses and consequential losses. Direct losses refer to culled animals under depopulation and welfare slaughter programs. Consequential losses deal with business interruption from depopulation and movement restrictions. These are losses from welfare slaughter programs (pig farms), breeding prohibition (pig farms) and from not being able to deliver milk (dairy farms).

The goal of rate making is to determine rates which will, when applied to the exposures underlying the risk being written, provide sufficient funds to pay expected losses and maintain an adequate margin for adverse deviation. The rate is the product of the frequency of an outbreak (number of observed losses divided by the number of exposure units) and the severity (total losses divided by the number of observed losses). There are various actuarial methods for premium calculation, more or less sophisticated, that are able to make optimal use of the data and information available. The loss ratio approach indicates the rate changes whereby the rate change is calculated as the experience loss ratio divided by the target loss ratio times the current premium rate. This method requires existing rates and can thus not be used for a new line of insurance, such as epidemic livestock insurance. The risk premium approach indicates rates on the basis of the total loss and the number of exposure units (Green, Driscoll and Bruch, this volume).

Since epidemics occur irregularly in time and place, it is difficult to derive general properties and predictive values. Also, the probability distribution describing the possible spread of FMD and hog cholera is difficult to ascertain. The latter is highly dependent on the control strategy applied. Another complicating factor is that the environment is dynamic, with respect to the control measures applied, the herd intensity and structure and the liberalization of markets, which may affect the trade opportunities for vaccinated animals and their products. Because of this lack of adequate historical data, the use of the expertise of those working in the epidemic disease risk area can complement the more standard rating methodologies. Elicitation of subjective expert knowledge can provide the necessary quantitative information for modelling purposes. Epidemiological models use expert information for parameters describing the spread of diseases. An example can be found in the model developed by Jalvingh *et al.* (1999) and Mourits, Nielen and Léon (2002). Their spatial, dynamic and stochastic model simulates the spread of virus between farms through local spread and various types of contacts, given a specific control strategy. In models with a more economic interest, expert opinions are generally used for more aggregated data such as the expected number of farms infected and the estimated duration of an epidemic, see for instance Meuwissen, Van Asseldonk and Huirne (2003a) and Van Asseldonk, Meuwissen and Huirne (this volume). In this chapter we follow this “aggregated approach”.

A Monte Carlo simulation model is constructed in order to obtain insight into annual loss distributions. Monte Carlo simulation is considered an appropriate and flexible method of investigating aspects that are stochastic in nature, such as livestock epidemics. Risks are thereby incorporated by using probability distributions, which can be manipulated by input modification. At each iteration randomly drawn numbers from specified distributions are used, representing a possible combination of values that could occur. Combining the results of all iterations will lead to a distribution of output values, for instance the annual direct losses of livestock epidemics. In the Monte Carlo simulation model, the uncertainty about the introduction of an epidemic in a specific year is reflected by a Poisson distribution. The region in which an epidemic occurs is simulated by a discrete probability distribution.

Table 10.2. Number of epidemics of Foot and Mouth Disease and hog cholera in the Netherlands for the five-year period 2002-2007 (most likely, minimum and maximum value) and probability of occurrence per region.[1]

Disease	Number			Probability of Outbreak	
	Most Likely	Minimum	Maximum	South	North
Foot and Mouth Disease	1	0	2	0.75	0.25
Hog cholera	1	0	2	0.80	0.20

[1] With respect to the number of epidemics of FMD in the Netherlands, Horst *et al.* (1999) found a most likely value of 1 per 5 years, with a range of 0.5 (25% percentile) to 2 (75% percentile). For hog cholera these numbers were 2.7, 1 and 5. Previous expert consultation by Meuwissen *et al.* (2003a) resulted in most likely values for FMD and hog cholera of 1 and 2.

Epidemics, and ultimately, economic consequences are reflected by triangular distributions, with parameters referring to the most likely, minimum and maximum scenarios. Results are based on 1,000 iterations.

Description of Regions

The country under consideration is the Netherlands and is subdivided into two regions. "NL-south" includes the four provinces in the southern and eastern part of the country with high farm and animal densities, especially for pigs. "NL-north" includes the eight other provinces.

Frequency of Epidemics

Per five-year period, the expected number of epidemics is one for FMD and one for hog cholera (Table 10.2). The chance of outbreaks varies across the two regions as a result of the number and type of farms, in combination with subjective expectations about the risk exposure of each region.

Size of Epidemics

When interviewing experts on the expected size of epidemics they were first asked to define the package of control measures for each of the regions under consideration. For FMD the control measures mentioned *in addition* to the minimum EU requirements (i.e. stamping-out of infected herds and contact herds and the implementation of restriction zones of 3 and 10-km) included: 1) a complete movement standstill of three days for the Netherlands as a whole, including transport of feed, milk and manure; 2) ring vaccination in 2-km zones around infected herds in "NL-south", 1-km for "NL-north"; and 3) pre-emptive slaughter of all susceptible herds within a radius of 1 km of infected herds. For hog cholera, the specified additional control measures included: 1) a complete movement standstill for live pigs of three days; and 2) pre-emptive slaughter of all pig herds within a radius of 1-km of infected herds. Additional measures for hog cholera do not differ among regions.

Given these additional control measures, Table 10.3 shows the most likely, minimum and maximum scenarios with respect to the size of epidemics. Included are the number of

Table 10.3. Expected size of Foot and Mouth Disease and hog cholera epidemics for the five-year period 2002-2007 (most likely, minimum and maximum scenario).

	NL-south[1]			NL-north[2]		
	M.L.	Min.	Max.	M.L.	Min.	Max.
Foot and Mouth Disease						
Number of pig and cattle farms infected	45	20	100	15	2	50
Ratio pig farms infected : cattle farms infected	1:3			1:10		
Duration of epidemic (days)	45	25	100	40	25	60
Radius of restriction zones (km)	20	10	40	15	10	30
Hog cholera						
Number of pig farms infected	30	4	200	10	1	40
Duration of epidemic (days)	60	30	100	45	30	80
Radius of restriction zones (km)	15	10	30	12	10	25

[1] Considering similar control strategies, simulations with the epidemiological model by Jalvingh *et al.* (1999) and Mourits *et al.* (2002) of an epidemic in a *densely* populated livestock area resulted in 28 and 111 infected cattle, pig, goat and sheep herds (50% and 95%) with a duration of 49 and 88 days.

[2] Considering similar control strategies, simulations with the epidemiological model developed by Jalvingh *et al.* (1999) and Mourits *et al.* (2002) of an epidemic in a *sparsely* populated livestock area resulted in 3 and 13 infected cattle, pig, goat and sheep herds (50% and 95%) with a duration of 33 and 48 days.

farms infected, the duration of an epidemic (in days) and the radius of restriction zones (in km). For the number of farms infected, the table also shows the expected ratio between the number of pig farms infected and the number of cattle farms infected. The radius of restriction zones refers to the radius of the total area that is expected to be confronted with restrictive measures.

With respect to the number of infected farms, it is generally expected that there are more cattle farms involved than pig farms. In the most likely scenario, FMD epidemics are expected to be largest in "NL-south" (45 infected farms, 45 days). The same is true for epidemics of hog cholera (30 infected farms, 60 days).

Financial Consequences

In order to calculate the losses for each scenario the financial consequences need to be assessed. Table 10.4 gives an overview of the financial parameters by farm type. Direct losses are reflected by the monetary values of animals destroyed. Consequential losses are derived from the costs of idle production factors.

Direct and consequential losses per scenario (see Table 10.5) are obtained by merging the aggregated herd characteristics, epidemiological parameters and financial parameters. Additional scenario assumptions are identified in Appendix I.

Table 10.4. Financial parameters by farm type.

	Direct losses (€/animal)	Consequential losses (€/day)			
		Business interruption: depopulation	Business interruption: breeding prohibition	Business interruption: welfare slaughter	Business interruption: from not delivering milk
Dairy cows	688	6.66/cow[1]	–	–	6.66/cow[1]
Farrowing sows	450	1.02/sow	1.80/sow	0[2]	–
Finishing pigs	83	0.18/place	–	0.18/place	–

[1] Assuming that leasing out milk quotum or catching up after an epidemic is not possible.
[2] If piglets are compensated at cost of production.

Sources: Agricultural Information and Knowledge Center and Research Station for Animal Husbandry (2002); Hogeveen *et al.* (2002); Meuwissen *et al.* (2003a).

Annual Loss Distribution for the Netherlands

The annual loss distributions for the Netherlands as a whole is the result of 1) the number of epidemics in a certain year, if any (Table 10.2); 2) the region of occurrence (Table 10.2); and 3) the losses per epidemic (Table 10.5). Table 10.6 shows the annual loss distributions for direct losses and consequential losses, expressed in per mille ‰ (per thousand) of the total animal value. Distributions are presented by their average and fractile values. For example, considering the annual distribution of direct losses there is a 90% probability that losses in the dairy sector are less than 2.93‰ of the animal value. However, there is a five percent probability that losses exceed 4.07‰ of the animal value. There is a 75% probability that an epidemic is absent.

Direct losses are particularly high for the farrowing sow and the finishing pig sectors. This is mainly the result of having two perils, i.e. FMD and hog cholera. The opposite holds true with respect to consequential losses, where dairy farms face high losses from not being able to deliver their milk.

Risk Financing

Taking the estimated loss distributions as a starting point, alternative risk-financing models and related levels of risk premiums are analyzed. These are a simplification of the proposed model in Fig. 10.1, i.e. in the default situation there is a 50% share for the EU budget (only for direct losses) but no share for the NL budget, and we do not consider the impact of a common pool. In the alternative models the NL budget is considered on the basis of various thresholds. For the direct losses (Fig. 10.2) these thresholds are losses in excess of 2.5‰, 7.5‰ and 10‰, respectively, of the total animal value. Given the relatively low losses in the dairy sector, applying only the 2.5‰ threshold leads to a small reduction in the total loss borne by the private sector. For the farrowing pig and finishing pig sectors, the NL budget takes on average the largest portion of the loss, even at the highest threshold.

Table 10.5. Losses of Foot and Mouth Disease and hog cholera epidemics in most likely, minimum and maximum scenarios.

	Direct losses (€)			Consequential losses (€)		
	Most Likely	Minimum	Maximum	Most Likely	Minimum	Maximum
Foot and Mouth Disease						
NL-south						
Dairy cows	3,144,476	1,387,269	6,936,344	22,952,586	3,201,703	203,376,030
Farrowing sows	6,801,075	2,043,604	24,262,619	2,394,862	350,299	20,649,303
Finishing pigs	9,433,090	2,641,538	35,273,713	1,936,664	283,278	16,698,567
NL-north						
Dairy cows	1,241,249	177,321	3,989,728	4,718,704	1,309,644	28,205,991
Farrowing sows	310,157	90,594	1,348,549	89,185	22,001	551,016
Finishing pigs	410,949	142,335	1,737,005	72,122	17,792	445,593
Hog cholera						
NL-south						
Farrowing sows	7,561,293	1,893,417	42,810,106	2,304,683	411,631	18,226,179
Finishing pigs	8,464,357	2,519,424	44,490,724	1,863,739	332,876	14,739,048
NL-north						
Farrowing sows	1,181,482	194,627	4,776,946	163,984	33,220	1,205,754
Finishing pigs	1,092,769	229,435	4,451,229	132,610	26,864	975,063

Table 10.6. Annual direct and consequential losses of Foot and Mouth Disease and hog cholera epidemics in per mille (‰) of the total animal value (fractile values and average value).[1]

	50%	75%	90%	95%	100%	Average
Direct losses						
Dairy cows	0.00	0.00	2.93	4.07	9.29	0.60
Farrowing sows	0.00	0.00	34.61	52.36	96.05	7.16
Finishing pigs	0.00	0.00	48.61	74.69	161.86	10.51
Consequential losses						
Dairy cows	0.00	0.00	44.37	81.04	229.83	10.52
Farrowing sows	0.00	0.00	18.05	26.36	67.12	3.66
Finishing pigs	0.00	0.00	17.38	27.40	62.77	3.72

[1] 1,000 @RISK-iterations.

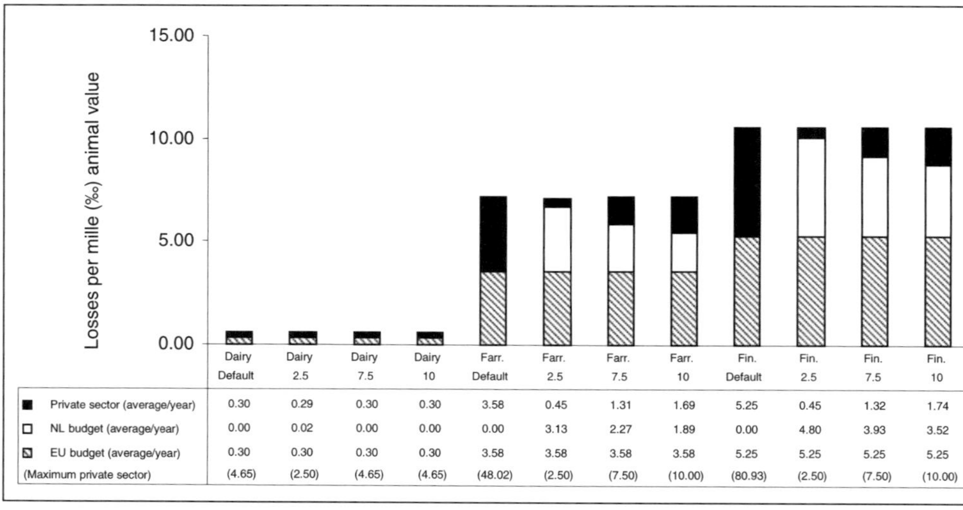

	Dairy Default	Dairy 2.5	Dairy 7.5	Dairy 10	Farr. Default	Farr. 2.5	Farr. 7.5	Farr. 10	Fin. Default	Fin. 2.5	Fin. 7.5	Fin. 10
■ Private sector (average/year)	0.30	0.29	0.30	0.30	3.58	0.45	1.31	1.69	5.25	0.45	1.32	1.74
□ NL budget (average/year)	0.00	0.02	0.00	0.00	0.00	3.13	2.27	1.89	0.00	4.80	3.93	3.52
▨ EU budget (average/year)	0.30	0.30	0.30	0.30	3.58	3.58	3.58	3.58	5.25	5.25	5.25	5.25
(Maximum private sector)	(4.65)	(2.50)	(4.65)	(4.65)	(48.02)	(2.50)	(7.50)	(10.00)	(80.93)	(2.50)	(7.50)	(10.00)

Fig. 10.2. Average annual share of private sector, NL budget and EU budget in financing *direct* losses of dairy, farrowing pig and finishing pig sectors. Estimated maximum share for private sector between brackets. Default scenario without NL budget involvement, alternative scenarios with loss-based thresholds for NL budget, i.e. for losses exceeding 2.5‰, 7.5‰ and 10‰ of the animal value.

For consequential losses, the default agreement implies that the private sector takes up 100% of the losses, i.e. there is zero co-finance from the EU and NL budgets. In the alternative models, we consider (again) a non-proportional coverage by the NL budget with similar thresholds as before. Now the largest effect of the NL budget involvement arises in the dairy sector (Fig. 10.3).

Instead of basing participation on a loss-based threshold, governments could also consider providing reinsurance coverage based on the duration of an epidemic. Fig. 10.4 shows the effect of a day-based threshold, i.e. the NL budget provides non-proportional coverage for consequential losses if epidemics exceed 14, 28, 42 or 56 days. Given the relatively short expected duration of epidemics (Table 10.3) a threshold of 56 days puts the largest burden of the consequential losses on the private sector, while at lower thresholds the largest part of the losses is on average taken up by the NL budget.

Discussion and Conclusions

Financing epidemic livestock diseases is a new field of risk financing that starts to develop mainly because governments are withdrawing and private market solutions need to be found. In this chapter we demonstrated that the three major difficulties in this arena, i.e. the large influence of human behaviour on the course of an epidemic, the systemic character of the risk and the lack of adequate data, can be addressed. We therefore conclude that epidemic livestock insurance is a feasible line of insurance. Crucial aspects in the design of epidemic livestock insurance are:

1. The incorporation of incentives for good risk management in insurance contracts and organizational structures. We therefore recommend obligatory coverage for direct losses with no inclusion of deductibles (other than not compensating sick and already dead animals), a well-developed risk classification of farms according to their risk for livestock epidemics, and the organization of the scheme through some mutual insurance pool.
2. A risk-financing model that anticipates the changing role of governments and that thus can be self-sufficient over time. We propose a financial role for national and European governments until reinsurance and capital markets have enlarged their capacity for epidemic disease risks. However, from the start, the role of public budgets is limited by a common pool organized by the private sector. In addition, for private markets to be able to supplant the role of public budgets once additional capacity is organized, public-private risk-sharing agreements are recommended 1) not to be free of charge; and 2) to be fully transparent right from the beginning. Transparency relates to both the costs at which coverage is provided and the type of threshold used for public budget involvement. Thresholds based on some objective index, such as animal mortality rates, are preferred to loss-based thresholds.
3. Some level of flexibility in the risk-financing model. Since losses of livestock epidemics are dynamic and, to some extent, uncertain, because of changing control measures and farm characteristics, we encourage a risk-financing system that partly operates through additional assessments. This prevents large amounts of farming capital from being idled as reserves in case a big calamity were to hit. From an incentive point of view it is important that additional assessments also apply to farmers who quit their business after an epidemic.

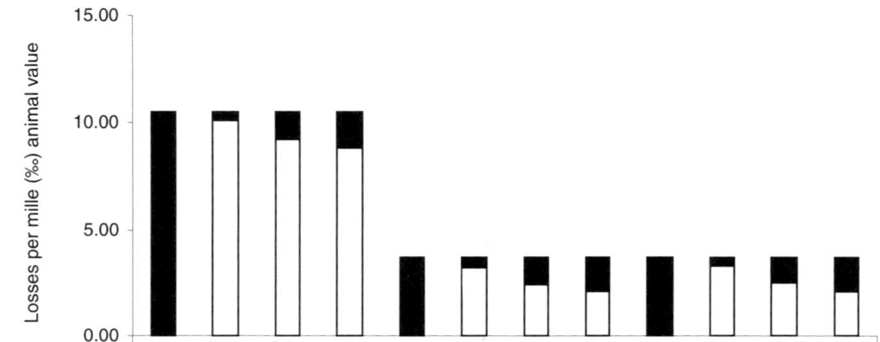

	Dairy Default	Dairy 2.5	Dairy 7.5	Dairy 10	Farr. Default	Farr. 2.5	Farr. 7.5	Farr. 10	Fin. Default	Fin. 2.5	Fin. 7.5	Fin. 10
■ Private sector (average/year)	10.52	0.45	1.32	1.72	3.66	0.44	1.25	1.61	3.72	0.44	1.27	1.63
□ NL budget (average/year)	0.00	10.07	9.20	8.80	0.00	3.21	2.41	2.05	0.00	3.27	2.45	2.09
(Maximum private sector)	(229.83)	(2.50)	(7.50)	(10.00)	(67.12)	(2.50)	(7.50)	(10.00)	(62.77)	(2.50)	(7.50)	(10.00)

Fig. 10.3. Average annual share of private sector and NL budget in financing *consequential* losses of dairy, farrowing pig and finishing pig sectors. Estimated maximum share for private sector between brackets. Default scenario without NL budget involvement, alternative scenarios with loss-based thresholds for NL budget, i.e. for losses exceeding 2.5‰, 7.5‰ and 10‰, respectively, of the animal value.

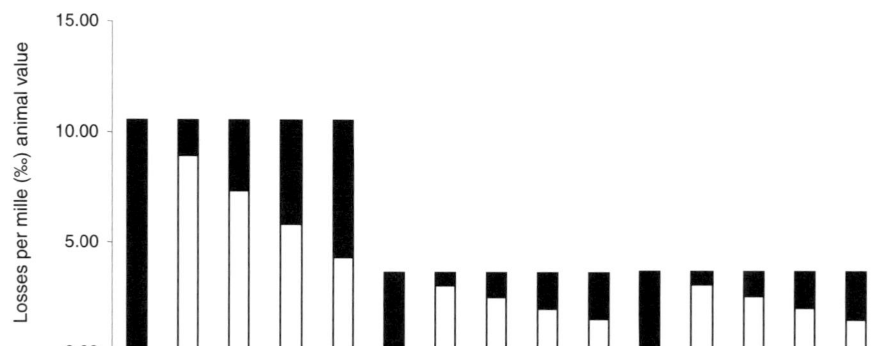

	Dairy Default	Dairy 14	Dairy 28	Dairy 42	Dairy 56	Farr. Default	Farr. 14	Farr. 28	Farr. 42	Farr. 56	Fin. Default	Fin. 14	Fin. 28	Fin. 42	Fin. 56
■ Private sector (average/year)	10.52	1.61	3.19	4.66	6.14	3.66	0.55	1.10	1.60	2.10	3.72	0.56	1.11	1.62	2.13
□ NL budget (average/year)	0.00	8.91	7.33	5.86	4.39	0.00	3.10	2.55	2.05	1.55	0.00	3.16	2.61	2.10	1.59
(Maximum private sector)	(229.83)	(32.18)	(64.35)	(96.53)	(128.70)	(67.12)	(9.42)	(18.84)	(28.26)	(37.68)	(62.77)	(8.84)	(17.67)	(25.89)	(35.34)

Fig. 10.4. Average annual share of private sector and NL budget in financing *consequential* losses of dairy, farrowing pig and finishing pig sectors. Estimated maximum share for private sector between brackets. Default scenario without NL budget involvement, alternative scenarios with day-based thresholds for NL budget, i.e. for epidemics exceeding 14, 28, 42 or 56 days, respectively.

Acknowledgements

The authors would like to acknowledge Dr Wilkens, Agricultural Committee of Comité Europėan des Assurances, Brussels, Belgium, for his ideas about the common pool.

References

Agricultural Information and Knowledge Center and Research Station for Animal Husbandry (2002) *Quantitative Information on Livestock Industry 2000-2001* Karmac Lelystad, The Netherlands.

Hogeveen, H., Meuwissen, M.P.M. and Huirne, R.B.M. (2002) *Animal health in the dairy sector; risk and risk financing.* Institute for Risk Management in Agriculture, Wageningen University, Wageningen, The Netherlands.

Horst, H.S., Dijkhuizen, A.A., Huirne, R.B.M. and Meuwissen, M.P.M. (1999) Monte Carlo simulation of virus introduction into the Netherlands. *Preventive Veterinary Medicine* 41, 209-229.

Jalvingh, A.W., Nielen, M., Maurice, H., Stegeman, A.J., Elbers, A.R.W. and Dijkhuizen, A.A. (1999) Spatial and stochastic simulation to evaluate the impact of events and control measures on the 1997/98 CSF-epidemic in the Netherlands. *Preventive Veterinary Medicines* 42, 271-295.

Meuwissen, M.P.M., Van Asseldonk, M.A.P.M. and Huirne, R.B.M. (2003a) Alternative risk financing instruments for swine epidemics. *Agricultural Systems* 75, 305-322.

Meuwissen, M.P.M., Van Asseldonk, M.A.P.M. and Huirne, R.B.M. (2003b) *Options on epidemic livestock insurance. Proceedings 10th Conference of the International Society for Veterinarians, Epidemiologists and Economists,* Vina del Mar, Chile, pp. 95-98.

Meuwissen, M.P.M., Velthuis, A.G.J., Hogeveen, H., and Huirne, R.B.M. (2003c) Traceability and certification in meat supply chains. *Journal of Agribusiness* 21, 167-181.

Michel, I. and Windhorst, W.-H. (2003) Densely populated livestock areas: definition and spatial distribution in the European Union – a case study for five member states. *Development of Prevention and Control Strategies to Address Animal Health and Related Problems in Densely Populated Livestock Areas of the Community.* In: Huirne, R.B.M. and Windhorst, H.-W. (eds) Final Report-FAIR5-CT97-3666, European Commission, Brussels, pp. 10-25.

Mourits, M.C.M., Nielen, M. and Léon, C.D. (2002) Effect of control measures on the course of simulated Foot and Mouth Disease epidemics that started on different farm types in various Dutch areas. *Proceedings of the Society for Veterinary Epidemiology and Preventive Medicine*, Cambridge, UK, pp. 190 - 200.

Skees, J.R. and Barnett, B.J. (1999) Conceptual and practical considerations for sharing catastrophic/ systemic risks. *Review of Agricultural Economics* 21, 424-441.

Van Asseldonk, M.A.P.M., Meuwissen, M.P.M. and Huirne, R.B.M. (2004) Risk financing strategies to manage epidemic animal disease costs in the UK. *Farm Management* 11, 686-697.

Appendix I: Additional Scenario Assumptions

For calculating the financial consequences per epidemiological scenario, the following additional assumptions were made:

1. For each infected farm, three farms are slaughtered pre-emptively. We need to make this assumption since only the pre-emptive slaughter policy considered by the experts is known, i.e. 1 km, but not the exact number of pre-emptively slaughtered farms. The ratio coincides with the ratio from both the 1997/98 hog cholera epidemic in the Netherlands and the 2001 FMD epidemic in the UK. The ratio from the 2001 FMD epidemic in the

Netherlands was approximately 1:40, but this was an exceptionally stringent policy and included vaccinated animals.

2. All affected farms, i.e. all farms that are infected, pre-emptively slaughtered and/or located in a restriction zone, face restrictions for the whole duration of the epidemic, so there are no temporary removals of restrictions.
3. A welfare slaughter program for animals that are ready to be delivered (25-kg pigs on farrowing farms and 110-kg hogs on finishing farms) is enforced right from the beginning of an epidemic. Welfare slaughter of very young piglets is not considered at all.
4. With respect to the enforcement of a breeding prohibition on pig farms, we assume a breeding prohibition for both FMD and hog cholera, starting from the beginning of an epidemic.
5. For depopulated dairy farms, it is assumed that there are no possibilities for leasing and/or "catching up" after the epidemic. Similarly, for dairy farms in restriction zones, we assume that milk is not collected or paid for and that there are also no possibilities for leasing or catching up.
6. Herds that have been vaccinated are not destroyed.

Chapter 11

The German System of Compensating Animal Keepers in Cases of Outbreaks of Animal Diseases

H.-J. Bätza

Introduction

The experiences of Germany in compensating animal owners, as well as funding the institutions to provide such compensation, represent an interesting case for other countries considering such mechanisms. This chapter gives the specific details of Germany's compensation strategy, as well as a specific regional organizational design.

In cases of animal disease outbreaks in Germany, animal keepers are compensated under the Animal Disease Act in the version promulgated on April 11, 2001. The Act states that pecuniary compensation shall be paid for animals whose destruction had been officially ordered or which have died after the destruction had been ordered, as well as for animals in which a notifiable disease has been detected after their death. The compensation is based on the animal's market value. Market value is calculated without taking account of the loss of value resulting from the disease, any measure prescribed under animal disease legislation or any officially ordered measure. The Animal Disease Act establishes compensation limits. Compensation rates may not exceed €5,113 for horses, €3,068 for bovine animals, €1,278 for pigs, €767 for sheep, €307 for goats, €51 for poultry and €102 per bee colony.

The value of parts of the animal that are usable in accordance with animal disease provisions or official decrees is deducted from the compensation. Costs resulting from the rendering or destruction of the animal are not included in the compensation payment. Taxes are not taken into account when it comes to fixing compensation rates. There is no compensation for consequential economic damage, e.g. due to infected area restrictions and the corresponding marketing bans.

Animal keepers are not entitled to compensation if they culpably failed to abide by statutory provisions in conjunction with the case triggering compensation payments. In addition, it is also possible to reduce the compensation payment in cases of minor guilt.

The Animal Disease Act

The Animal Disease Act prescribes who grants compensation, the *Länder* (federal states), and how they pay it. In principle, it is the *Land* which has to make compensation payments.

However, if animal keepers pay a mandatory contribution for certain animal species, entitling them to compensation payments, the *Land* has to pay only 50% of the compensation rate, thus the total compensation is 50% from public funds (that is *Land* funds) and 50% from farmers' contributions. According to the Animal Disease Act, contributions are charged to owners of horses, bovine animals, pigs, sheep, poultry and freshwater fish. Animal censuses may be conducted for the purpose of charging contributions. Contributions do not have to be charged for poultry and freshwater fish if they result in an unacceptable burden to contributories, particularly due to the low number of animal keepers affected or if the disease situation does not require these contributions for funding.

The Animal Disease Act also prescribes that contributions have to be separately charged for each animal species. However, it is possible to relate them to herd sizes and infection risk, particularly with regard to the farm's organizational structure, and also to age, weight or type of use of the animal. The risk run by a farm should not, as in the past, result in a reduction of the compensation payment, but only be taken into account when it comes to fixing contribution payments to the Animal Disease Funds. Yet this rule is not binding, it is at the discretion of the *Länder* and their Animal Disease Funds to use this framework for a case-by-case evaluation. Experience has shown that the size of the herds alone does not result in increased risk of a disease outbreak. Disease cases involving large herds do burden the farmers' solidarity more than cases involving herds with fewer animals, as the total compensation will be higher for large herds. However, there are many other factors influencing the overall risk of a farm with regard to hygiene and sanitation influencing disease control. Therefore, on the whole, more account must be taken of the risks of infection when it comes to fixing compensation payments. In doing so, the type of farm organization has a decisive role to play (e.g. disposal of slurry – accumulated in separate housing units or several farms – in a joint or a separate plant, joint vehicle fleet, delineation vis-à-vis other farms, in-house clothing, stocking according to the "all in-all out" procedure, and type of animal purchasing). Thus, farms that continuously buy animals from many different livestock dealers, without completely emptying their housing units before the new animals arrive, run considerably higher risks than farms that buy only from certain farms monitored by veterinarians and with known health status.

Such structures, which have now been developed in practice, were created in part by the possibility of paying lower contributions to the Animal Disease Fund. Until 1998, the infection situation of a farm was taken into account (in an unjustified manner) by reducing the compensation payment for individual animals by a certain percentage a priori, provided that the animals for which a compensation is to be paid were kept on farms with a certain number of animals. Thus, the compensation payment was reduced

- by 20% for
    - bovine animals kept on farms with more than 500 bovines
    - pigs kept on farms with more than 1,250 pigs
    - poultry kept on farms with a minimum of 20,000 laying hens or 30,000 head of fattening poultry
- by 40% for
    - pigs kept on farms with more than 2,500 pigs
    - poultry kept on farms with a minimum of 50,000 laying hens or 100,000 head of fattening poultry.

The compensation rules laid down by the Animal Disease Act are embedded in the overall system of public animal disease control, fulfilling special functions. The compensation payments serve to support the animal keeper's cooperation in disease control and to mitigate economic losses resulting from the death or destruction of diseased or suspect animals. It represents an indirect disease control measure which is not less effective than direct control measures. Thus, compensating for animal losses forms an integral part of the measures the Government has to take to exercise its duty in regard to the protection of the community's legal assets against health hazards and national economic losses caused by animal diseases. These public measures are aimed at disturbers as defined by police law. In all cases, the animals concerned pose an actual or considerable potential risk, albeit to a varying degree, to other animals susceptible to the disease. Therefore, animal disease compensation does not represent a compensation for expropriation, but a stand-alone right granted by legislators on both grounds of equity and as a matter of convenience.

Structure of the Animal Disease Fund

Since the Animal Disease Act obliges the *Länder* to regulate compensation, let us now specify what their compensation arrangements look like. There is an Animal Disease Fund in every *Land.* The Animal Disease Funds are enshrined in the respective – *Länder* Acts implementing the Animal Disease Act. As an example of the Animal Disease Fund of the *Länder,* rules governing the Lower Saxony Animal Disease Fund will be discussed.

The Implementing Act of the Animal Disease Act of the *Land* of Lower Saxony lays down that the Lower Saxony Animal Disease Fund is an institution under public law with legal capacity. The Animal Disease Fund manages its affairs under its own responsibility. In addition, in accordance with the Implementing Act, the Animal Disease Fund must take responsibility for animal losses caused by animal diseases or infectious diseases, costs accrued through the control of animal or infectious diseases and/or damage resulting from these diseases and costs of setting up and operating vaccine banks in which the *Land* has a contractual share. Furthermore, a share of the Animal Disease Funds are used in preventive animal health programmes to lessen damage caused by animal diseases.

A Governing Board and a Board of Directors organize the Animal Disease Fund and both bodies have a term of office of six years. The Governing Board consists of 13 members, including nine members nominated on the suggestion of agricultural organizations, two members on the suggestion of Lower Saxony's rural district parliament and two members designated by the relevant Ministry. In Lower Saxony, these final two members are designated by the federal state's Ministry of Agriculture. The governing board decides on amendments to the statutes, budget, contributions by animal keepers, appointing auditors, formal approval of the director's actions, and on payments of the Animal Disease Funds that are not based on statutory obligations. The governing board holds its first meeting within two months after its term of office has started and meets at least twice annually during this term. It elects from its members a chairperson for the duration of its term.

The Board of Directors consists of: 1) four members elected by the Governing Board for the duration of the term of office; 2) two other members designated by the competent Minister; and 3) the managing director. The Board of Directors executes the decisions taken by the Governing Board. The board chairperson represents the Animal Disease Fund externally in all legal and administrative matters. The board chairperson is the superior of all the officials of the Animal Disease Fund. He or she represents employer interests vis-à-vis the employees and workers of the Animal Disease Fund. The managing director manages day-to-day administrative business of the Animal Disease Fund. He or she is the superior of officials, employees and workers at the Animal Disease Fund. According to Lower Saxony law, the managing director must be a veterinarian.

The Animal Disease Fund is subject to supervision exercised by the *Land,* the relevant Ministry being its supervisory authority. Supervision is limited to the tasks transferred to the Animal Disease Funds in accordance with the respective Implementing Acts and to questions of legality. Statutes of the Animal Disease Funds require the relevant Minister's approval, who must publish them in the respective Official Journal.

The *Länder*'s Implementing Acts also assure that the Animal Disease Fund must make compensation payments for animal losses to beneficiaries that are prescribed under the Animal Disease Act. It is also stated that the official veterinarian must estimate, either before culling, if possible, or immediately after culling, the value of an animal or its parts to calculate the compensation payment. Upon request of the animal keeper, the official veterinarian must consult with two estimators; in this case the above value consists of the mean value determined by the veterinarian and the estimators. If the Animal Disease Fund has doubts about the results of the estimate, it can request an advisory opinion from an expert to be designated by the competent Chamber of Agriculture. The outcome must then be taken as a basis for calculating the payments of the Animal Disease Fund.

The *Länder*'s Implementing Acts also include information on how to raise funds for compensating payments. Thus, the Animal Disease Funds must raise funds for administrative costs and necessary reserves through the animal keepers' contributions. There are specific statutes for the collection of contribution payments. The Acts implementing the *Land* Animal Disease Act include the legal basis for these statutes. For instance, the statutes regulate that calculation of contributions must be based on the number of animals that were kept the day before the Animal Disease Fund conducted its official survey and that animal keepers must inform the Animal Disease Fund within two weeks after the closing date, including names and addresses, of the animal species they kept at the closing date. Furthermore, animal keepers must inform the Animal Disease Fund not later than two weeks after the incident about any increase in herd numbers, new business or re-entrances following the closing date. In 2002, animal keeper contributions to the Animal Disease Fund in Lower Saxony were €5.00 for bovine animals and €0.80 for pigs.

Irrespective of the compensation rules under the Animal Disease Act, farmers may insure themselves against certain animal disease risks. There are various insurance companies offering these types of coverage. Subject to the insurance animal keepers may purchase additional underwriting to augment the compensation for animal losses granted by the state as insurance against this risk, particularly against consequential damage resulting from the establishment of isolation or surveillance zones and the related marketing ban. However, the premiums probably constitute a limiting factor, since there is usually a certain deductible. Hence, animal health insurances generally do not represent an alternative to Animal Disease Funds as they are also used within the framework of preventive animal health programmes, e.g. for measures which minimize the risk of disease introduction and thus risk of loss. These measures are not covered by private insurance programmes. What is more, experience gathered in conjunction with Foot and Mouth Disease events has shown that, in the case of crisis, insurance companies tend to exclude certain risks from their catalogue of benefits.

The system of animal disease compensation in place in Germany has stood the test through past outbreaks. Thus, the basic principles and the aim of the current rules will be maintained.

Chapter 12

Managing the Risks and Impacts of Animal Diseases in the Australian Livestock Sector

G.B. Neumann and R.C. Keogh

Introduction

Management of risks associated with animal diseases impacting commercial livestock industries in Australia has increasingly been a matter for close cooperation between government and livestock industry sectors.

Natural barriers to entry by major animal diseases have endowed Australia with high animal health status. The importance of livestock industries to Australia's economy and the importance of exports to Australia's livestock industries have driven the commitment of significant resources over many years to maintaining the natural advantages Australia has enjoyed in regard to its animal health status.

Strong border protection has been complemented with successful campaigns to eradicate significant diseases that had become endemic. Further, there have been coordinated efforts to enhance disease response preparedness, capability and funding arrangements.

Most recently Australia's governments and livestock industries have concluded a comprehensive and innovative agreement, the Emergency Animal Disease (EAD) Response Agreement, to further reinforce cooperative animal disease preparedness, response and funding arrangements. This agreement institutionalizes concepts of government, industry, enterprise biosecurity planning, auditable animal health service performance standards, and industry and government participation in managing and funding animal disease response.

The direction of such resources to largely successful risk-mitigation activities has coincided with, if not given rise to, very limited use of "conventional" insurance mechanisms to manage livestock disease risks.

With the establishment of the EAD Response Agreement to secure strong minimum animal disease cost recovery arrangements as well as the risks of natural, intentional or inadvertent introduction of new animal diseases to Australia, the case for supplementing the current risk management arrangements with "conventional" livestock disease insurance is being assessed.

Background

Management of Animal Health in Australia

Australia is a federation with a central Commonwealth Government with powers set out in the Australian Constitution and six State and two Territory governments that have exclusive authority in areas not defined in the Constitution as being within the Commonwealth authority. For example, quarantine and international trade are within the authority of the Commonwealth, while the delivery of other animal health services within Australia is a State and Territory responsibility. Thus the Australian States and Territories are responsible for disease surveillance, control and eradication within their own boundaries.

Development of national agricultural policies for Australia needs to be consistent with the objectives of both the Commonwealth and the State/Territory governments and needs to take account of the interests of the industries involved. To facilitate this, a set of key policy development committees bring together the nine jurisdictions that make up the Australian federation to provide a focus for resolution of national issues such as animal health.

The animal health role of the Commonwealth Government is delivered through the Office of Product Integrity Animal and Plant Health and the Australian Chief Veterinary Officer, the Australian Quarantine and Inspection Service (AQIS), and Biosecurity Australia. The Commonwealth Scientific Industrial Research Organisation (CSIRO) operates the Australian Animal Health Laboratory (AAHL) at Geelong, which provides diagnostic services, exotic and emergency disease support, and independent scientific advice.

The Australian Quarantine and Inspection Service (AQIS) is responsible for export health certification and regulation of the import of animals and plants, their genetic material and their products. AQIS also has responsibility for veterinary public health inspection of meat for export and domestic consumption, through a national inspection service, animal quarantine involving imports of live animals and animal products, and the health certification of exports of live animals and animal reproductive material.

Biosecurity Australia is responsible for consulting with the livestock industry and the community, conducting research and developing policy to protect Australia's animal and plant health status and natural environment, undertaking import risk analyses to determine whether a product may be allowed into the country and if so, under what quarantine conditions, and assisting Australia's export market programme by negotiating with other countries to allow the importation of Australian agricultural products.

State and Territory governments are responsible for monitoring and improving the health of livestock in their region and managing disease control programmes. This is delivered via an animal health administrative unit headed by a veterinarian designated as the State's or Territory's Chief Veterinary Officer (CVO).

The State and Territory services administer the application of relevant State or Territory Acts and Regulations, maintain records of the animal health status of farms in their region, contribute to the prevention and control of livestock diseases and conditions and implement designated disease control programmes. The States and Territories also have animal health laboratories that provide disease diagnosis and investigation services, undertake applied research, and work closely with their field veterinary staff.

Australia's Animal Health Advantages

Largely due to its geographic isolation, Australia has enjoyed a favorable animal health status since the introduction of commercial livestock with the first European colonists in the late 18th century.

As animal health policy has evolved since the 1850s, a consistent priority has been to prevent the establishment of new diseases with significant resources devoted at all levels of government to preserving the advantages that have arisen from natural circumstances. In large part the priority given to maintaining a favourable animal health status has been driven by the importance of livestock industries to the economic well-being of the country throughout Australia's history since colonization. Within a few decades of its founding, European society and commerce in Australia had become largely dependent on exports of fine wool. The predominance of this industry within the Australian economy continued until the 1950s. The development of refrigerated trans-oceanic transport in the latter part of the 19th century resulted in beef and dairy production as well as mutton and lamb joining wool as important exports originating from the country's livestock industries.

The recognition of the importance and the potential of the cattle industries to Australia's international trade is illustrated by the investment in the ultimately successful campaigns conducted between the 1920s and 1990s to eradicate bovine pleuropneumonia, and then bovine tuberculosis and bovine brucellosis from Australian herds. These initiatives involved the coordinated efforts of the Commonwealth, State and Territory governments as well as the beef and dairy industries in all States and Territories.

The key driver for the initiatives that resulted in the elimination of the latter two diseases was the desire to satisfy the developing requirements of importing markets. The improvements to productivity that also arose from the elimination of these diseases were important secondary drivers. Currently, Australia's livestock industries are worth some AU$13 billion (US$7 billion) annually. Approximately 70% of this income is derived from exports. Australia remains the world's largest producer and exporter of wool and the world's largest exporter of beef.

Australia continues to both enjoy, and place a high value on, a favourable animal health status. While there are currently no internal initiatives to match the commitment and ambition of the national campaigns to eradicate bovine brucellosis and tuberculosis, there continues a massive commitment to border protection and quarantine to maintain disease freedom and a considerable emphasis on surveillance to demonstrate this freedom.

A third priority, in recent decades in particular, has been the development of the capability and systems to respond to incursions of exotic livestock diseases. A cornerstone of the response system is a national commitment to ensuring that disease responses are carried out according to a set of nationally agreed principles and processes described in the *Australian Veterinary Emergency Plan (AUSVETPLAN).* This is a series of 52 manuals that provide guidance on the conduct and management of emergency disease responses in Australia for OIE List A diseases and for List B diseases of particular concern to Australia. AUSVETPLAN also includes manuals addressing generic issues involved in animal disease response management and control.

AUSVETPLAN is supported with systems to maintain the currency of the information in the manuals and to train personnel in the roles described in the plan. The training includes the conduct of simulation exercises, usually on a modest scale, to test components of the response process in confined areas. An exercise conducted in 2002, "Exercise Minotaur", was an exception in that it tested a significant range of management processes at a national level. Where a specific strategy to control a disease has not been prepared and agreed upon (because of low priority), a set of "response policy briefs" have been prepared as the basis for the response.

Responses to animal disease emergencies in Australia are generally geared to stamping out disease incursions when feasible. There is an emphasis on protecting the national interest and having prosperous livestock industries competing successfully in international markets. This approach, necessarily, requires the reconciliation of the tension between the burden of the costs of a response being confined to relatively small areas and few enterprises and being borne by individual combat authorities over short, unpredictable time-frames on one hand and the benefits of eradication applying nationally, for prolonged periods and extending throughout value and production chains, on the other. For many years there operated a formal agreement between the State and Territory governments and the Commonwealth Government that specified arrangements under which costs incurred by one or more of the combat jurisdictions in responding to a disease incursion would be shared by all.

The commitment to including the livestock industries in the maintenance of Australia's advantageous animal health status, which has a history beginning before the brucellosis and tuberculosis eradication and which was institutionalized in the establishment of Animal Health Australia, has now been extended to sharing in the costs associated with emergency animal disease responses. In April 2002 the "EAD Response Agreement", that includes the livestock industries as well as government, replaced the cost-sharing agreement between governments.

Animal Health Australia

Animal Health Australia is a not-for-profit company established under Australian corporations law in 1996 to develop national priorities in animal health and to protect and enhance Australia's animal health status via a partnership of governments and the livestock industries. The 23 Members (shareholders) of the Company include the Commonwealth Government, the eight State and Territory governments, twelve national animal/livestock industry bodies, as well as the Australian Veterinary Association and the Australian Animal Health Laboratory at Geelong. Members contribute to the funding of activities by means of annual subscriptions based on the value of production of the livestock industries. A seven-person skills-based independent Board of Directors determines the strategic direction of the Company in consultation with Member organizations. Collectively, Directors have experience in animal health services; major export markets for livestock and their products; industry processing and marketing; extensive and intensive livestock production; government and industry organizational arrangements, legislation and policy development processes; as well as strategic, economic and financial management skills.

A small management team combining skills in animal health, business, financial and programme management, policy development, planning, negotiation, accounting, information technology and administration is responsible for the day-to-day operations of the company.

A wide range of services is delivered to members, extending from advice on national policy and strategy development to the coordination and management of national programmes and projects where both governments and the affected livestock industries are engaged. Thus, while some services are particular to a subset of Members (and funded by them), the core functions of the company are funded by members' subscriptions.

The "core" work of the company is directed at the establishment and management of a range of national animal health programmes directed at services accruing benefits to all members.

- **Animal Health Services** (targeted at improving the standards and capability of Australia's animal health human resources and infrastructure).
- **Animal Disease Surveillance** (targeted at improving Australia's surveillance and reporting systems to enhance Australian trade and includes the National Animal Health Information System (NAHIS) that delivers Australia's annual status report in areas such as food safety, residues, meat inspection (domestic and export), quality assurance and processed foods).
- **Emergency Animal Disease Preparedness** (targeted at improving Australia's preparedness and response capability for disease outbreaks).
- **Future Directions** (targeted at the early identification and analysis of strategic issues for stakeholders in Australia's animal health system).

In addition, the Company assists sub-groups of government and livestock industry members to establish national animal health programmes funded by beneficiaries including:

- **Tuberculosis Freedom Assurance Programme** (surveillance and control activities to maintain Australia's disease-free status for bovine tuberculosis).
- **National Johne's Disease Control Programme** (targeted at developing control strategies, disease information, zoning policies and market assurance programmes for all forms of Johne's disease).
- **National Ovine Johne's Disease Control and Evaluation Programme** (evaluating the best way to manage Ovine Johne's Disease in Australia in the long term and in the meantime controlling further spread of the disease).
- **National Arbovirus Monitoring Programme** (identification of areas free of arboviruses to underpin the international trade of cattle from northern and eastern Australia).
- **National Transmissible Spongiform Encephalopathies Surveillance Programme** (testing programme to meet international requirements to justify Australia's freedom from TSEs).
- **Screw-Worm Fly Project** (to develop a cooperative government and livestock industry suite of activities to maintain Australia's freedom from the Old World SWF).

Other work recently undertaken on behalf of sub-groups of Members includes:

- Development of a national strategy to address Foot-and-Mouth Disease (FMD) and Bovine Spongiform Encephalopathy (BSE) in Australia (governments and the beef, dairy, feedlot, sheep, wool, pigs and goat industries).
- Bovine Johne's Disease Control and Evaluation Project (for the beef, dairy and feedlot industries).
- Survey of Bovine Johne's Disease in South-East Australian Beef Cattle Project (for the beef industry).
- Cattle Buy-Back Scheme (designed to minimize risks of BSE in the beef industry).

By virtue of its independence and peak-body status Animal Health Australia is uniquely placed to identify strategic priorities and marshal the support necessary to take action. Examples of areas where the company has taken a leadership position include:

- Development of national animal health services performance standards.
- Development of a competency-based training programme for personnel involved in animal health emergencies.
- Management of a national programme for accrediting veterinarians to participate in government programmes.
- A national programme of awareness building among livestock producers and other personnel about the importance of early reporting of unusual disease symptoms.
- A national initiative to improve on-farm biosecurity provisions in all livestock industries.
- Negotiation of a new agreement on the management of emergency animal diseases.

The unique arrangement outlined here has proved successful at brokering outcomes that build partnerships, generate cost-efficiencies for stakeholders and secure equitable funding arrangements that assist in advancing Australia's animal health status at home and abroad with direct spin-offs in terms of facilitating trade. Formal consultative mechanisms ensure that Members' views are heard, that their priorities are pursued and that they have a direct input into the development of operational imperatives, thus assisting the Board to develop longer-term directions. To put this investment in perspective, the collective livestock industry's funding of the one-third share of core funding to enable the above suite of activities to be undertaken by the Company in 2000/01 ($932,907) equates to 0.007% of forecast national GVP for livestock and livestock products. More information on Animal Health Australia can be gleaned from the company website: www.aahc.com.au

Animal Disease Risk Management

The EAD Response Agreement

Animal Health Australia has worked with its members for more than four years to develop a new national agreement for managing responses to outbreaks of serious diseases. This agreement, the *Government and Livestock Industry Cost Sharing Deed in Respect of Emergency Animal Disease Responses,* or "EAD Response Agreement", binds together governments and the livestock industries in an agreed response process to serious animal diseases and is a significant achievement given the diverse interests involved and the major enhancement it provides to Australia's mitigation of risks associated with disease outbreaks.

The previous agreement, known as the "Commonwealth/States Cost Sharing Agreement" covered only twelve animal diseases, meaning that funding and management arrangements of outbreaks of the many other diseases exotic to Australia as well as emerging diseases were uncertain. Experience has indicated that such uncertainty results in procrastination over the direction of a disease response and exposes both governments and livestock industries to considerable risk should an outbreak of an unlisted disease occur.

The inclusion of industry organizations in the Agreement is consistent with a "beneficiary pays" policy principle that has developed in many aspects of Australian life. This principle became a central consideration in agreeing the funding criteria for the EAD Response Agreement. Inclusion of the livestock industries in funding responses has guaranteed their inclusion in decisions relating to proposed responses to disease outbreaks.

The Diseases and their Management

In the context of this agreement, emergency animal diseases are defined as those animal diseases likely to have a significant effect on livestock with potential for mortalities, production loss and, in some cases, impacts on human health and the environment, and that meet one or more of the following specific criteria:

- It is a known disease that does not occur in endemic form in Australia and it is considered to be in the national interest for the country to be free of the disease.
- It is a variant form of an endemic disease, caused by a strain or type of the causal agent that can be distinguished by appropriate diagnostic methods, and which if established in Australia would have a national impact.
- It is a serious infectious disease of unknown or uncertain cause, which may, on the evidence available at the time, be an entirely new disease, or one not included in the categorised disease list.
- It is a known endemic disease, but is occurring in such a severe outbreak form, that an emergency response is required to ensure that there is neither a large-scale epidemic of national significance nor serious loss of market access.

A list of such diseases was compiled and each assigned to one of four categories according to the following classification criteria. The category of a disease determines the proportions of government and industry funding in response to that disease.

Category 1 diseases *(funded 100% by government)* are those that predominantly and seriously affect human health and/or the environment (depletion of native fauna) but may only have minimal direct consequences to the livestock industries. This includes diseases, such as rabies and Nipah virus.

Category 2 diseases *(funded 80% by government and 20% by the applicable industry[s])* have the potential to cause major national socio-economic consequences through very serious international trade losses, national market disruptions, and very severe production losses in the livestock industries that are involved. This category includes diseases that may have slightly lower national socio-economic consequences, but also have significant public health and/or environmental consequences. This category includes BSE, Hendra virus (formerly called equine morbillivirus) and FMD.

Category 3 diseases *(funded 50% by government and 50% by the applicable industry[s])* are of moderate public impact and have the potential to cause significant (but generally moderate) national socio-economic consequences through international trade losses, market disruptions involving two or more states, and severe production losses to affected industries, but have minimal or no effect on human health or the environment. This category includes highly pathogenic avian influenza, classical swine fever and Newcastle disease.

Category 4 diseases *(funded 20% by government and 80% by the applicable industry[s])* are those that could be classified as being mainly production loss diseases. While there may be international trade losses and local market disruptions, these would not be of a magnitude that would be expected to significantly affect the national economy. The main beneficiaries of the successful emergency response to an outbreak of such a disease would be the affected livestock industry(s). Included in this category are Aujeszky's disease, contagious equine metritis and equine influenza.

Provisions exist in the Agreement to review the classification of a disease and the classification of new diseases.

Government and Livestock Industry Consultative Processes

Under the EAD Response Agreement, governments and the livestock industries are represented on two key committees. The Emergency Animal Disease National Management Group (NMG) is a high-level committee that carries responsibility for decision making on policy and resource allocation issues during an emergency animal disease response. This group comprises the chief executives of government parties and presidents of the livestock industry parties affected by the particular disease outbreak. It approves response plans and budgets and monitors expenditures. While the NMG reports to the Primary Industries Ministerial Council (comprising the Ministers responsible for Agriculture of each of the nine government parties), the joint nature of the Agreement means that ultimate accountability for the cost-sharing arrangements remains with all the parties.

Industry representatives are also included on the Consultative Committee on Emergency Animal Diseases (CCEAD). This technical advisory committee comprises the Chief Veterinary Officers from each of the States, Territories, and the Commonwealth and the head of the Australian Animal Health Laboratory. The views of each of the affected livestock industries are represented by a technical expert (usually a veterinarian) with wide experience in the animals affected by the particular disease, and an additional technical expert carries the combined interests of the non-affected industries.

Each industry party authorizes appropriate "Industry Representatives" who are accredited to represent that industry in meetings of the CCEAD or the NMG and to provide an industry liaison at disease control centers at local and state levels. To ensure they are equipped for these roles, they are required to undertake competency-based training (Livestock Industry Leader Training) conducted by Animal Health Australia as part of a competency-based National EAD Training Programme.

Managing the Response to a Disease Outbreak

The Agreement contains a series of provisions to provide assurance that the respective State or Territory agency (that has legal authority under local legislation) handles a disease outbreak appropriately.

In the event of a disease event, the Chief Veterinary Officer of the State or Territory develops (in consultation with the CCEAD and consistent with the relevant AUSVETPLAN disease strategy) a Response Plan and budget to deal with the outbreak. When satisfied with the technical aspects of this plan, the CCEAD recommends it to the NMG. The CCEAD may also recommend to the NMG variations to the content of any AUSVETPLAN manual that will assist the particular response. Application of the content of the following Management Manuals is specifically required.

- Control Centre Management Parts 1 and 2
- Destruction of Animals
- Disposal procedures
- Public relations
- Valuation and Compensation
- Decontamination
- Laboratory preparedness
- Mapping

(Full information on all AUSVETPLAN Management and Disease Strategies is available at http://www.aahc.com.au/ausvetplan/index.html.)

Once endorsed by the NMG, the Response Plan commits the affected jurisdiction to the key strategies and core operational activities, subject to any variations that may be subsequently advised by the CCEAD and agreed to by the NMG. Endorsement of the Response Plan also brings into operation the cost-sharing provisions of the Agreement to the limit of the budget agreed upon by the NMG.

To reassure parties that a disease response is being conducted efficiently, the NMG is required to obtain advice from an independent source. The efficiency audit is a systematic examination to determine whether the eradication/containment activities comply with the approved Response Plan, and whether the Plan itself is being implemented effectively and is suitable to achieve the objectives. There are also provisions in the Agreement for a financial audit to ensure that the costs to be shared are those allowed by the Agreement and that appropriate cost recording and management consistent with the agreed budget are practiced.

Because States and Territories have differing systems for the management and delivery of animal health services, national standards of performance for all aspects of Australia's animal health system have been developed and are referred to in the Agreement. These performance standards will form the basis for measuring the effectiveness of each party's animal health management capabilities. The Agreement requires the parties, wherever possible, to use personnel for key roles who are accredited under the competency-based National EAD Training Programme conducted by Animal Health Australia for its members.

Sharing the Costs of Disease Responses

The government and livestock industry parties have agreed to share costs for an outbreak of a disease that falls within one of the previously described four categories of disease. The costs of salaries and wages, operating expenses, capital costs incurred by parties responding to the disease, and compensation to affected owners are covered. It does not cover consequential losses.

Government and industry parties share the agreed proportions that vary with the disease and the parties involved. For example, where a disease affects only one species, that industry alone bears the livestock industry proportion of costs to be shared. Where more than one animal species is affected by a disease, the contributions from the affected livestock industry parties take account of both the gross value of production (GVP) of each industry and the importance of that particular disease for that industry. For example, in the case of FMD, the weighting is 50% cattle, 30% sheep/goats and 20% pigs. For Surra, the agreed weighting is 50% cattle and 50% horses.

Where more than one industry represents a single animal species, such as is the case with the beef and dairy industries, they have agreed to share costs on the basis of the GVP of each sector. For example, the agreed split for cattle diseases are beef grazing (52.94%), beef feedlots (5.88%), and dairy (41.18%). Comparable arrangements exist in sheep industries (wool and sheepmeats) and poultry industries (meat chickens and eggs).

Of course, the costs of responding to a disease outbreak will depend on the nature of the disease and the specific circumstances of the outbreak. In a worst-case scenario of a major outbreak of FMD, an agreed limit is to cost sharing of 1% of the gross value of production (GVP) of the industries involved. This provides a basis to calculate maximum liabilities for all parties. The three-year average total GVP of the industries concerned (for the period to 2001) is $11,235 million, of which 1% is $112.35 million. For FMD, a Category 2 disease, costs are split 80:20 between government and industry, making the collective industry liability 20% of $112.35 million, or $22.47 million. Applying the agreed division between the affected industries, the maximum industry liabilities are: cattle $17.14 million, sheep/goat $4.79 million and pork $0.54 million. Applying the division of costs between Australia's governments, per the Agreement for the same FMD scenario, the shares for the various jurisdictions are: Commonwealth $44.94 million, New South Wales $11.69 million, Victoria $10.51 million, Queensland $11.82 million, South Australia $3.10 million, Western Australia $5.38 million, Tasmania $1.28 million, Northern Territory $1.17 million and Australian Capital Territory $0.02 million.

Livestock industry parties pay their share of the costs of a disease response through industry statutory levy arrangements or voluntary means. In most cases, the industry parties have put in place a levy set at $0.00, to be activated at the time of a disease response to raise sufficient funds to cover that industry's liability. That is, the levy is in place but does not collect funds until needed. Some industries have also put in place arrangements to accumulate a contingency fund to cover part or all of their liability.

Claims for expenditure to be compensated by cost sharing are managed by Animal Health Australia. This involves each party submitting claims that are summed to determine the aggregate amount of expenditure and to determine whether the upper limit on expenditure agreed upon in the NMG, which may be less than 1% of GVP of the collective industries involved, is likely to be breached. If the NMG believes that the cost of a particular response will exceed the agreed limit, it will determine whether:

- the agreed limit should be increased;
- the emergency response should be continued;
- the proportional shares of costs should be altered;
- the emergency response should be transformed into a long term control programme; or
- any other appropriate alterations should be made to the Response Plan.

Thus it is the responsibility of the NMG, where all appropriate parties are represented at the most senior level, to make decisions acceptable to all parties.

Actions to Minimize the Likelihood of Outbreaks of Emergency Animal Diseases

The importance of improving biosecurity as a step towards reducing the overall risk to all parties was recognized early in negotiations for a new Agreement. The implementation of improved biosecurity practices within each livestock industry was recognised as being linked to wider government programmes aimed at minimizing the risks of disease establishment and spread. Feral animal control is such a programme.

A timetable for the development and implementation of the biosecurity programme is included in the Agreement, with a national communications programme to raise community awareness of the importance of biosecurity measures forming part of the *Protect Australian Livestock Campaign,* conducted by Animal Health Australia on behalf of its government and livestock industry members.

The focus of an agreed risk reduction programme included in the Agreement is the development of individual on-farm biosecurity[1] plans. These plans are designed to provide a simple vehicle for reinforcing on-farm management practices that will reduce the likelihood of disease spread. The agreed plans are one-to-two page documents that identify simple actions producers can take to reduce the chances of a disease entering their property or of spreading. In many cases such actions will be part of existing management.

Given the substantial variation in the nature of livestock enterprises between and within industries, generally biosecurity plans are a guide. Some industries such as the chicken, egg, pork and beef feedlot industries, which operate with high concentrations of stock, may experience more rapid and more catastrophic losses should a serious disease enter their enterprise. Thus, they will have developed biosecurity plans of a relatively high standard.

The key principles of enterprise biosecurity plans are to identify and reduce/manage the risks of introduction and spread of disease, both within the enterprise in question and also to other enterprises. In order to achieve this, the main (highest risk and/or highest impact) diseases of concern and the key features of those diseases are identified and addressed.

The EAD Response Agreement and Animal Disease Insurance

For many years Australia's agricultural producers have funded a variety of activities including research and development, marketing and risk mitigation by agreement to levies on the commodities they produce. The direction of a small proportion of the livestock industry levies to fund industry membership of Animal Health Australia and the industry obligations under the EAD Response Agreement are two of the more recent applications of livestock levy funds.

The EAD Response Agreement as Insurance

In many respects the payment provisions of the EAD Response Agreement mean that the Agreement is itself a basic insurance policy covering the livestock industries for the direct costs of a response to an animal health emergency. On the one hand, it confines the exposure of the livestock industries to between zero and 80% of the direct costs of a response, up to

---

[1] Biosecurity is taking steps in everyday management of discrete livestock populations that will eliminate or minimize the possibility of selected disease agents entering or being disseminated from such populations.

a limit of that proportion of 1% of the annual Gross Value of Production of the particular livestock industry and, on the other it provides that the insurance "premium" is payable retrospectively over a period of up to ten years upon a loss being incurred and at a generous interest rate.

Livestock Insurance in the Light of the EAD Response Agreement

While Australian agricultural producers have historically been regular users of insurance to manage some risks associated with major activities (insuring crops against hail damage is a particularly well-established commercial practice), insuring livestock against the risks of animal diseases is not widely practiced and few major agricultural insurers have "off-the-shelf" products for this type of risk. The reason for this position is probably a combination of the very low frequency of disease outbreaks, the cost of the insurance, and the level of the excess the producer must fund before a claim is paid. These factors may be complemented by an expectation that in the event of a very serious outbreak, public funds would be directed to alleviating extreme hardship, as has been the case with the worst flood and fire events.

Three factors may cause the current practice to change. First, the expectation that public funds will be applied to alleviating the impacts of manageable risks is progressively being reduced. Second, recent events in other parts of the word are changing the perception that Australia's isolation immunizes it from accidental or intentional incursions of high-consequence adverse-impact events. Third, the EAD Response Agreement could well make livestock disease insurance more attractive to both buyer and seller because it effectively funds a very significant "excess" before a claim on a commercial insurer would be made.

During 2002, Animal Health Australia held preliminary discussions with elements of the agricultural risk management industry, to ascertain whether, in the light of the EAD Response Agreement, attractive livestock disease insurance products might be devised for the Australian market. The indications at that time were that possibilities occur in two areas. The first was to insure livestock industry associations who are parties to the Agreement against the risk that an incursion of certain infectious diseases, such as FMD, could result in direct costs exceeding the 1% of GVP non-discretionary limit. The rationale was that an industry could be so decimated by an outbreak that the obligation to pay the share of costs, even within the ten years available, would necessitate such a high levy upon the reduced numbers of stock remaining that it would severely hamper the industry's recovery. The second was in providing protection for individual producers against the consequential losses they might incur in the event of a disease incursion. Such losses are specifically excluded from the EAD Response Agreement, but in the case of an FMD outbreak, indications are that consequential losses are likely to be in the range of AU$4.0 billion (US$2.0 billion) to AU$14 billion (US$7 billion) while direct costs, recoverable under the EAD Response Agreement, are likely to be of the order of AU$500 million (US$250 million).

Little substantive progress has been made in developing such products although the potential to do so remains. It is interesting to note that the condition precedent delaying the progress of Animal Health Australia's engagement with the insurance industry is the difficulty of a robust probability analysis of a FMD or any other emergency animal disease outbreak occurring in Australia.

Next Steps

The EAD Response Agreement

As events have transpired in the first six months since signing, the cost-sharing aspects of the EAD Response Agreement have been applied in two real and one simulated event. These occasions have shown the Agreement to be robust, flexible and effective. They have also confirmed that, in regard to the cost-sharing aspects at least, the spirit of the parties has matched the intent and letter of the formal instrument. This experience has also demonstrated the value of the effort put into familiarizing the parties with the processes of the Agreement and in training representatives for their roles in these processes.

The underlying intent of the entire agreement, however, is that a party "qualifies" for the benefits and protection of the cost-sharing aspects of the agreement by compliance with the risk mitigation (biosecurity) and preparedness enhancement (Performance Standards) components. While all parties have initiated activities to address these issues, the real test of whether or not the spirit matches the letter and whether the Australia of the future matches the commitment of the past to maintaining its advantageous animal health status, will come when, as will inevitably be the case, parties are faced with the need to commit discretionary resources to honoring obligations arising from the biosecurity and performance standards aspects of the agreement.

Livestock Disease Insurance

Animal Health Australia will continue to investigate options for livestock disease insurance to supplement the protection afforded under the EAD Response Agreement. This will require the implementation of a risk analysis process to establish a credible risk profile, followed by closer engagement with the commercial agricultural insurance industry and the development of product concepts to be canvassed with industry associations and producers and, if attractive to them, to be commercialized and deployed.

Conclusion

Historically, Australian governments and livestock industries have sought to manage the risks associated with incursions of serious livestock diseases by collectively funding activities to prevent incursion, preparing to stamp out incursions, and to eradicate established livestock diseases of national concern – rather than applying the available funds to insurance arrangements to recover the costs associated with the same diseases.

The recent extension of collective and cooperative arrangements between governments and the livestock industries to mitigate EAD risks by means of the EAD Response Agreement, combined with apparent increases in the risk of incursions by emergency animal diseases, may eventually see livestock industry associations and individual livestock producers add livestock disease insurance to their risk management portfolios.

Chapter 13

Livestock Industry Insurance: Canada

Bruce Stephen and Terri Epps

Introduction

Agricultural producers in Canada have identified a number of gaps in available farm-level risk management coverage, despite a wide array of programmes and services offered by the public and private sectors. Historically, these gaps have been characterized by the lack of insurance coverage offered to livestock producers, although gaps have also been identified in the crops sector. In the development and implementation of Canada's Agricultural Policy Framework (see below), the Federal, provincial and territorial governments have moved to significantly expand risk management coverage to the agricultural sector in those areas where the public sector has a defined role in sharing risk management with farmers (e.g. production insurance, income stabilization, etc.). However, government programmes are not necessarily the appropriate answer to all farm-level risk management issues. Governments recognize the important role of the private sector in developing and delivering risk management products and services to the agriculture sector.

In providing for and encouraging an increasing role for the private sector, the Federal government has an opportunity to assist livestock organizations in addressing the remaining gaps, as well as emerging gaps which may be identified in the wider agricultural industry (environmental and/or food safety liabilities, etc.), for the long-term benefit of producers.

Agriculture and Agri-Food Canada's (AAFC) Private Sector Risk Management Partnerships (PSRMP) programming is designed to help the agricultural industry find risk management solutions for these gaps using products and services developed and delivered by the private sector. Without adequate coverage for ongoing and emerging agricultural risks, the Canadian agricultural industry may be hampered in its efforts to maintain a sustainable and innovative industry. The PSRMP will complement the Federal government's ongoing investment in Canadian agriculture by assisting producers in managing business risks while moving the industry beyond crisis management to long-term profitability and competitiveness.

The PSRMP represents an alternative approach for governments to address gaps in farm-level agricultural risk management coverage. By facilitating an expanding role for private sector providers of risk management products and services in areas where there is no clear rationale for public sector programming, the PSRMP approach allows for a limited investment of public resources in industry-led projects where perceived farm-level risk management gaps can be narrowed or closed. The result of these specified investments will be an increased capacity within the agriculture sector to manage farm business risks, supported by an increasing range of private sector risk management products and services designed to meet the needs of Canadian farmers.

Canada's Agricultural Policy Framework

Since 2001, Federal, provincial, and territorial governments have been working with the agriculture and agri-food industry to help strengthen and revitalize the industry through the Agricultural Policy Framework (APF). Over a five-year period, which started on 1 April 2003, Canadian producers will be given access to new tools, services and options to strengthen their businesses, increase prosperity and meet the demands of consumers at home and abroad.

The goal of the APF is to foster an economic and social advantage for Canada to help ensure that Canadian producers and agri-food companies meet today's agricultural production and marketing challenges. To assist the industry to reach this goal, federal, provincial and territorial Ministers of Agriculture have jointly developed "... an agricultural policy that is comprehensive, integrated and ensures that producers have the tools to address issues, be competitive and capture the opportunities these challenges present in the areas of science, food safety and environmental stewardship".

The APF brings together five key elements:

- Business Risk Management (BRM);
- Environment;
- Food Safety and Food Quality;
- Renewal; and
- Science and Innovation.

Together, these five elements form a single, solid platform that will help Canadian agriculture maximize new opportunities in world markets. Few other countries can claim such a cohesive and integrated policy approach to agriculture. Canada will strive for international recognition as the world leader in food safety and food quality, environmentally responsible production and innovative products.

Business Risk Management Programming

One of the objectives of the APF is to move from "safety nets" to "risk management". This means that BRM programming will be focused on growth and improving the income prospects in agriculture. It also means taking a comprehensive approach to risk management that encourages producers to make decisions to manage risk and improve the viability of their farm business through change and innovation. BRM programming will contribute to the success of the APF by creating an environment where producers actively manage their business risks utilizing an appropriate combination of private and public sector risk management tools.

BRM programming is a critical foundation of the APF. It will promote more rapid adoption of food safety and environmental practices, support other elements of the framework, enhance the expansion and diversification of farm businesses through new applications of science, and enhance managerial and strategic planning skills through (industry) renewal programming.

The BRM programme includes the following elements:

- The Canadian Agricultural Income Stabilization (CAIS) programme, (see http://www.agr.gc.ca/caisprogram);
- Production Insurance (PI), (see http://www.agr.gc.ca/progser/ci_e.phtml);
- Cash advance programming:
  - Advance Payments Programme (APP); and
  - Spring Credit Advance Programme (SCAP), (see http://www.agr.gc.ca/nmp/app).
- Private Sector Risk Management Partnerships (PSRMP), (see http://www.agr.gc.ca/psrmp).

Existing province-based programmes will continue through a transition period ending on 31 March 2006.

Notwithstanding the coverage offered by BRM programming, there remain small but significant risk management challenges affecting Canadian producers. Examples of these include:

- Existing as well as emerging concerns related to producer liability resulting from environmental and food safety practices;
- Business interruption losses associated with a plant or animal disease;
- Animal care issues; or
- The loss of market access.

Traditionally, these challenges would have been addressed with insurance coverage provided by the private sector or via self-insurance by individual producers.

However, over the past couple of decades, the agricultural industry has expressed concerns regarding private sector insurance market trends in Canada. For example, the narrowing range of coverage insurance companies are willing to provide, as well as significant exclusions for the coverage they do offer. Two drivers of these trends have been the emergence of new and more complicated risks (e.g. terrorism, third party liability and disease) and an increasingly precarious financial picture (e.g. higher claims and awards, increased competition and lower profit margins). Private sector insurance coverage providers in the agricultural industry are perceived as increasingly focused on mainstream coverage (barns, equipment, crops, transportation, etc.) while shying away from innovative – and likely more risky – areas of activity. The APF specifically recognizes and promotes the role the private sector shares with producers in Canada and the challenges the industry faces. Products and services in the areas of insurance, banking and investments continue to support and contribute to the success of agriculture in Canada, and BRM programming has been enhanced to complement these activities. It aims to expand the role of the private sector in agricultural risk management wherever feasible.

The BRM programme Private Sector Risk Management Partnerships is designed to assist the agriculture industry in cooperating with the private sector with the goal of developing private sector risk management solutions to fill the gaps in available risk management tools. To do this, the PSRMP will provide a portion of the necessary funding for the research and development activities required in developing these solutions, thus reducing the up-front financial investment required from potential private sector providers and reversing the perceived trend away from exploring innovative coverage solutions. In short, it will invest in the public good of animal health management through research on appropriate mechanisms, pricing and implementation of insurance and other risk management programmes.

Private Sector Risk Management Partnerships

The PSRMP is part of the Federal Government's policy towards agricultural business risk management. The PSRMP is designed to help the agricultural industry find risk management solutions for risks or perils which are predictable and represent a significant threat to an affected farm when the overall impact on the industry or the market is expected to be insignificant. The PSRMP will also help the financial services industry (banking, insurance, investment, etc.) to identify and develop new products and services for producers.

The PSRMP is not designed to provide risk management solutions for larger, uninsurable disasters such as drought or disease epidemics since the private sector is unlikely to get involved with such risks. Plus, the objective of a PSRMP project is to facilitate a private sector solution. These more financially intensive risk management needs may be addressed by other BRM programming, such as the CAIS programme, PI or the cash advance programmes.

Funding for the PSRMP will be delivered through Contribution Agreements between AAFC and producer organizations, who will be the clients of PSRMP programming, and will, in turn, represent the interests of their member-producers. The PSRMP will provide up to $15 million in funding for approved risk management projects over the five-year period of the APF.

The PSRMP – Objectives and Approach

The PSRMP is designed to achieve two APF-BRM objectives:

1. To enhance the capacity of the agricultural industry to manage risks traditionally not covered by public sector programmes and for which private sector coverage is not currently available; and
2. To increase the participation of the private sector financial services industry in providing risk management solutions (products and services) to the agricultural industry.

The PSRMP: a Two-Pronged Approach

One prong facilitates relationships between producer organizations and the financial services industry. These relationships will identify opportunities for the private sector to expand its role in agricultural risk management. These relationships will also provide private sector expertise (e.g. data collection and analysis, risk and actuarial assessments, legal and underwriting advice, etc.) to PSRMP projects.

The second prong provides financial and technical assistance to agricultural industry-led projects to develop a comprehensive business case for negotiating a private sector risk management solution. The business case will demonstrate that the identified risks/perils are quantifiable, predictable and have a measurable financial impact, thereby presenting an opportunity for a risk-transfer (i.e. insurance) solution.

The process of developing a business case is structured in four stages:

1. Identifying the risks to be addressed and developing an objective description of the risks;
2. Outlining the risk control process in place to mitigate these risks;
3. Presenting a history of representative risk data; and
4. Analyzing the data and developing viable coverage options for the identified risks.

Based on the four stages of developing a business case and the stage at which the prospective PSRMP client will start their project, three types of projects can be considered for PSRMP assistance:

1. A comprehensive business case for securing a private sector risk management solution;
2. An interim risk management solution, with the expectation that the data collected during this period would support the future development of a business case for insurance-based coverage; or
3. A risk assessment to identify and rank the business risks facing the member-producers of the producer organization.

In drafting a detailed project work plan, PSRMP staff will work with the client in identifying project activities to be carried out in each of these stages, as appropriate. Or, for those project applications ineligible for PSRMP support, staff will help the applicant to identify other potential sources for assistance from either the public or private sector.

Financial Support for PSRMP Projects

Funding from the PSRMP is available for up to 100 percent of the eligible costs of a client's project activities. While costs will vary significantly from project to project, it is anticipated that a PSRMP project involving data collection, analysis, development of alternatives and negotiations with private sector providers could take 18-24 months to complete. On the other hand, a risk assessment could reasonably be completed within six months.

Financial assistance for PSRMP will be available for eligible project activities including, but not limited to:

- Research and development costs;
- Data collection and/or analysis expenses;
- Legal costs directly related to project objective(s);
- Actuarial costs and other professional services;
- Consultations; and
- Travel related to project activities.

PSRMP Project Results

The expected results of a PSRMP project are:

- The initiation of partnerships or expansion of existing relationships between producer organzations and private-sector providers of financial products and services;

- The implementation of a new risk management product or an interim risk management solution in the private sector; and/or
- A basic research contribution to the risk profile of the client's commodity through the analysis of the risks identified in a PSRMP project.

PSRMP Staff Activities

The PSRMP staff will work closely with clients in the following areas:

- Consultation with
  - Agricultural industry stakeholders;
  - Representatives of the domestic and international financial services industry; and
  - Other APF programmes.
- Provision of technical advice and expertise;
- Support of clients in fulfilling their responsibilities with respect to project management; and
- Reporting and wrap-up.

In addition, the PSRMP staff will provide clients with contact information on potential providers of professional services to support project activities. These professional services may include, but are not limited to:

- Risk assessments, including data collection and assembly;
- Ranking of identified risk/peril(s);
- Risk/peril(s) specific analysis, including actuarial and financial analysis;
- Development of coverage options, including ranking and testing of options; and
- Promotion of selected option(s) to member-producers, assessing their acceptance and determining their intention to participate.

Support for Other Agricultural Policy Framework Elements

The PSRMP will contribute to Canada's Agricultural Policy Framework by helping to build a strong risk management model for producer organizations. The PSRMP will help create an environment where producers can better understand and actively manage their business risks and encourage the participation of producers in other APF programmes. In addition, information collected through PSRMP projects may contribute to AAFC's understanding of the risks facing individual commodities.

Chapter 14

The Current State of US Federally Supported Livestock Insurance

Chad Hart

Introduction

The US Agricultural Risk Protection Act (ARPA) of 2000 substantially changed the risk management landscape for livestock. Before the passage of this act, livestock was explicitly excluded from coverage under federally supported agricultural insurance programmes. Federal agricultural insurance has been available in some form for some commodities for the past seventy years, but for many of those years, participation was very limited. Strong interest in insurance programmes did not surface until the government began to provide significant premium subsidies. Over the last decade, federal agricultural insurance programmes have expanded tremendously. Revenue insurance coverage did not exist in the early 1990s; now revenue insurance is the most popular form of coverage. Federally supported agricultural insurance now exists for over 100 commodities. Roughly 70% of crop insurance-eligible agricultural production has been covered since 1995. The entire federal crop insurance programme has grown into a $3-4 billion programme.

All of this growth occurred in the crop sector, as livestock could not be covered. The livestock sector in the USA represents roughly half of the total value of agricultural production in the USA. The move to allow federal livestock insurance provides another avenue for expansion of the federal agricultural insurance programme. As the US Department of Agriculture (USDA) Risk Management Agency (RMA) proceeds in fostering livestock insurance products, it will need to examine the similarities and differences between livestock and crop farming to provide adequate and appropriate types of insurance coverage. Also, there are insurance products sold by private companies that cover some livestock risks (usually death losses due to certain named perils). In adding to the list of federally supported agricultural insurance products, RMA typically avoids additions that would directly compete against such private sector insurance products.

The risks involved in livestock farming are not exactly the same as those for crop farming. As part of an RMA-funded study on livestock insurance, researchers at Iowa State University held several listening sessions with livestock producers across the USA and discussed the various risks livestock producers face. Most of the producers indicated that price risk, on both the output and input markets, is their biggest concern. Figure 14.1 shows the relative price movements for maize, beef cattle and hogs in Iowa since June 1994. In all of these markets, prices have, at one time, shifted at least 60% from their June 1994 levels. As the figure shows,

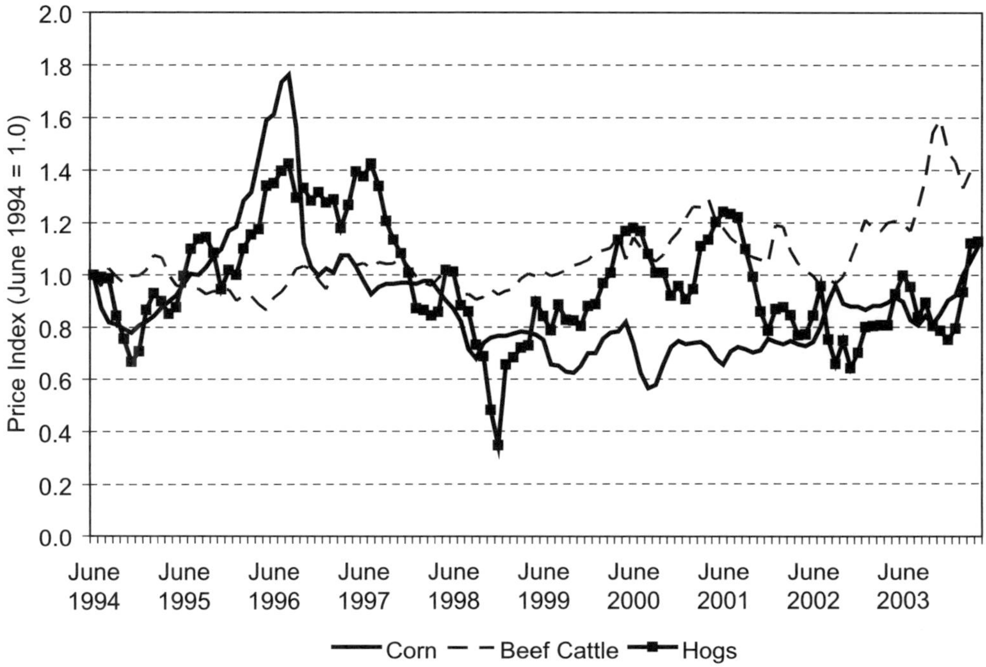

Fig. 14.1. Price Indices for Maize, Beef Cattle and Hogs.

livestock producers have experienced dramatic price swings for their outputs and their major input, feed.

However, livestock producers have not traditionally used available price risk management tools such as futures and options. In the 1996 Agricultural Resource Management Study, conducted by the USDA's Economic Research Service, agricultural producers were asked about their usage of various risk management tools. The results showed that only 15 to 25% of cattle and hog producers utilized futures and options (USGAO, 1999). Increasing livestock producers' usage of risk management tools was a goal of RMA even before the passage of ARPA, as RMA sponsors several risk management education programmes.

In the listening sessions, livestock producers indicated that production risk was of lesser importance and could be handled through management practices, while crop farmers faced significant yield risk. Disease risk was also not a major concern of the producers in the sessions. But the listening sessions were conducted before the bovine spongiform encephalopathy (BSE) cases in Canada and the USA. If the listening sessions were conducted today, disease risk would likely be high on the list of risk management concerns.

This chapter will briefly outline the types of insurance typically provided by federally supported agricultural insurance, discuss ARPA and its specific livestock provisions, provide detailed descriptions of approved livestock insurance products, discuss the usage and performance of these products thus far, and preview product proposals that are currently being discussed for implementation. Currently there are six federally supported insurance products that can cover livestock. They are Livestock Risk Protection (LRP) for hogs, fed cattle and feeder cattle; Livestock Gross Margin (LGM) for hogs; Adjusted Gross Revenue (AGR); and Adjusted Gross Revenue-Lite (AGR-Lite). The LGM and LRP products are specifically targeted at livestock and cover price risk. The LRP products cover livestock price risks, while LGM covers both livestock and feed price risks. The AGR products cover livestock as part of a whole-farm revenue policy. The differences among these products will highlight several issues in providing livestock insurance: what risks are covered, what levels of coverage are allowed, and how government involvement may spur both the development of risk management tools for livestock producers and the usage of these tools and additional tools provided by the private sector, such as futures and options.

Various Types of Agricultural Insurance

Federally supported agricultural insurance can be categorized into three types: yield or production, price and revenue policies. Traditional crop insurance, the only type of crop insurance available throughout most of the history of the Federal Crop Insurance Corporation (FCIC), is based on the average yield or production on the farm or insurance area. Typically, producers provide evidence on their yields or production over the last four to ten years to create an average yield or production level. This average yield is called an APH yield (i.e. Actual Production History yield). Yield insurance allows producers to insure against their current yield or production falling below a set percentage of the APH yield. The set percentage is equal to the coverage level of the insurance policy. The prices for these policies are set at the beginning of the insurance period and are held constant over the life of the policy. Per-unit indemnity payments are given by the following equation:

$$\text{Per-unit Indemnity} = \text{Max}(0, \text{YCL} \times \text{APH} - \text{AY}) \times \text{PCL} \times \text{Price}$$

where Max is the maximum function (choosing the larger of the two values), YCL is the producer's chosen yield coverage level, APH is the producer's APH yield, AY is the producer's actual current yield, PCL is the producer's chosen price percentage and Price is RMA's expected market price for the commodity. Once the producer has signed up for yield insurance, the only unknown is the actual yield, as all of the other variables have been set beforehand or chosen by the producer.

Yield insurance is available on over 75 different commodities. Producers can choose among coverage levels, in five percent increments, between 50% and 85% of their APH yields. Higher coverage levels are not available in all states for all crops. Producers may also choose their insured price, in one percent increments, from 55% to 100% of RMA's expected market price (RMA, 2003b). An area-based yield insurance (insuring against shortfalls in area yields, as opposed to farm yields) is also available. The area yield insurance has different coverage levels for both prices and yields.

Price insurance policies are very recent additions to the federal agricultural insurance portfolio and work much the same way as yield insurance policies. Producers can insure against prices falling below a set percentage of an insurable price. The set percentage is equal to the coverage level of the insurance policy. The yields or production for these policies are set at the beginning of the insurance period and are held constant over the life of the policy. Per-unit indemnity payments are given by the following equation:

Per-unit Indemnity = Max(0, PCL × Price – AP) × YCL × Yield

where Max is the maximum function, PCL is the producer's chosen price coverage level, Price is the maximum insurable price for the commodity, AP is the actual current price for the commodity, YCL is the producer's chosen yield percentage and Yield is the maximum insurable yield for the commodity. Once the producer has signed up for price insurance, the only unknown is the actual price, as all of the other variables have been set beforehand or chosen by the producer.

Most of the livestock insurance products discussed later are price insurance products, albeit some are more complicated than the form shown above as they insure against price changes in several markets. Allowed coverage levels are higher for price insurance products, usually between 70 and 100% of the maximum insurable price.

Revenue insurance policies are also recent additions to the risk management menu for agricultural producers. These policies combine the first two types of policies as shortfalls in either yield or price may trigger insurance payments. There are two types of revenue insurance policies currently offered. We will refer to them as basic and enhanced policies. With basic revenue insurance, producers insure against revenues below a set percentage of insurable revenue. The insurable revenue is determined in a variety of ways, depending on the policy. Again, the set percentage represents the coverage level of the policy. The insurable revenue is held constant over the life of the policy. Per-unit indemnity payments are given by the following equation:

Per-unit Indemnity = Max(0, PCL × Price × YCL × Yield – AP × AY)

where Max is the maximum function, PCL is the producer's chosen price coverage level, Price is the maximum insurable price for the commodity, AP is the actual current price for the commodity, YCL is the producer's chosen yield percentage, Yield is the maximum insurable yield for the commodity and AY is the producer's actual current yield for the commodity. Once the producer has signed up for revenue insurance, the only unknowns are the actual price and yield, as all of the other variables have been set beforehand or chosen by the producer. In many cases, PCL is set to one and the producer only needs to choose a yield coverage level, which is by default the revenue coverage level. Income Protection and Revenue Assurance without the Harvest Price Option are two examples of basic revenue insurance programmes.

Enhanced revenue insurance policies cover the same risks as basic revenue insurance policies and add some additional coverage for low-yield events. Crop Revenue Coverage and Revenue Assurance with the Harvest Price Option are two examples of enhanced revenue insurance policies. For both policies, the enhancement is that the insurable revenue can move up with the current actual price. Per-unit indemnity payments are given by the following equation:

Per-unit Indemnity = Max(0, PCL × Max(Price, AP) × YCL × Yield – AP × AY)

where Max is the maximum function, PCL is the producer's chosen price coverage level, Price is the insurable price for the commodity, AP is the actual current price for the commodity, YCL is the producer's chosen yield percentage, Yield is the maximum insurable yield for the commodity and AY is the producer's actual current yield for the commodity. Once the producer has signed up for revenue insurance, the only unknowns are the actual price and yield, as all of the other variables have been set beforehand or chosen by the producer, but now the actual price affects both the insurable revenue and the actual revenue. The effect this has on the policy is that an enhanced revenue insurance policy covers all of the revenue losses that a basic revenue insurance policy would, but also covers all of the yield losses that would trigger payments under a yield insurance policy with the same coverage level (assuming PCL is equal to one).

Figure 14.2 shows the coverage differences among the insurance types for a representative maize farm. For this example, let's assume an APH yield of 100 bushels per acre, an insurable price of $2.00 per bushel and a coverage level of 75%. Yield insurance would pay when yields are below 75 bushels per acre (below the line with square markers). Price insurance would pay when prices fall below $1.50 per bushel (to the left of the line with no markers). Basic revenue insurance would pay when revenues fall below $150 per acre (75% × 100 bushels per acre × $2.00 per bushel); this is represented by the area to the left and below the line with triangle markers. Enhanced revenue insurance would pay when revenues fall below $150 per acre or yields are below 75 bushels per acre (below and to the left of the line with circle markers).

There is an extensive literature covering crop insurance. Below is but a brief summary of the literature. For a more extensive listing, see Goodwin and Smith (1995), Knight and Coble (1997) or Coble and Knight (2002). Crop insurance demand has been investigated at both the aggregate (Barnett and Skees, 1995; Richards, 2000) and the disaggregate (Wu, 1999; Makki

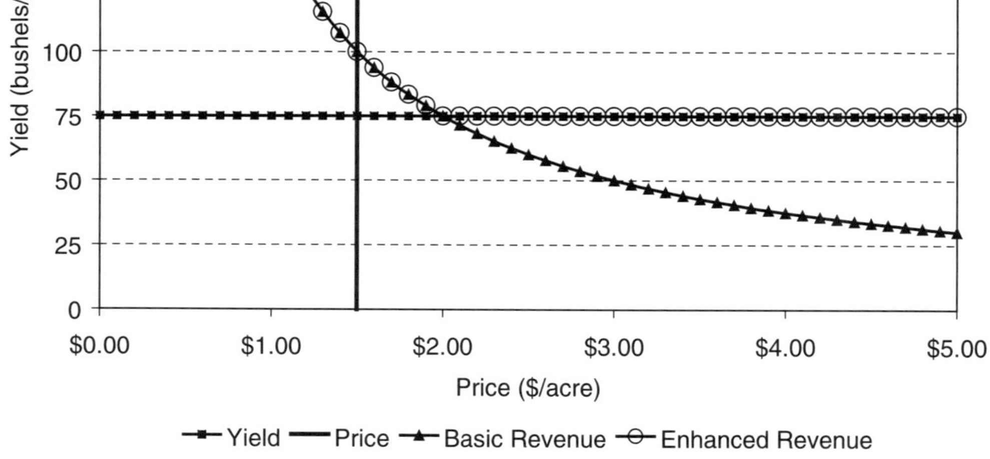

Fig. 14.2. Example Graph of Various Insurance Types.

and Somwaru, 2001; Sherrick, *et al.*, 2004) levels. Many of these studies have found demand for crop insurance to be inelastic. A myriad of papers have proposed alternative insurance designs. Mahul and Wright (2003), Hennessy, Babcock and Hayes (1997) and Stokes (2000) study crop revenue insurance structure. Skees, Barnett and Black (1997), Mahul (1999) and Ker and Goodwin (2000) study area yield insurance policies. Turvey and Islam (1995) compare the equity and efficiency of area versus individual crop yield insurance coverage. Hart, Babcock and Hayes (2001) examine several livestock insurance schemes and compare them to strategies involving existing futures and options.

Moral hazard describes the effect insurance has on the insured's choices. In the case of agricultural insurance, actions taken by producers that may increase the likelihood that insurance will pay out are referred to as moral hazard. Adverse selection describes the self-selection problem in insurance. The most likely customers for an insurance policy are those that are most likely to collect on the insurance. Just, Calvin and Quiggin (1999) examine reasons for crop insurance participation. They segment the participation decision into risk aversion, subsidy and adverse selection incentives. Their results suggest crop insurance participation is primarily based on subsidy and adverse selection incentives. Ker and McGowan (2000) examine adverse selection on the reinsurance side, that crop insurance companies may exploit their reinsurance relationship with the federal government via adverse selection. Coble, Knight, Pope and Williams (1997) look at moral hazard issues within crop insurance. Their results suggest moral hazard occurs in poor production years, but is not significant in good production years.

ARPA and Livestock

The federal government first began its venture into agricultural insurance in the 1930s. The FCIC was formed in 1938 to administer the programme. Due to severe losses in the first few years, the programme was scaled down to an experimental level, covering only a few crops within a limited geographic area. Federal crop insurance continued in this manner until 1980. The Federal Crop Insurance Act of 1980 established federal crop insurance as the primary type of government disaster aid. The crop insurance programme was extended to all counties that had significant agricultural production and to all crops for which sufficient actuarial data could be obtained. The programme was designed to be delivered by private insurance companies reinsured by FCIC (RMA, 2003a). To induce both farmers and private insurers, the government subsidized the premium cost, paid for both the delivery and servicing costs of private insurers, and assumed any excess indemnities over premiums. Thus, the federal government assumed most of the risk associated with the programme.

Congress passed the Agricultural Risk Protection Act (ARPA) in the summer of 2000. The act made several substantial changes to the creation and implementation of federally supported agricultural insurance. The private sector's role in the creation of insurance products was strengthened. Research and development of new insurance products was to be conducted by the private sector, with RMA serving in an advisory role. Partnership agreements between the private sector and RMA were to target increasing agricultural insurance coverage in under-served areas and commodities. Additional premium subsidies were provided to entice new participants into the programme and to spur current participants to choose higher levels of coverage. Regulations were put in place to limit waste and fraud within the programme (United States Congress, 2000).

But the portion of the act that is likely to have the largest impact on the future of agricultural insurance is the inclusion of livestock in the list of commodities eligible for federally supported agricultural insurance. Before the passage of ARPA, livestock had been specifically excluded from federally supported agricultural insurance. The only exception to this had been the limited coverage for livestock revenues under AGR policies. The justification for this exception was that the livestock revenue represented the value of the crop production fed to the livestock. With livestock no longer excluded, insurance policies can now be tailored specifically for livestock. The possible impacts are tremendous. In terms of gross value of production in the USA, livestock accounts for roughly half of the value of agriculture. So before ARPA, half of agriculture did not have access to federally supported agricultural insurance. With this restriction gone, federally supported agricultural insurance has the potential for astounding growth.

However, the language in ARPA is fairly specific on how livestock insurance coverage can be expanded. Section 132 of ARPA details the livestock provisions. Pilot programmes may be created that protect against losses from livestock poisoning and disease (this book is an example of the research ongoing to support such pilot programmes). FCIC shall conduct at least two livestock pilot programmes that may cover price, income, or production losses. The total amount of government expenditures on all livestock programmes was limited to $10 million in fiscal years 2001 and 2002, $15 million in fiscal year 2003, and $20 million in fiscal year 2004 and thereafter (United States Congress, 2000).

While at the outset the limits on expenditures would seem to constrain growth in livestock insurance products, such has not been the case. Several pilot programmes are currently under way, covering a variety of livestock species in several states. Several research partnerships on livestock insurance issues have been funded. With continued success from these partnerships and pilot programmes, Congress may be moved to loosen the purse strings and allow livestock insurance to expand greatly.

Current Federally Supported Livestock Insurance Products

Livestock Risk Protection

Livestock Risk Protection for hogs, also known as LRP-Swine, was first offered to livestock producers in the summer of 2002. LRP-Swine covers downside price risk in hogs. The insurance coverage is based on prices derived from the Chicago Mercantile Exchange (CME) futures prices for lean hogs. Under an LRP-Swine contract, a producer selects to insure against hog prices falling below an insured price. The producer chooses from a menu of choices on the length of the coverage period, the coverage level, and the target weight for the insured animals. At the end of the coverage period, the actual ending value is seen and if it is below the insured price, the producer receives an indemnity.

There are four coverage period lengths for LRP-Swine: 13, 17, 21 and 26 weeks. These allow producers to select a coverage period that most closely matches their marketing pattern. Insurable prices (called expected ending values) are derived from current CME lean hog futures prices. Coverage levels range between 70% and 95% of the expected ending values. Target weights are stated for lean hogs using a constant 0.74 lean weight-to-live weight conversion factor. Target weights are restricted to be between 150 and 225 pounds per hog.

LRP-Swine is sold on a continuous basis throughout the year; there is no set sales period for the product. This is a common feature of all of the LRP products. There is an annual limit of 32,000 hogs that can be covered for an individual producer during the insurance year. Producers can obtain coverage multiple times throughout the year; each time is called an endorsement. There is a limit of 10,000 hogs per endorsement. The actual ending values are derived from the USDA-Agricultural Marketing Service cash index for hogs (RMA, 2003h).

The premium is subsidized; the government pays 13% of the premium and, as is the case for all federally supported insurance products, all administrative and operating expenses associated with providing and servicing the policy. Since the product is subsidized and is closely related to the commodity futures and options market, there are restrictions on certain transactions by the producer to limit the possibility of having off-setting positions in the market and allowing the producer to capture the premium subsidy. Examples of off-setting transactions to the insurance contract are the selling of a CME lean hog put option and the buying of a CME lean hog futures contract. These transactions are blocked because they have the potential to offset the benefits from the insurance while allowing users to partially capture the insurance premium subsidy over a wide range of prices. Other types of market transactions, such as traditional hedging strategies of selling futures or buying put options, are still allowed.

To show how the LRP-Swine product works, let us work through an example based on price and insurance information for January 15, 2003. The information for this example can be gathered from the RMA website. A pork producer wants to insure 100 hogs that they intend to market in three months. They expect to market the hogs at 270 pounds each, which translates into a lean weight of 200 pounds. The producer is the sole owner of the hogs, so their share of the hog operation is 100%. The expected ending value is $59.30 per hundredweight. The producer chooses a 13-week policy at 90% coverage, implying an insured price of $53.33 per hundredweight (LRP coverage levels are not exact; they are calculated from the insurance prices). The premium rate for this coverage is 3.4% of the liability. The producer's premium is equal to the product of the number of head insured, the target weight, the insured price, the producer's share of the hog operation, the premium rate and 100% less the premium subsidy rate (13%)

$$\text{Producer Premium} = 100 \text{ head} \times 2 \text{ cwt/head} \times \$53.33\text{/cwt} \times 100\% \times 3.4\% \times (100 - 13)\%$$
$$= \$316 \text{ or } \$3.16 \text{ per head insured.}$$

After 13 weeks, the actual ending value equals $49.62 per hundredweight. The indemnity is equal to the product of the number of head insured, the target weight, the maximum of zero or the difference between the insured price and the actual ending value, and the producer's share of the hog operation

$$\text{Indemnity} = 100 \text{ head} \times 2 \text{ cwt/head} \times \text{Max}(\$0, \$53.33 - \$49.62)\text{/cwt} \times 100\% = \$736$$
$$\text{or } \$7.36 \text{ per head insured.}$$

If the actual ending value had been greater than $53.33 per hundredweight, then the indemnity would have been zero. Figure 14.3 shows the indemnity schedule for various actual ending values in this example.

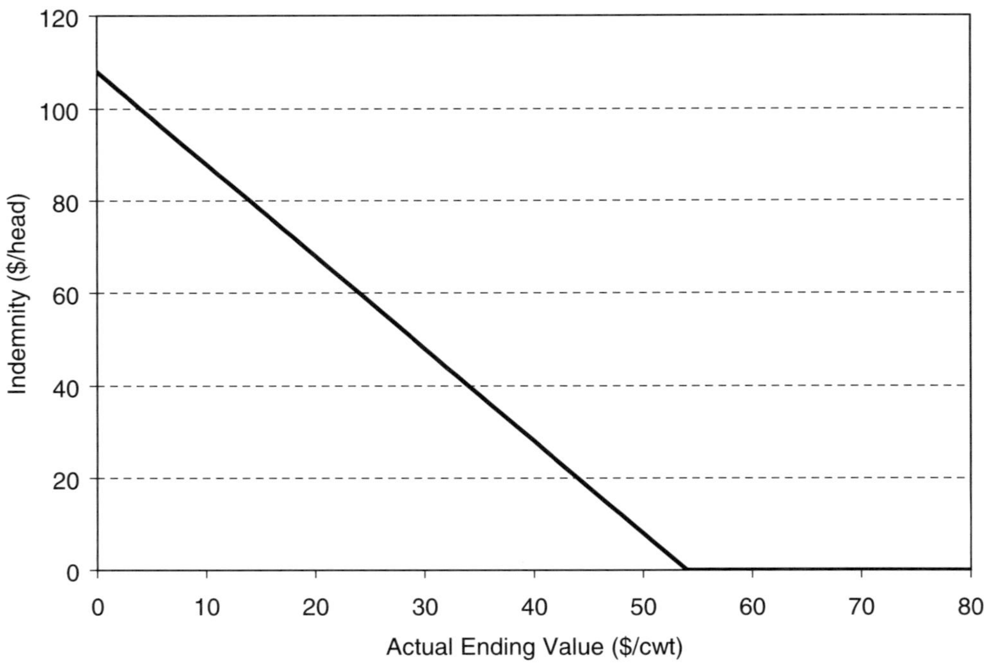

Fig. 14.3. Example of LRP Indemnities.

The LRP-Swine policy does not require actual sales values, sales weights or the number of sold head. The coverage remains in effect whether the producer markets the hogs at that time or not. The only limitation comes if the hogs are marketed more than 30 days before the end of the insurance period. Production and marketing losses, such as death loss due to disease or local price movements, are not covered by the policy. When LRP-Swine was first released, only Iowa hog producers could obtain coverage. LRP-Swine has been expanded to Colorado, Illinois, Indiana, Kansas, Michigan, Minnesota, Missouri, Nebraska, Nevada, North Dakota, Ohio, Oklahoma, South Dakota, Texas, Utah, West Virginia, Wisconsin and Wyoming.

The other LRP products (LRP-Fed Cattle and LRP-Feeder Cattle) are similar to LRP-Swine. LRP-Fed Cattle covers downside price risk in fed cattle; LRP-Feeder Cattle does the same for feeder cattle prices. The insurance coverage is based on prices derived from the CME futures prices. Under the products, a producer selects to insure against prices falling below an insured price. The producer chooses from a menu of choices on the length of the coverage period, the coverage level, and the target weight for the insured animals. At the end of the coverage period, the actual ending value is seen and if it is below the insured price, the producer receives an indemnity. Both of these products were introduced in early 2003 and are now available in the same states as LRP-Swine. Table 14.1 provides some of the policy settings for the cattle LRP products.

Insurable prices for LRP-Fed Cattle are derived from current CME live cattle futures prices. The actual ending values are derived from the USDA-Agricultural Marketing Service

Table 14.1. Livestock Risk Protection Settings for Cattle.

	Fed Cattle	Feeder Cattle
Insurance Period	13, 17, 21, 26, 30, 34, 39, 43, 47 and 52 weeks	21, 26, 30, 34, 39, 43, 47 and 52 weeks
Coverage Levels	70 - 95%	70 - 95%
Target Live Weights	1,000 - 1,400 pounds	650 - 900 pounds
Annual Limits	4,000 head	2,000 head
Endorsement Limits	2,000 head	1,000 head

cash prices for slaughter cattle (RMA, 2003f). Insurable prices for LRP-Feeder Cattle are derived from current CME feeder cattle futures prices. The actual ending values are derived from the CME feeder cattle index (RMA, 2003g). Since both LRP cattle products are subsidized and are closely related to the commodity futures and options markets, there are restrictions on certain transactions by the producer to limit the possibility of having off-setting positions in the market and allowing the producer to capture premium subsidies. Examples of off-setting transactions to the insurance contract are the selling of a CME live or feeder cattle put option and the buying of a CME live or feeder cattle futures contract.

Premium and indemnity calculations are very similar to those for LRP-Swine. The producer's premium is equal to the product of the number of head insured, the target weight, the insured price, the producer's share of the operation, the premium rate and 100% less the premium subsidy rate. The indemnity is equal to the product of the number of head insured, the target weight, the maximum of zero or the difference between the insured price and the actual ending value, and the producer's share of the operation.

The policies do not require actual sales values, sales weights or the number of head sold. Similar restrictions on marketings apply for the cattle LRP policies as for LRP-Swine. Production and marketing losses are not covered by the policy.

Livestock Gross Margin

LGM was first offered to livestock producers in the summer of 2002. LRP-Swine covers downside price risk in hogs and upside price risk in feed. The insurance coverage is based on prices derived from CME futures prices for lean hogs and Chicago Board of Trade (CBOT) futures prices for maize and soybean meal. Under an LGM contract, a producer selects to insure against the margin between hog prices and feed prices falling below an insured level. The producer chooses the coverage level and the spread of marketings across the insurance period. At the end of the coverage period, the actual margin between hog prices and feed prices is seen and if it is below the insured level, the producer receives an indemnity.

Originally, LGM had two sales periods, at the end of January and July. Coverage extends six months after the sales periods, so policies cover the February–July period and the August–January period. Within each period, producers project the number of hogs they will market in each month. Expected gross margins are derived from current CME lean hog futures and current CBOT maize and soybean meal futures prices. Insurable hog, maize, and soybean meal prices are determined for each month. For futures contract months for each commodity, the insurable price is equal to the average futures price in the last three trading days prior to the 15th of the sales month. For non-futures contract months, the insurable prices are calculated as the weighted average of surrounding prices from futures contract months. For example, lean hog futures trade for the months of February and April, but not March. The insurable hog price for March is the average of the insurable prices for February and April. Coverage levels are from 80 to 100%, in five percent increments.

The gross margin calculation is based on a specific marketing weight, 260 pounds live weight, a given lean hog conversion factor, 0.74 and specific feed rations that depend on the production practice, farrow-to-finish versus finishing. For farrow-to-finish operations, it is assumed that 12.95 bushels of maize and 184.89 pounds of soybean meal are required to raise a hog to a market weight of 260 pounds. For finishing operations, the rations are 10.41 bushels of maize and 149.46 pounds of soybean meal. Since the feed is consumed during the months preceding the sale of the animal, the prices for the feed are lagged in the gross margin to represent the period of time when roughly half of the feed has been consumed by the hog. The lag of farrow-to-finish operations is three months (the hog price for August is paired with feed prices for May), while the lag for finishing operations is two months (the hog price for August is paired with feed prices for June). For a given month t, the expected gross margin is given by

$$(\text{Hog Price}_t \times 0.74 \times 2.6 \text{ cwt}) - (\text{Maize Price}_{t-3} \times 12.95 \text{ bu}) - (\text{Soybean Meal Price}_{t-3} \times 184.89 \text{ lb}/2000 \text{ lb})$$ for farrow-to-finish operations and

$$(\text{Hog Price}_t \times 0.74 \times 2.6 \text{ cwt}) - (\text{Maize Price}_{t-2} \times 10.41 \text{ bu}) - (\text{Soybean Meal Price}_{t-2} \times 149.46 \text{ lb}/2000 \text{ lb})$$ for finishing operations.

There is a limit of 15,000 hogs that can be covered for an individual producer during each six-month period. Actual gross margins are derived from the same futures markets, using the average futures prices for the contracts over the last three trading days prior to the contract expiration. The same lag structure, feed ration weights and marketing weights are used in the calculations.

Table 14.2 shows the expected hog, maize and soybean meal prices and the expected gross margins for the January 2003 sales period for farrow-to-finish operations. The lag in feed prices is already taken into account in the table, so the formulas above can be used to calculate the expected gross margins.

Table 14.3 shows the actual hog, maize and soybean meal prices and the actual gross margins for the January 2003 sales period for farrow-to-finish operations. The lag in feed prices is already taken into account in the table, so the formulas above can be used to calculate the actual gross margins.

When a producer signs up for LGM coverage, they provide target marketings for each month of the insurance period. Their total expected gross margin is the sum of the products of the monthly target marketings and the monthly expected gross margins. Given Table 14.1, if a producer wanted to insure 100 hogs in each month of the insurance period, the total expected gross margin for the policy would be equal to $41,418. If the producer only wanted to insure 100 hogs in April, then the total expected gross margin for the policy would be $6,840. The total actual gross margin is calculated the same way. For the producer covering all six months, their total actual gross margin would be $36,781. For the producers insuring only in April, the total actual gross margin would be $4,995.

Table 14.2. Insurable prices and expected gross margins for the January 2003 insurance period.

Month	Lean Hog Price ($/cwt)	Maize Price ($/bu)	Soybean Meal Price ($/ton)	Expected Gross Margin ($/hog)
February	52.81	2.50	162.88	54.17
March	56.06	2.36	163.33	62.20
April	59.31	2.36	163.87	68.40
May	63.79	2.36	164.60	76.95
June	64.82	2.36	165.33	78.87
July	62.19	2.38	164.90	73.59

Table 14.3. Actual prices and gross margins for the January 2003 insurance period.

Month	Lean Hog Price ($/cwt)	Maize Price ($/bu)	Soybean Meal Price ($/ton)	Expected Gross Margin ($/hog)
February	48.95	2.50	162.88	46.75
March	49.34	2.36	163.33	49.27
April	49.72	2.36	163.87	49.95
May	59.78	2.36	169.52	68.78
June	66.62	2.36	175.17	81.42
July	62.67	2.45	186.15	71.64

The premium and indemnity calculations are based on the total gross margins. The producer's premium is based on a Monte Carlo simulation of the total expected gross margin, the coverage level and the producer's share of the hog operation. The indemnity is equal to the producer's share of the operation times the difference between the product of the coverage level and the total expected gross margin and the actual gross margin. The premium for LGM is not subsidized, but the government pays all administrative and operating expenses associated with providing and servicing the policy.

To compare coverage across LRP and LGM, let us look at an LGM example that parallels the LRP example. The producer purchases LGM at the 90% coverage level and insures 100 hogs for an April sales date. The total insured gross margin is $6,156. The total and producer premium for the coverage is $383 or $3.83 per head. Given the actual gross margins reported in Table 14.3, this producer would have received an indemnity of $1,161 or $11.61 per head. These examples are only meant to be illustrative; depending the market conditions in the hog, maize and soybean meal markets, LGM premiums and indemnities could be higher or lower than those for LRP over the same period.

LGM, like LRP-Swine, does not require actual sales values, sales weights or the number of sold head. Production and marketing losses, such as death loss due to disease or local price movements, are not covered by LGM. LGM is currently available to Iowa hog producers.

FCIC has approved several changes in LGM. The number of sales periods will expand to 12, one in each month. An additional production practice for segregated early-weaned pigs will be added. The insurance coverage will shift from a six-month policy with the insurance starting in the month immediately after the sales closing date to a five-month policy with the insurance starting in the second month after the sales closing date. The price discovery will be shortened to the last three trading days of each month and the sales period will be condensed into a one-day period. Restrictions on offsetting futures and/or options positions will be put in place (RMA, 2003e).

Adjusted Gross Revenue Products

AGR and AGR-Lite are revenue-based products that cover whole farm incomes. AGR was first offered in 1999; AGR-Lite is based on the structure of AGR and followed in 2003. The policies are targeted at protecting agricultural producers from low-income events due to unavoidable natural disaster and severe market fluctuations. Livestock and aquaculture revenues can be covered under the policies, but their value is meant to represent the value of crop production fed to animals. The coverage is annual in nature, smoothing year-to-year variations in income.

A producer's insurable income is determined with data from the producer's past federal tax forms and annual farm reports. Since the incomes are taken from tax forms, they also incorporate on-farm expenses. The insurable income can be derived from many commodities, so one policy can cover many commodities. Using income as a basis for insurance creates a common base across commodities for an insurance product. AGR and AGR-Lite were designed to complement existing agricultural insurance plans in that producers can obtain commodity-specific insurance where available and add AGR or AGR-Lite to cover all other agricultural activity on the farm at a reduced cost.

Under the AGR products, the producer chooses a coverage level and a payment rate. The payment rate is the percentage of the income shortfall that will be reimbursed to the producer. At the end of the coverage period, the actual revenue and expenses for the year are seen and if income is below the insured level, the producer receives an indemnity.

Coverage levels range from 65 to 80% of the expected income. The payment rate can be equal to 75 or 90%. There are requirements on the number of commodities and the size of the commodity revenues for the higher coverage levels and payment rates. AGR limits the proportion of income that can come from animal or animal products to 35% or less (RMA, 2003c). AGR-Lite allows any proportion of income from animal or animal products, but limits the total liability on individual policies to $250,000 (RMA, 2003d). Premiums are a function of the producer's average and trend income, the number of commodities covered by the policy, and the share of income attributed to each covered commodity. The premiums are subsidized and the government pays all administrative and operating expenses associated with providing and servicing the policy. If the current year's expenses are less than 70% of average past expenses, then the insured income will be reduced accordingly.

The following example holds for both AGR and AGR-Lite. A producer has purchased an AGR-type policy at the 75% coverage level and a 90% payment rate. Their average historical income from agriculture is $100,000. Drought conditions impact the farm and income for the current year falls to $20,000. The producer would receive an indemnity payment of $49,500 (90% × [75% × $100,000 – $20,000]). If the current year's income had exceeded $75,000, then there would be no indemnity payment.

Table 14.4. LRP and LGM statistics.

Year	Insurance Plan	Policies	Number of Head	Liabilities	Total Premium	Premium Subsidy	Indemnity	Loss Ratio
2002	LRP-Swine	186	129,762	$8,550,085	$527,077	$68,528	$1,313	0.00
2002	LGM	68	48,817	1,909,985	218,992	0	3,104	0.01
2003	LRP-Fed Cattle	43	12,382	10,179,406	259,993	33,804	9,314	0.00
2003	LRP-Feeder Cattle	175	19,903	12,079,987	253,576	32,965	0	0.00
2003	LRP-Swine	94	64,110	5,849,219	201,098	26,145	30,069	0.03
2003	LGM	276	680,710	51,082,942	2,757,296	0	6,159,451	2.23
2004	LRP-Fed Cattle	175	84,793	73,146,706	1,960,974	254,923	201,397	0.10
2004	LRP-Feeder Cattle	227	51,839	34,060,325	954,823	124,134	31,986	0.03
2004	LRP-Swine	97	91,861	9,481,256	372,289	48,402	4,265	0.01
2004	LGM	111	179,477	12,792,359	879,090	0	1,017,935	1.16

Source: Data from RMA's online Summary of Business Reports, September 9, 2004.

AGR-type policies require actual sales values and expenses. Losses due to negligence, mismanagement, abandonment or failure to follow good farming practices are not covered by the policies. AGR is available in selected counties in California, Connecticut, Delaware, Florida, Idaho, Maine, Maryland, Massachusetts, Michigan, New Hampshire, New Jersey, New York, Oregon, Pennsylvania, Rhode Island, Virginia, Vermont and Washington. AGR-Lite is available in selected counties in Alaska, Connecticut, Delaware, Idaho, Maine, Maryland, Massachusetts, New Hampshire, New Jersey, New York, North Carolina, Oregon, Pennsylvania, Rhode Island, Vermont, Washington and West Virginia.

The Results Thus Far

Given the short history of these insurance products, the figures in Tables 14.4 and 14.5 show that agricultural producers have begun to utilize these products for their risk management needs. Table 14.4 contains the sales and performance figures for the LRP family of products and LGM. The 2004 figures are preliminary. According to the September 2003 Quarterly Hogs and Pigs report, Iowa had roughly 14.5 million market hogs in 2002 and 14.68 million market hogs in 2003 (NASS, 2003). Based on the sales figures below, LRP-Swine and LGM covered 1.3% of those in 2002 and 5.8% in 2003. Hog producers have insured over 1 million animals, paying roughly $4 per head for the insurance. Fed cattle producers have insured nearly 100,000 head at nearly $20 per head. Feeder cattle producers have insured 72,000 cattle at $14.65 per animal. Producers have received nearly $7.5 million in indemnities.

Two issues that often come up when looking at new insurance products are moral hazard and adverse selection. RMA and the private insurance companies that promote and service agricultural insurance continue to work on insurance plans to limit these insurance problems. For example, the recent changes in LGM with regards to the insurance period, price discovery period and sales period are targeted at reducing possible adverse selection in LGM.

Table 14.5. AGR and AGR-Lite statistics.

Year	Insurance Plan	Policies	Liabilities	Total Premium	Premium Subsidy	Indemnity	Loss Ratio
1999	AGR	86	$12,617,938	$624,640	$312,310	$851,497	1.36
2000	AGR	79	9,330,632	387,820	193,896	389,989	1.01
2001	AGR	565	188,748,088	6,822,528	4,117,511	12,046,110	1.77
2002	AGR	785	244,797,134	8,966,153	5,229,179	10,831,181	1.21
2003	AGR	971	318,925,056	12,160,375	6,884,931	8,233,741	0.68
2003	AGR-Lite	73	2,633,229	126,038	96,511	9,631	0.08
2004	AGR	966	301,884,296	11,688,886	6,346,165	0	0.00
2004	AGR-Lite	95	3,285,690	155,907	84,039	0	0.00

Source: Data from RMA's online Summary of Business Reports, September 9, 2004.

Table 14.5 contains the sales and performance figures for the AGR family of products. The 2004 figures are preliminary. Figures on the number of policies that covered livestock could not be obtained. Again, as the data show, agricultural producers have begun to utilize these insurance products for their risk management needs. Since 1999, over $1 billion in agricultural income has been insured under AGR. On average, the total premium per dollar of coverage has been four cents. With the premium subsidies and some additional state support, producers have paid less than two cents for each dollar of insurance coverage. Over $32 million in indemnities have been paid out.

The results thus far are encouraging. Six insurance products have been developed to cover some livestock risks since the passage of ARPA. Usage of these products has been growing over time. RMA has been able to expand the number of producers eligible to use these products. The products address price risk – the main risk livestock producers indicated they faced. The products also are structured to address various issues livestock producers may have faced with futures and options. All of these insurance products are scalable to the size of the operation. The LRP products are available on a near continuous basis, being sold every business day. LGM and the AGR products offer risk protection across several markets.

However, the livestock insurance products are still a work in progress. RMA suspended sales of LRP-Fed Cattle and LRP-Feeder Cattle after the announcement of the US BSE case in December 2003. LRP-Swine and LGM were suspended at the end of the 2004 insurance year (June 30, 2004) to implement changes in price discovery and sales periods and to set up procedures for suspension of sales due to catastrophic events. Sales of all four products resumed October 1, 2004 (RMA, 2004). Early sales reports show that producers have also resumed their usage of LRP and LGM. Producers insured over 22,000 animals with LRP in the first 20 days since sales restarted, while LGM covered nearly 172,000 hogs within its first sales period under the new rules.

Possible Areas of Expansion for Livestock Insurance

RMA continues to look at further expansions of agricultural insurance for livestock producers. RMA has considered expansions that cover price risks for dairy, Brahman, cow-calf and heifer operations. RMA also has several partnership agreements targeted at constructing risk management tools for livestock and crop price risk; forage, pasture, and hay production risk; and stock reductions due to drought. LRP- and LGM-type products could be expanded and/or developed for other regions of the country and other livestock species. AGR and AGR-Lite could also continue to be expanded. But there are also other livestock risks that could be addressed by insurance. The current array of insurance products mainly targets price risk in livestock. Production risks are mostly left uninsured. This may be partially due to the wealth of data available on livestock prices and the relative lack of data on some livestock production risks. The area of livestock disease is one area where this relative lack of data may inhibit insurance development. ARPA specifically listed livestock disease as one area for possible expansion. Recent episodes in the USA, Canada, Japan and Europe have highlighted the possibility that a livestock disease outbreak can have significant impacts on the livestock sector. As we examine the possibility of covering livestock disease outbreaks with insurance, we will need to answer numerous questions dealing with the likelihood and the severity of outbreaks, the government's response to disease episodes, the direct and indirect costs associated with outbreaks, etc. Moral hazard and adverse selection issues will need to be addressed as well. Insurance may, or may not, be the best mechanism for providing financial security against certain livestock risks.

References

Barnett, B.J. and Skees, J.R. (1995) Region and crop specific models of the demand for federal multiple peril crop insurance. *Journal of Insurance Issues* 19, 47-65.

Coble, K.H. and Knight, T.O. (2002) Crop insurance as a tool for price and yield risk management. In: Just, R.E., and Pope, R.D. (eds). *A Comprehensive Assessment of the Role of Risk in US Agriculture.* Kluwer Academic, Norwell, Massachusetts.

Coble, K.H., Knight, T.O., Pope, R.D. and Williams, J.R. (1997) An expected indemnity approach to the measurement of moral hazard in crop insurance. *American Journal of Agricultural Economics* 79, 216-226.

Goodwin, B.K. and Smith, V.H. (1995) *The Economics of Crop Insurance and Disaster Aid.* The American Enterprise Institute Press, Washington, D.C.

Hart, C.E., Babcock, B.A. and Hayes, D.J. (2001) Livestock revenue insurance. *Journal of Futures Markets* 21, 553-580.

Hennessy, D.A., Babcock, B.A. and Hayes, D.J. (1997) The budgetary and producer welfare effects of revenue assurance. *American Journal of Agricultural Economics* 79, 1024-1034.

Just, R.E., Calvin, L. and Quiggin, J. (1999) Adverse selection in crop insurance: actuarial and asymmetric information incentives. *American Journal of Agricultural Economics* 81, 834-849.

Ker, A.P. and Goodwin, B.K. (2000) A nonparametric estimation of crop insurance rates revisited. *American Journal of Agricultural Economics* 83, 463-478.

Ker, A.P. and McGowan, P. (2000) Weather-based adverse selection and the US crop insurance program: the private insurance company perspective. *Journal of Agricultural and Resource Economics* 25, 386-410.

Knight, T.O. and Coble, K.H. (1997) A survey of literature on US multiple peril crop insurance since 1980. *Review of Agricultural Economics* 19, 128-156.

Mahul, O. (1999) Optimum area yield crop insurance. *American Journal of Agricultural Economics* 81, 75-82.

Mahul, O. and Wright, B.D. (2003) Designing optimal crop revenue insurance. *American Journal of Agricultural Economics* 85, 580-589.

Makki, S.S. and Somwaru, A. (2001) Farmers' participation in crop insurance markets: creating the right incentives. *American Journal of Agricultural Economics* 83, 662-667.

NASS (2003) *Hogs and Pigs.* US Department of Agriculture, National Agricultural Statistics Service. [Accessed 2003.] Available at http://jan.mannlib.cornell.edu/reports/nassr/livestock/php-bb/2003/hgpg0903.pdf.

Richards, T.J. (2000) A two-stage model of the demand for specialty crop insurance. *Journal of Agricultural and Resource Economics* 25, 177-194.

RMA (2003a) A history of the crop insurance program. US Department of Agriculture, Risk Management Agency. [Accessed 2003.] Available from http://www.rma.usda.gov/aboutrma/history.html.

RMA (2003b) Multiple Peril Crop Insurance. US Department of Agriculture, Risk Management Agency. [Accessed 2003.] Available from http://www.rma.usda.gov/pubs/rme/fsh_6.html.

RMA (2003c) Adjusted Gross Revenue (AGR). US Department of Agriculture, Risk Management Agency, Fact Sheet, Program Aid Number 1667-06, [Accessed 2003.] Available from http://www.rma.usda.gov/pubs/2003/PAN-1667-06rev.pdf.

RMA (2003d) Adjusted Gross Revenue-Lite (AGR-Lite). US Department of Agriculture, Risk Management Agency, Fact Sheet, Program Aid Number 1667-07. [Accessed 2004] Available from http://www.rma.usda.gov/pubs/2003/PAN-1667-07.pdf.

RMA (2003e) Board Memorandum No. 709, Docket No. CI-LGM-03-01, Livestock Gross Margin Proposed Changes. US Department of Agriculture, Risk Management Agency. Board Memorandum No. 709. [Accessed 2003.] Available from http://www.rma.usda.gov/aboutrma/fcic/2003/731lgmresolution.pdf.

RMA (2003f) Livestock Risk Protection Insurance Pilot Program, LRP-Fed Cattle. US Department of Agriculture, Risk Management Agency, Fact Sheet, Program Aid Number 1667-10. [Accessed 2003.] Available from http://www.rma.usda.gov/pubs/2003/PAN-1667-10.pdf.

RMA (2003g) Livestock Risk Protection Insurance Pilot Program, LRP-Feeder Cattle. US Department of Agriculture, Risk Management Agency, Fact Sheet, Program Aid Number 1667-09. [Accessed 2003.] Available from http://www.rma.usda.gov/pubs/2003/PAN-1667-09.pdf.

RMA (2003h) Livestock Risk Protection Insurance Pilot Program, LRP-Swine. US Department of Agriculture, Risk Management Agency, Fact Sheet, Program Aid Number 1667-08. [Accessed 2003.] Available from http://www.rma.usda.gov/pubs/2003/PAN-1667-08.pdf.

RMA (2004) USDA to Resume Sales of Livestock Protection Programs. US Department of Agriculture, Risk Management Agency, Press Release. [Accessed 2004.] Available from http://www.rma.usda.gov/news/pr/2004/08/812livestock.html.

Sherrick, B.J., Barry, P.J., Ellinger, P.N. and Schnitkey, G.D. (2004) Factors influencing farmers' crop insurance decisions. *American Journal of Agricultural Economics* 86, 103-114.

Skees, J.R., Barnett, B.J. and Black, J.R. Designing and Rating an Area Yield Crop Insurance Contract. *American Journal of Agricultural Economics* 79, 430-438.

Stokes, J.R. (2000) A derivative security approach to setting crop revenue coverage insurance premiums. *Journal of Agricultural and Resource Economics* 25, 159-176.

Turvey, C. G. and Z. Islam. (1995) Equity and efficiency considerations in area versus individual yield insurance. *Agricultural Economics* 12, 23-35.

United States Congress, 106th. (2000) Agricultural Risk Protection Act of 2000. [Accessed 2003.] Available from http://frwebgate.access.gpo.gov/cgi-bin/getdoc.cgi?dbname=106_cong_bills&docid=f:h2559enr.txt.pdf.

USGAO. (1999) Agriculture in transition: farmers' use of risk management strategies. US General Accounting Office, Report # GAO/RCED-99-90.

Wu, J. (1999) Crop insurance, acreage decisions, and nonpoint-source pollution. *American Journal of Agricultural Economics* 81, 305-320.

Chapter 15

Livestock Disease Eradication Programmes and Farm Incentives: the Case of Bovine Tuberculosis

Christopher A. Wolf

Introduction

Government-controlled eradication of diseases that pose a human health risk or that are a tremendous food security/supply concern is a staple of US livestock disease control policy. Understanding the incidence of costs, from biosecurity measures and other avoidance costs as well as expected losses from a disease outbreak, allows for both an improved understanding of farmer reaction and a foundation for examining policy alternatives. Bovine tuberculosis has recently become a problem in Michigan and statewide testing of all cattle, bison and captive cervids was mandated when Michigan lost "tuberculosis free" accreditation in June 2000. The financial consequences of depopulating a livestock herd and the resulting political pressure have resulted in the State allowing farmers to choose between depopulation and a continuous test-and-slaughter positive animals protocol. Thus far beef farms have uniformly chosen to depopulate but dairy farms seem to have a much larger business interruption cost, resulting in a socially sub-optimal decision to follow the test-and-slaughter protocol. This chapter examines the farm-level decision on whether to depopulate or test-and-slaughter, and explores potential societal welfare improving policies including subsidized insurance.

Background

Livestock disease and related government control programmes have significant economic implications for both farmers and governments. Disease effects at the farm and industry level are potentially devastating (witness BSE in the British beef industry). Disease control at the private and public levels is a trade-off between control costs and losses due to disease (e.g. mortality and productivity losses).

Economic justifications for public intervention in disease control include externalities, public good aspects, coordination failures, information failures and income distribution considerations (Ramsay, Philip and Riethmuller, 1999). To facilitate an understanding of livestock disease public intervention, we make a simple disease typology: those diseases with public control or eradication programmes and those without. Diseases where human health is at risk or which have large potential economic effects, which can include trade effects or impacts in industries up and down the supply chain from livestock operations, are in the domain of public control. The potential to pose a large economic cost depends on many factors including trade laws and level of infectiousness. Public policies range from bounties for infected livestock to required herd depopulation and farm decontamination. Some programmes are legal mandates and others are explicit assistance programmes. These programmes are often partnerships between the federal and state governments and may involve several agencies depending on the species affected (e.g. Department of Agriculture and Department of Natural Resources).

Diseases that are not human health risks and do not pose a significant potential economic impact are often left to private control. In this case, the private result does not vary enough from the publicly desired result to cause implementation of public control measures. Public resources may be allocated to programmes for education as well as research into vaccinations and treatments. In some cases, industry leaders or groups may facilitate programmes intended to foster private disease prevention and control. However, without government programmes, the absence of insurance markets for disease outbreaks necessarily means that farmers are self-insured. In many cases, these diseases – or their control – possess externalities that have economic effects on public resources.

This chapter considers livestock disease eradication programmes, incidence of costs related to these programmes, and resulting farm-level incentives and responses. The application is bovine tuberculosis in cattle herds. The focus is largely on the disease in Michigan but information and lessons from the experiences in other states are also considered. The objectives of this chapter are to: understand farm incentives to prevent and control livestock disease; understand the role of public policy in affecting farm incentives and behaviour; discuss economic incentives in disease control and impact; and to apply these concepts to farm decisions regarding bovine tuberculosis.

The chapter is organized as follows. The next section briefly outlines public policy parameters, farm responses and incentives for livestock disease management. The following section reviews the history of bovine tuberculosis and the eradication programme. Public costs, as well as farm effects and responses to the programmes, are then examined for representative Michigan dairy and beef case farms. A summary and policy implications conclude.

Public Policy and Private Disease Control

The primary government objective relative to livestock disease is welfare maximization considering public health, cost to consumers, cost to producers (including longer-run investment and structural adjustment behaviour) and cost to taxpayers. The rationale for disease eradication is often founded on a benefit-cost calculation. Eradication costs are viewed as an up-front investment for a long stream of benefits as production losses, deaths and trade sanctions are avoided – sometimes in perpetuity.[1] To accomplish eradication, the disease reservoir and vectors must be eliminated. Movement restrictions/quarantines, testing,

---

[1] Assuming the benefit stream in perpetuity is a brave assumption as the disease in question could often return tomorrow if it exists elsewhere. An accurate framework would incorporate uncertainty and optimize expected net benefits.

slaughtering (usually with indemnity payments to the owners), vaccinations and education programmes are often involved.

Farmers can manage livestock disease in several ways. Livestock disease management is a trade-off between *ex ante* prevention and *ex post* control (McInerney, 1996; Chi *et al.*, 2001). In general, additional resources allocated to prevention should result in lower disease incidence and, thus, lower control costs. With respect to prevention, farmers can allocate operating resources or investments to biosecurity programmes. These programmes often involve limiting livestock contacts (e.g. through quarantines and controlled movement of livestock new to the operation) as well as preventing other potential vectors from entering farms (e.g. wildlife or human contacts). In addition, biosecurity may involve testing livestock and feed prior to purchase, strict sanitation of people and vehicles entering the farm, separation of newborns from infected dams and protecting feed supplies from wildlife. Farmers control disease by monitoring and testing their own herd and reporting relevant infections to authorities. Treating disease is possible in some cases and may involve vaccines or antibiotics as well as veterinary visits. In situations where recovery from the disease is impossible or treatment expense is economically prohibitive, culling animals may be the only recourse. Early involuntary culling has long-term implications for livestock capital stock and significant costs related to animal replacement.

Farmer motivations to control disease are to avoid livestock mortality and related replacement expenses; livestock morbidity and related production losses; increased veterinary and medicine expenses; and potential business interruption losses when government programmes mandate slaughtering animals. Farmer motivations to shirk disease control include time, labour, management and capital constraints; increased cost of replacement animals; and that only the value of the animal is reimbursed in most governmental indemnity programmes.

The farm maximizes expected net profits by choosing appropriate prevention and control management practices. A standard optimization results in first-order conditions that equate the decline in marginal expected loss from disease (the marginal benefit from prevention measures) to the marginal cost of control measures (McInerney 1996; Chi *et al.*, 2001). The decline in expected losses is a benefit from farm biosecurity. This decline in probability of disease may apply to many diseases or pests, it may spill over benefiting neighbours and associates by lowering their incidence of disease, it may lower the possibility of wildlife disease, and it may lower the incidence of human health risk. However, without legal requirements or subsidies, farmers can only be expected to react to the incentives provided by their direct, on-farm benefits. Therefore, we must have estimates of farm losses and the probability of these losses occurring. We must also know the cost of mitigating practices and the change in probability associated with these practices.

Public policies affect farm incentives through indemnity payments for infected animals (or herds), movement and testing restrictions (including required vaccinations and identification), sanitary licensing requirements (e.g. grade A milk) and output grades and standards. Government policies in some instances have sought to prevent and control disease by implementing testing, quarantines and slaughter in reaction to infectious disease. In some instances, slaughter of the infected animal or herd is mandatory and the farmer is reimbursed with an indemnity payment. In other cases, and relevant to the example discussed below, farmers may have a choice in the course of disease control. As we will show, when the farmer has a choice, the financial consequences associated with business interruption are crucially important. These losses vary across farms but also systematically across livestock enterprises.

Bovine tuberculosis appears to be a situation where both society and farmers could be better off by compensating farmers for business interruption losses to depopulate infected herds.

Bovine Tuberculosis Eradication Programmes

Bovine tuberculosis (*Mycobacterium bovis*), hereafter TB, is a contagious, infectious, bacterial disease so named because it apparently originated in cattle. However, the disease can infect livestock (e.g. cattle, bison, goats), wildlife (e.g, deer and elk) and humans. When left unmitigated, the disease is slowly degenerative and can be fatal. If detected early, TB is easily treatable in humans with low-cost antibiotics. Eradication programmes in the past century have brought the disease close to eradication in developed countries. The National Cooperative State/Federal Bovine Tuberculosis Eradication Programme began in 1917. At that time, five percent of cattle nationally tested reacted positive and tuberculosis was the most common cause of death in the human population (Frye, 1995). Consumption of raw milk from tuberculous cattle was a common method of transmission (Hickling, 2002).

The US programme consisted of test-and-slaughter as well as movement restrictions and was effective in drastically reducing the prevalence rate. For example, in the first 13 years of the programme, the national reactor rate declined 64% (Frye, 1995). Michigan has been a traditional hot spot of TB, producing more than 30% of the TB-positive reactor cattle in the 1950s (Hickling, 2002). However, by 1979, Michigan had achieved TB accredited-free status, reflecting that the state had no positive cattle for the previous four years. Accredited-free status is desirable as it prevents other states from placing testing, movement and quarantine restrictions on interstate animal exports. While the US Department of Agriculture has jurisdiction to award or withdraw TB accredited-free status, the USDA does not mandate interstate requirements or restrictions in response to lost status. In the case of TB, individual State Veterinarians determine these restrictions.

Indemnity payments for condemned infected livestock are paid using federal and state funds. Federal indemnity payments may be subsidized by state funds. State indemnities range from zero to amounts that exceed the federal payment (National Research Council, 1994).[2] Over the 80-year period ending in 2001, the cooperative state/federal programme spent a total of $666 million: $291 million in federal funds and $375 million in nonfederal funds (US Animal Health Association 2001). The benefits of the eradication programme are in increased productivity and competitiveness in international markets facilitated by livestock product quality and credibility.

In 1975 and again in 1994, TB-positive deer were found by hunters in northeast lower Michigan. Subsequent investigation following the 1994 positive deer revealed further cases of TB in wild deer, captive deer and cattle. Prior to this outbreak, conventional wisdom held that TB was not self-sustaining in wildlife populations. The first positive cow was found in a beef herd in the same region as the infected deer in 1998. Subsequent testing revealed several additional infected beef and dairy herds resulting in Michigan's status being changed to "modified" accredited free in June 2000. The loss of accredited-free status meant that other states could place testing and movement restrictions on livestock originating from Michigan. Further, in order to regain accredited-free status, Michigan was required to institute

[2] In Michigan, the indemnity payment is currently set to a value up to 90% of market value (although market value is not explicitly defined).

statewide testing of all cattle, bison and captive cervid (deer and elk) operations. Michigan has approximately one million head of cattle within its borders. As of 2003, 31 total cattle herds had been found to be positive – 29 beef and two dairy herds. The struggle to eradicate the disease is likely to be long and arduous because of the potential for deer and cattle to pass the disease back and forth.

The presence of an apparently self-sustaining disease in the wild whitetail deer herd makes the current situation unique and difficult for policymakers. Several policy complications arise including the fact that the Michigan Department of Agriculture has jurisdiction over the livestock and captive deer and elk herds while the Michigan Department of Natural Resources manages the wild deer herd through hunting levels and activities associated with hunting. Deer hunting is a highly valued activity in the infected region. Indeed it is possible that the presence of hunt clubs and the rigorous deer feeding programmes are a primary reason that the disease has been self-sustaining (Hickling, 2002). These feeding practices, along with the presence of crops to consume and lack of natural predators (other than cars), have resulted in high deer densities. In some cases, these densities are estimated at four or more times their historic values. Hickling (2002) found that deer density and TB prevalence rate were positively related. Deer TB management is important to cattle disease management as epidemiological studies have strongly linked TB in the deer and cattle populations (Kaneene *et al.,* 2002). However, we concentrate on the disease management in cattle herds and the incentives provided by current cattle eradication policies.

Public Policies and TB Eradication

The stated goal of the US Department of Agriculture is to eradicate bovine tuberculosis from the domestic livestock population (USDA-APHIS, 1999). As is often the case with livestock disease, one of the primary motivations for eradicating TB is the trade benefit that accompanies freedom from restrictions. The USDA's Animal and Plant Health Inspection Service (APHIS) sets the parameters for achieving modified accredited-free status. To achieve that desirable status again, Michigan is required to have zero herds infected for the past five years or zero infected herds for the past two consecutive years if all positive herds are depopulated. The last condition exists because herds are currently given a choice of two eradication programmes: depopulation or test-and-slaughter.

Depopulation refers to removing all cattle from the farm. A mandatory period without cattle on the farm follows depopulation while the farm is cleaned and testing protocols are tailored to the individual farm situation (Michigan Legislative Council, 1988). Alternatively, farms may choose to remove only test-reactor cattle and continue in business – termed a "test-and-slaughter" protocol. Subsequent tests follow frequently if other test positive animals are found and less frequently when no test reactors are found. During the period of testing, which can occur for an indefinite amount of time, there are movement restrictions on the herd.

Public costs of the disease can be separated into direct and indirect costs (also called consequential losses) (Meuwissen *et al.,* 2000). Direct costs include those involved with testing all cattle, bison, goat and captive cervid herds. All grade "A" dairy herds were required to be tested to maintain status under the Pasteurized Milk Ordinance. All beef, bison, goat and captive cervid herds were also located and tested. The state incurs costs to purchase equipment, pay veterinarians and other workers involved in testing, and all laboratory expenses. Any indemnified animals are reimbursed using state and federal funds and must be disposed of properly. In addition, the state changed research and monitoring programmes for both cattle and deer and incurred costs in managing the wild deer herd. Indirect costs were incurred by the directly affected operations and landowners, as well as related industries.

The state covers the direct testing costs (e.g. lab tests and veterinary visits), while farmers incur the incidental testing costs (e.g. labour and lost performance) as well as increased transportation and trade requirements. The total testing costs for livestock agriculture were estimated at around $12 million per year if all herds are tested annually (Wolf and Ferris, 2000). Annually testing only the herds in the core area substantially lowers the state testing cost. Another state cost is the indemnity payment for purchasing test positive animals. While the current law mandates that farmers be compensated for 90% of the fair market value up to $3,000 per animal, many of the payments in practice likely exceed objective fair market value as the state has motivation to remove animals quickly rather than argue over a relatively small amount of money.

The logic for bestowing accredited-free status earlier, when all positive herds are depopulated, has support in past experience. The test-and-slaughter strategy to control TB has been in practice in other states with mixed results. Several TB-positive dairy herds in west Texas followed the strategy unsuccessfully in the 1990s before being coaxed into depopulation using large buy-out payments in 2000. Depopulation is likely to achieve the goal of eradication more quickly as TB can take a long period to manifest in cattle. Additionally, since the farm remains in business, thorough decontamination associated with depopulation is not performed. If the disease vector is still present then the test-and-slaughter strategy might be unsuccessful. However, in many circumstances the costs of business interruption associated with depopulation make test-and-slaughter the rational choice for individual farmers.

Public and Private TB Control Incentives in Michigan

When a TB-positive animal is identified in Michigan the owner of the infected herd has two choices. Either the entire herd is depopulated or the test-and-slaughter protocol is followed. In either scenario, all animals classified as "reactors" are slaughtered. Reactors are not required to be tissue culture positive – meaning that it is possible to slaughter animals that test false positive. The time between tests in the test-and-slaughter track increases with clean herd tests but the time horizon in the test-and-slaughter scenario is essentially indeterminate (in the Texas case, the involved herds were in a programme of this type for many years before the mandatory depopulation).

From an economic standpoint, it seems rational for a farmer with an infected herd to choose the less costly of depopulation or test-and-slaughter options. While each involves some uncertain costs, we can outline the basic farm implications for beef and dairy herds to assess the likely choices and determining factors. The basic result is that when business interruption losses are large during depopulation, test-and-slaughter is preferred. As is shown below, the business interruption losses are smaller for beef than dairy herds, making the decision less clear.

Depopulating cattle herds in an effort to eradicate disease has significant economic effects. When a herd is depopulated for TB control reasons, animals, feed, labour and net farm revenues are lost. There is also a period of time when animals must be kept off the farm. The Michigan Animal Industry Act of 1988 (PA 466) specifies that a "...premises that has been depopulated shall be cleaned and disinfected as prescribed by the director" (Michigan Legislative Council). Because TB is a hardy bacterium, the cleaning and disinfecting process is rigorous and time consuming. Each infected herd is assessed to determine when they may repopulate. Experience with operations in Michigan suggests that a one-year depopulation period can sometimes occur.[3] For our purposes the time at which the herd is dispersed and the period without cattle are referred to as the "depopulation period".

Assuming that the farm intends to resume operation after a period of depopulation, farm-level losses from depopulation may be divided into lost cattle, or replacement, value and forgone net revenues, also termed "business interruption" losses (Wolf, Harsh and Lloyd, 2000). Lost cattle value reflects any difference between the value of the cattle slaughtered and the indemnity payment. Michigan law stipulates that farmers can be paid up to 90% of fair market value. As officials are primarily interested in eradicating TB, payments have been generous to encourage cooperation. Farmer risks in the interim depopulation year prior to repopulation include the inability to replicate the herd quality, potential increases in cattle prices, and search and transportation costs that might not be adequately reimbursed. Any discrepancy between indemnity payment and true cattle value is also an issue for animals removed under the test-and-slaughter regime but is a larger consequence when the whole herd is removed.

The role of indemnity payments in farm incentives for disease control was examined by Kuchler and Hamm (2000) for the case of scrapie in sheep. They determined that farms found more infected animals when indemnity payments were increased. Because bovine TB testing is mandatory, the moral hazard problem of finding (or concealing) infected animals does not apply. With state and federal indemnity payments, the maintained hypothesis here is that any discrepancy in cattle value for depopulation is small in comparison to business interruption losses.

Business interruption losses include forgone revenues net of avoided costs. Economists discern between variable and fixed costs where fixed costs are unavoidable even when production does not occur. This concept is useful for understanding business interruption losses. When the cattle enterprise is removed, many (and perhaps most) variable costs of production are not incurred. For example, there is no need to purchase feed for cattle.

By definition, fixed costs cannot be avoided or varied when production ceases (over the time period considered). These include overhead expenses and other costs that accompany an operational farm. The standard list of fixed costs includes interest on investment, depreciation, property taxes and insurance. Because the depopulated farms are assumed to resume operation, the fixed costs must be covered during the interim period. For this reason, we do not subtract fixed costs from the farm revenues in order to allow the farm to stay current on these expenses.

There are also situations where it might be appropriate to treat labour as a "fixed cost" for business interruption loss calculations. Labour is often a scarce resource on farms. During the depopulation period, labour used for these enterprises is not required. However, it is often

---

[3] The costs below are calculated as annual costs of business interruption from depopulation. For dairy farms, because they are continuous revenue generators, it may be reasonable to simply multiply this annual cost by the fraction of a year depopulated for business interruption cases of less than a year. For beef operations, this question is more difficult and depends on the operation's characteristics. However, many beef farms operate on an annual cycle with respect to revenue generation.

Table 15.1. Example Annual Dairy and Beef Costs Associated with Business Interruption.

	Milk Production[1]	Milk Production[1]	Feeder Steer[2]
	$/cwt	$/cow	$/cow
Revenue	14.89	2,990	833
"Variable" costs[3]	9.11	1,829	705
"Fixed" costs[4]	4.67	917	79
Business Interruption Loss[5]	5.78	1,160	127

[1] Source: Wittenberg and Wolf (2003).
[2] Source: Wittenberg and Black (2003).
[3] Include costs that are not incurred without the cattle enterprise including feed, veterinary and marketing.
[4] Fixed costs include costs of empty facilities (taxes, insurance, depreciation) as well as hired labour. Hired labour was $2.42/cwt for dairy and $39/head for beef.
[5] Business interruption annual losses calculated as revenue less variable costs.

infeasible to "lay-off" hired labour and expect that labour will be available when the farm is repopulated. For this reason, business interruption losses might include payment of workers during the depopulation period. For farms that utilize only part-time labour, this may not be an issue. Similarly, for very large farms with many homogeneous workers, labour may be a strictly variable expense. In the example below, we subtract labour costs out of revenues allowing for the reimbursement of labour specific to the cattle-related enterprises (the case of hired labour as a strictly variable cost is noted in a footnote).

Consider example revenues and costs for calculating annual business interruption losses using average Michigan data from 2002 (see Table 15.1). Foregone revenues include milk, calves, cull cows and government payments tied to production. For 2002 in Michigan, the average herd received $14.89 per hundred pounds of milk (hundredweight abbreviated cwt). This translated to $2,990 per milk cow. Variable milk production costs that occur with milk production and raising replacements include feed, herd health, marketing, supplies, repairs, custom hire, fuel, replacement and interest on cattle or operating capital. Fixed costs for business interruption calculations included hired labour, taxes, insurance and depreciation on machinery and facilities. Because dairy operations often have a large amount of capital tied up in facilities, fixed costs are relatively large ($917 per cow which amounts to 33 percent of total costs). Business interruption losses, calculated as revenue less variable costs, totalled $1,160 per cow on an annual basis ($5.78/cwt).[4]

The example beef operation is a feeder steer operation. With $833/cow revenues and $705/cow in variable expenses, the business interruption losses are $127/cow.[5] This value is

[4] For the case in which labour is considered strictly a variable expense, the hired labour expense of $2.42/cwt becomes a variable cost and business interruption loss is $3.36/cwt or $675 per cow annually.

[5] Hired labour costs averaged $39 per head. If these costs were considered variable, the business interruption loss would be $89 per head.

much smaller than the dairy business interruption on a per-head basis. To be an equivalent amount on a per-farm basis, the beef farm would need nine times as many animals.

Implications of Test-and-Slaughter

When a herd chooses the test-and-slaughter strategy, the costs may be expressed as the incidental costs and losses associated with each test along with any shortfall in cattle reimbursement from indemnity payments. Costs related to TB testing include labour and inconvenience to support each herd test as well as any production losses associated with animal stress from the test. The state of Michigan continues to pay for the laboratory and veterinary costs associated with testing.

Producer costs associated with herd tests were estimated by the National Research Council (1994) to be $300 per herd plus five percent of production for a week. The average dairy herd lost $476 per test using these values. Wolf and Ferris (2000) assumed that milk production would be curtailed two-tenths of one percent of the annual production by testing stress (for a single test). In addition, Wolf and Ferris (2000) assumed that cattle prices would be discounted by one percent to account for potential discrimination in the market. The discounted cattle price affected dairy returns as well as sales from beef cow herds and feedlots.

Other financial implications of the test-and-slaughter protocol are restrictions on cattle movement and sale. In the case of Michigan, the state agrees to purchase cull animals sold off the farm and takes them to slaughter resulting in small financial consequences relative to the milking herd value. However, movement restrictions constrain activities, such as utilizing custom heifer raisers, which may ultimately constrain resources put into the herd. In addition, the ability to sell breeding stock is curtailed.

Also of concern in the case of test-and-slaughter is the potential for future forced depopulation. With respect to bovine TB in Texas, a test-and-slaughter protocol did not succeed in eliminating the disease, which resulted in a long-term depopulation.

Farm Decisions

The expected business interruption losses on dairy farms are large because the facilities investment, interest, taxes, and other fixed costs are a larger portion of total cost, relative to beef cow-calf operations. Dairy farms are also almost always full-time occupations for at least one person while many beef enterprises are often part-time occupations. Job search costs mean that it is likely to be more costly for a full-time manager to be unemployed for a year. Finally, milk production is a flow good that generates a revenue stream throughout the year. Accordingly, expenses – such as investment in facilities – may be set up to include regular payments. Beef production often includes the sale of a single crop of animals, and sales to the government may not entail serious cash-flow difficulties.

The business interruption losses are key to understanding farm decisions in the presence of a positive cow. Dairy farms tend to have relatively more serious financial implications from business interruption costs than beef operations as they tend to have higher fixed costs, depend on regular sales for cash flow requirements, and require full-time commitments for the farm manager with relatively less income available from off-farm sources.

To date in Michigan, all infected beef herd owners have chosen to depopulate while dairy farmers chose the test-and-slaughter strategy. The reason for this choice seems clear, as the business interruption losses are larger for the dairy farms. As was mentioned above, the state of Michigan has clear incentives to depopulate and acquire TB accredited-free status as soon as possible, thus saving additional testing and control costs.

A producer might choose depopulation if the business interruption losses were expected to be small, if test-and-slaughter was expected to take a large portion of the herd, or if cash flow constraints would not be oppressive (Nott and Wolf, 2000). A beef producer who focuses on stocker calves might experience a relatively small business interruption cost if the government purchases their animals. In the case of beef steers, the producer is essentially losing the gain on pasture or home-grown feed. The pasture is not storable while the feed may be; both may have alternative uses. Investment in facilities on beef operations is relatively smaller than a typical dairy operation. This has importance relative to the size of depreciation, repairs and maintenance, property taxes and interest expenses.

A producer might also choose depopulation if the farm enterprise is relatively unimportant to family income, for example a part-time undertaking, or if the producer is at a point where retirement is viable or preferable. In many cases, the land-owner purchases beef cattle to utilize pasture but does not depend on the enterprise to provide substantial family income. In these part-time farming situations, the financial ramifications of depopulation may be relatively small, thus encouraging the adoption of this practice.

One Complicating Factor: a Wildlife Disease Reservoir

Disease in the deer population is important to the livestock disease control decisions – both at the farm and state levels. Wild deer disease control is under the direction of the Michigan Department of Natural Resources.[6] Control instruments to eradicate TB in deer include hunting as well as practices such as feeding and baiting (Horan and Wolf, 2003). The stated goal in wild deer, as in livestock, is disease eradication. To this end, the deer herd has been thinned considerably in the region. Hickling (2002) found that prevalence rate was positively correlated to population density supporting this policy. In addition, the practice of large-scale feeding, once common by the hunt clubs in the region, has been limited. Baiting for hunting purposes, defined as five gallons or less of small grain, is still allowed (over objections by agricultural interests). However, Hickling (2002) estimated that the current policies would not likely eradicate the disease. TB is difficult to eradicate because the natural mortality rate is quite low – even lower than the transmission rate. Horan and Wolf (2003) found that with a valuable deer herd, and disease controlled by hunting and feeding, it was not economically optimal to eradicate the disease.

Eradication Policy Implications

This analysis points to a number of issues and potential actions for policy. Given the large public costs associated with allowing farms to follow test-and-slaughter protocols, namely years without accreditation, the potential exists for the state to spend more to encourage farms to choose depopulation. One way to encourage this outcome is to pay at least a portion of

---

[6] Captive deer and elk herds are considered livestock and so fall under the direction of the Michigan Department of Agriculture.

business interruption costs. At the same time, completely reimbursing farm losses may lead to a moral hazard problem where the farm has no incentive to prevent or control disease.

Other steps that policymakers might consider are paying indemnity only one time and leaving farms responsible for future infections. To the extent that farms can prevent infection, this would provide incentive to adopt biosecurity measures. Alternatively, indemnity payments could be tied to best management biosecurity practices. In the same manner that protection from litigation with respect to environmental measures is tied to best management practices, tying indemnity payments (or even business interruption losses) to biosecurity practices would be a movement towards compatibility between public and private incentives. Finally, it might also be economically justifiable to provide public subsidies for biosecurity measures, especially those that involve capital investments. In the case of TB, there has been cost sharing of fences to prevent deer-cow contact. The cost-sharing programmes could operate similar to the conservation cost-sharing programmes with eligibility and targeting of resources.

The Michigan TB case also has implications regarding business interruption insurance. It was shown that business interruption losses can be forecasted on a farm-by-farm basis using previous records to divide costs into fixed and variable components. This allows one to project operation-specific business interruption losses. Meuwissen *et al.* (2000) showed that actuarially fair business interruption insurance premiums were prohibitively expensive in the case of classic swine fever in Europe. This result might prove true for the majority of livestock diseases, however; it is worth considering subsidized insurance that might also provide the appropriate incentives for farms to prevent and control infectious disease, depending on some of the factors addressed in this chapter.

When designing policies for livestock disease control, it is critical to understand that farmers react to private incentives. The best course of action for an individual farm will depend on the farm type and size as well as on individual solvency and cash-flow considerations. The option to choose the test-and-slaughter method of control means that dairy farms will avoid the potentially large business interruption losses of depopulation by choosing this avenue. However, the state costs of this are those associated with more years without accredited TB-free status. Thus, the state could afford to pay more to encourage depopulation and still have a financial improvement. One critical aspect of reimbursing business interruption losses is avoiding the incentive to produce infected herds.

References

Chi, J., Weersink, A., Van Leeuwen, J. and Keefe, G. (2001) The economics of controlling infectious diseases on dairy farms. Working Paper 02/03, Department of Agricultural Economics and Business, University of Guelph, Canada.

Frye, G.H. (1995) Bovine tuberculosis eradication: the program in the United States. In Thoen, C.O., and Steele, J.H. (eds) *Mycobacterium Bovis Infection in Animals and Humans.* Iowa State University Press, Ames, Iowa, pp. 119-129.

Hickling, G. (2002) Dynamics of bovine tuberculosis in wild white-tailed deer in Michigan. Michigan Department of Natural Resources Wildlife Division Report 3363, Lansing, Michigan.

Horan, R. and Wolf, C. (2003) The economics of managing wildlife disease. Selected Paper, American Agricultural Economics Association Annual Meetings, Montreal, Quebec, Canada.

Kaneene, J.B., Bruning-Fann, C.S., Granger, L.M., Miller, R. and Porter-Spalding, B.A. (2002) Environmental and farm management factors associated with tuberculosis on cattle farms in northeastern Michigan. *Journal of the American Veterinary Medical Association* 221, 837-842.

Kuchler, F. and Hamm, S. (2000) Animal disease incidence and indemnity eradication programs. *Agricultural Economics* 22, 299-308.

McInerney, J.P. (1996) Old economics for new problems. *Journal of Agricultural Economics* 47, 295-314.

Michigan Legislative Council. (1988) Public Act 466 of 1988, The Animal Industry Act. [Accessed 2004] Available at: http://www.MichiganLegislature.Org.

Meuwissen, M.P.M., Skees, J.R., Black, J.R., Huirne, R.B.M. and Dijkhuizen, A.A. (2000) An analytical framework for discussing farm business interruption insurance for classical swine fever. Selected Paper, American Agricultural Economics Association Annual Meeting, Tampa, Florida.

National Research Council (1994) *Livestock Disease Eradication: Evaluation of the Cooperative State-Federal Bovine Tuberculosis Eradication Program.* National Academy Press. Washington, DC.

Nott, S. and Wolf, C. (2000) Dairy farm decisions on how to proceed in the face of TB. Michigan State University, Agricultural Economics Staff Paper 00-39, East Lansing, Michigan.

Ramsay, G.C., Philip, P. and Riethmuller, P. (1999) The economic implications of animal diseases and disease control at the national level. In Perry, B.D. (ed.) *The Economics of Animal Disease Control.* Office International Des Epizooties, Scientific and Technical Review Vol 18, 343-354.

U.S. Animal Health Association. (2001) Report of the Committee on Tuberculosis [Accessed 2004] Available at http://usaha.org/reports/reports01/r01tb.html.

U.S. Department of Agriculture – Animal and Plant Health Inspection Services – Veterinary Services. (1999) Bovine Tuberculosis Eradication Uniform Methods and Rules, Effective January 22, 1999. [Accessed 2004] Available at http://www.ncagr.com/vet/FedTBUMRTxt.htm.

Wittenberg, E. and Black, J.R. (2003) 2002 Michigan feeder steer business analysis summary. Michigan State University, Agricultural Economics Staff Paper 03-21, East Lansing, Michigan.

Wittenberg, E. and Wolf, C. (2003) 2002 Michigan dairy farm business analysis summary. Michigan State University, Agricultural Economics Staff Paper 03-14, East Lansing, Michigan.

Wolf, C. and Ferris, J. (2000) Economic consequences of bovine tuberculosis for Michigan livestock agriculture. Report to the Michigan Department of Agriculture, Lansing, Michigan.

Wolf, C., Harsh, S. and Lloyd, J. (2000) Valuing losses from depopulating Michigan dairy herds. Michigan State University, Agricultural Economics Staff Paper 00-10, East Lansing, Michigan.

Chapter 16

Economic Impacts of Eradicating Scrapie, Ovine Progressive Pneumonia and Johne's Disease on US Sheep, Lamb, Sheep Meat and Lamb Meat Markets

Ann Hillberg Seitzinger, Philip L. Paarlberg and John G. Lee

Introduction

Three diseases endemic to the US sheep flock are repeatedly cited as impediments to further development of the US sheep industry's export potential. Scrapie, ovine progressive pneumonia (OPP) and ovine Johne's disease (OJD) are slowly progressive diseases present at low levels in US adult sheep. Scrapie affects the central nervous system and has been identified as a fatal disease of sheep in the USA since 1947 (Kimberling, 1997). Also fatal, OPP may affect any of several organ systems and was described in Montana sheep as early as 1923 (Cutlip *et al.,* 1992). OJD affects the sheep's ability to absorb nutrients through the intestinal wall, and has been recognized in the US ovine population since the 1930s (Howarth, 1932). The dominant sheep exporters in world trade, Australia and New Zealand, are both considered free of scrapie and OPP but have dealt with OJD since the 1980s and the 1950s, respectively.

In 2000 the US Department of Agriculture (USDA) announced a plan to spend $100 million over ten years to eradicate scrapie from the US flock (US Federal Register, 2000). Key elements of the eradication plan are identification of pre-clinical infected animals via testing, tracing of identified animals and the provision of clean-up strategies. The primary clean-up strategy is genetic testing and removal of susceptible animals with producer indemnification. However, in some cases whole flock depopulation may occur, and live animal testing to allow the retention of some susceptible sheep is a further option. In addition to the long-held desires of breeding stock producers to gain expanded access to overseas markets, increased concern about the presence of a transmissible spongiform encephalopathy in the US livestock population has provided greater motivation to attempt elimination of scrapie from the USA.

In contrast to the eradication programme being pursued by the US government for scrapie, efforts to control OPP in the US sheep population are limited to producer-motivated initiatives (i.e. OPP Concerned Sheep Breeders Society – www.oppsociety.org). Control measures for Johne's disease in US sheep have taken a similar path. The USDA did institute a voluntary control programme for Johne's disease in bovines in the USA in 2002. In this case, concern over a possible link between Johne's disease in livestock and Crohn's disease in humans further motivated control efforts. However, these control efforts have not yet been extended to Johne's disease in US sheep, though such efforts have been initiated in the Australian sheep industry.

The objective of this chapter is to estimate the market and economic welfare effects of successfully eliminating scrapie, OPP and OJD from the USA. This objective is accomplished by constructing a model of the US lamb, lamb meat, sheep and sheep meat markets and simulating the consequences of eradication on prices, domestic output, domestic use and trade. Various measures of the changes in economic welfare for producers and consumers are presented. The next section presents the conceptual and empirical model developed. This section is followed by the results of the simulation analysis and the conclusions.

Modelling Framework

To determine the economic impacts from eradication of scrapie, OPP and OJD in US sheep and lambs, a model is required. This section presents both the modelling framework and the empirical model used.

Choice of Framework

A critical decision is the choice of framework. The commodities included in the model are sheep, lambs, sheep meat and lamb meat. This means a partial equilibrium model is used where all other prices, national income and consumer expenditure are unaffected by changes in variables for the four commodities included. This assumption is plausible. *Per capita* consumption of sheep and lamb meat is very low so price increases for consumers will have little effect on their level of spending. There is little evidence of large cross-price effects on other commodities so other meat and animal prices should be barely affected. Sheep and lambs are not large users of feedstuffs compared to cattle, swine and poultry, so feedstuff prices are not likely to be greatly affected.

The decision to model sheep, lambs, sheep meat and lamb meat also means there are horizontal and vertical linkages that must be explicitly considered if the model is to perform correctly. Two vertical channels are apparent – one from sheep to sheep meat; the other from lambs to lamb meat. These vertical channels are handled by explicitly modelling slaughter using complementarity conditions (Sanyal and Jones, 1982). Animal slaughter simultaneously generates the derived demand for animals and the supply of meat. The horizontal linkage is handled via ending inventory equations. The supply of lambs available at a point in time reflects previous inventory decisions. Ending inventory decisions are based on retention of lambs for breeding that depends on the relative profitability of marketing now versus the stream of benefits from holding the animal for the future.

Simple Theoretical Impacts

In order to give an intuitive idea of the impacts arising from the elimination of the three diseases from the US sheep population, a simple graphical model of the lamb and lamb meat markets is used (Fig. 16.1). Only the lamb and lamb meat markets are used since the analysis for sheep and sheep meat is analogous.

The initial situation is that scrapie, OPP and OJD are present in the US lamb market. The USA is an importer of lamb meat so faces an excess supply function, denoted ES, in the world lamb meat market. Domestic demand and domestic supply initially are D_0 and S_0, respectively. The difference between the US domestic demand and the US domestic supply is the US excess demand for lamb meat, denoted ED_0. Where excess supply intersects excess demand gives the initial price for lamb meat, P_0, and the initial quantity of lamb meat imports,

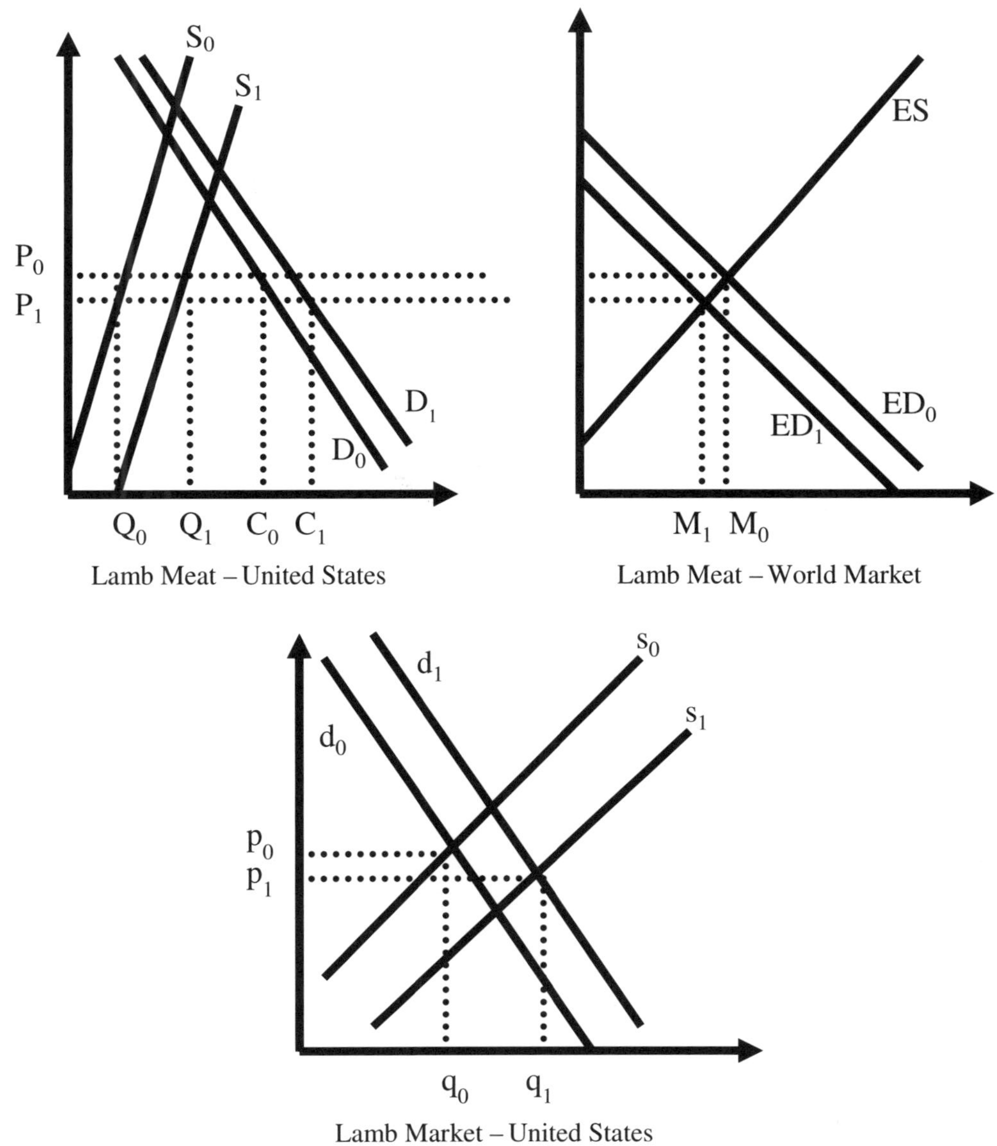

Fig. 16.1. Impact of Disease Eradication in the USA.

M_0. At P_0, the quantity of lamb meat demanded in the USA is C_0 and the quantity of lamb meat supplied is Q_0, with the difference equal to M_0.

The US domestic market for live lambs is shown in the lower panel of Fig 1. The assumption is that exports of lambs, x_0, are exogenous and included in the total US domestic demand for lambs, d_0. Note the vertical linkage as the US supply of lamb meat in the upper right is tied to the US demand for lambs in the lower panel. Imports of live lambs are also treated as exogenous and are included in the total US supply of live lambs, s_0. The lamb market clears at a price of p_0 and quantity, q_0.

Eradication of scrapie, OPP and OJD in the US lamb population causes several potential shifts. One potential shift is that some consumers who were afraid of eating lamb meat due to the diseases are now willing to consume the meat. This is depicted as a rightward shift in the demand for lamb meat in the USA from D_0 to D_1. Eliminating disease also means more lambs are supplied at every price so the US lamb supply function in the lower panel shifts rightward from s_0 to s_1. At the same time with the three diseases eradicated there may be more of an opportunity to export live lambs out of the USA so the demand for live lambs in the lower panel shifts from d_0 to d_1. Figure 16.1 is drawn such that the supply shift dominates the demand shift. The result is an increase in lambs marketed from q_0 to q_1 and a lower price for lambs, p_0 to p_1.

The changes in the lamb market are reflected in the lamb meat market as a rightward shift in the lamb meat supply from S_0 to S_1. The combined effect of the expansion in US demand for lamb meat and the US supply of lamb meat is to shift the US excess demand for lamb meat leftward from ED_0 to ED_1. As a result the price of lamb meat falls from P_0 to P_1 and the quantity of imports falls from M_0 to M_1. The extent of the price decline and the drop in imports depends on the elasticity of the excess supply for lamb meat facing the USA. The more elastic that relation is, the more the impact is felt in the trade volume and the less it is felt as a price decrease.

Structure of the Conceptual Model

The conceptual model is a richer form of the theoretical model outlined above. To facilitate applying the model to the base data, the model is developed in its logarithmic differential form (Sanyal and Jones, 1982). Thus, this gives the percentage changes from the base for a given set of shocks. In order to keep the presentation tractable, only the differential form is presented.

We begin with consumers. The model allows for two types of consumers. There may be a certain reluctance to consume lamb and sheep meat due to the presence of disease. Such consumers are assumed to be in the minority, with the majority of consumers unaffected by the presence of the diseases. To account for this possibility the changes in *per capita* demands for lamb meat, $\mathrm{dln}(PC_{LM})$, and for sheep meat, $\mathrm{dln}(PC_{SM})$, are modeled. In this partial equilibrium framework, these changes depend on the changes in the prices of lamb meat, $\mathrm{dln}P_{LM}$ and sheep meat, $\mathrm{dln}P_{SM}$:

(1) $$\mathrm{dln}(PC_{LM}) = \varepsilon_{LL}\mathrm{dln}P_{LM} + \varepsilon_{LS}\mathrm{dln}P_{SM}$$

(2) $$\mathrm{dln}(PC_{SM}) = \varepsilon_{SL}\mathrm{dln}P_{LM} + \varepsilon_{SS}\mathrm{dln}P_{SM}.$$

The own-price elasticities are ε_{LL} and ε_{SS} which are assumed to be non-positive. The price elasticity for sheep meat in the demand for lamb is ε_{LS}. The price elasticity for lamb meat in the demand for sheep meat is ε_{SL}. Both cross-price elastiticies are assumed to be non-negative. The changes in the total demands for lamb meat, $dlnC_{LM}$, and for sheep meat, $dlnC_{SM}$, depend on the change in the total population, dln(Pop), the change in the share of the population unafraid of eating the meats due to disease, $dln(\alpha)$, and the changes in *per capita* consumption, $dln(PC_{LM})$ and $dln(PC_{SM})$:

(3) $$dlnC_{LM} = dln(Pop) + dln(\alpha) + dln(PC_{LM})$$

(4) $$dlnC_{SM} = dln(Pop) + dln(\alpha) + dln(PC_{SM}).$$

Thus, one of the potential effects of eradicating disease appears with $dln(\alpha) > 0$.

The production of the meats and the derived demand for the animals emerge from Ricardo-Viner (i.e. fixed factor) models for meat production. Lamb and sheep slaughter are treated as separate production activities by profit-maximizing firms using constant returns to scale technologies. Slaughtering firms are treated as being perfectly competitive in that individual firms cannot affect output and inputs prices. This means meat output and animals killed are determined by supply and demand. Meat production uses three factors.

One factor, denoted by O, is mobile throughout the economy and in this partial equilibrium model, that factor's price, W, is exogenous to the lamb slaughter industry and to the sheep slaughter industry. Each industry uses some factors, capital and management, specific to that industry. These factors are assumed in fixed supply and are denoted, K_j, j = LM, SM, for the lamb meat and sheep meat industries, respectively. The unit returns to these specific factors are R_{LM} and R_{SM}, which indicate the per-unit rental rate to the specific factors (profits plus capital payments).

Next we consider that slaughter requires animals. The lamb meat industry uses lambs with a price of P_L to make lamb meat. The sheep meat industry uses sheep with a price of P_S to manufacture its meat. Under these assumptions each industry earns zero economic profit and the differentials of the zero profit conditions are:

(5) $$\theta_{O,LM}\, dlnW + \theta_{K,LM}\, dlnR_{LM} + \theta_{L,LM}\, dlnP_L = dlnP_{LM}$$

(6) $$\theta_{O,SM}\, dlnW + \theta_{K,SM}\, dlnR_{SM} + \theta_{S,SM}\, dlnP_S = dlnP_{SM}$$

where $\theta_{i,j}$ is the unit revenue share for factor i in output j, (i = O for other factors, i = K for the specific factor, i = L for lambs and i = S for sheep).

Inputs are allowed to be used in variable proportions as governed by elasticities of substitution. Let $\sigma_{i,h,j}$ be the elasticity of substitution between factors i and h in industry j. Define $a_{i,j}$ as the use of factor i per unit of output j. Using the envelope property, the percentage change in the per-unit factor use can be expressed as a unit cost share multiplied by an elasticity of substitution and the difference between factor price changes. Thus,

(7) $$dln(a_{K,LM}) = \theta_{L,LM}\, \sigma_{L,K,LM}\, (dlnP_L - dlnR_{LM})$$

(8) $$dln(a_{L,LM}) = -\theta_{K,LM}\, \sigma_{L.K,LM}\, (dlnP_L - dlnR_{LM})$$

(9) $\quad dln(a_{O,LM}) = -\theta_{K,LM}\, \sigma_{O,K,LM}\, (dlnW - dlnR_{LM})$

(10) $\quad dln(a_{K,SM}) = \theta_{S,SM}\, \sigma_{S,K,SM}\, (dlnP_S - dlnR_{SM})$

(11) $\quad dln(a_{S,SM}) = -\theta_{K,SM}\, \sigma_{S,K,SM}\, (dlnP_S - dlnR_{SM})$

(12) $\quad dln(a_{O,SM}) = -\theta_{K,SM}\, \sigma_{O,K,SM}\, (dlnW - dlnR_{SM})$.

The differential equations arising from the specific factor market clearing identities give the percent changes in the outputs of lamb meat, $dlnQ_{LM}$, and sheep meat, $dlnQ_{SM}$:

(13) $\quad dln(a_{K,LM}) + dlnQ_{LM} = dlnK_{LM}$

(14) $\quad dln(a_{K,SM}) + dlnQ_{SM} = dlnK_{SM}$.

The change in the derived demand for lambs for slaughter, dD_L, and the change in the derived demand for sheep for slaughter, dD_{SM}, come from the definition of the derived demand. Derived demand for animals is the per-unit use of animals in meat production multiplied by the meat output. Thus:

(15) $\quad dlnD_L = dln(a_{L,LM}) + dlnQ_{LM}$

(16) $\quad dlnD_S = dln(a_{S,SM}) + dlnQ_{SM}$.

The lamb meat and sheep meat markets are cleared using three relations each. The USA imports large quantities of both products relative to domestic use, so imports are modeled using excess supply functions. Exports of each product from the USA, X_{LM} and X_{SM}, are small and are treated as exogenous. Doing so allows US exports of lamb meat and sheep meat to be exogenously shocked as another effect of disease eradication. Import barriers consist of a negligible tariff. Thus, exports of lamb meat to the USA, M_{LM}, are a function of the US lamb meat price. Exports of sheep meat to the USA, M_{SM}, are a function of that good's price. Imports of meat by the USA equal exports to the USA. Thus, the differentials of the excess supply functions are:

(17) $\quad dlnM_{LM} = \xi_{LM} dlnP_{LM}$

(18) $\quad dlnM_{SM} = \xi_{SM} dlnP_{SM}$

where ξ_{LM} is the elasticity of excess supply of lamb meat to the USA and ξ_{SM} is the elasticity of excess supply of sheep meat to the USA. The excess supply elasticities are non-negative. Recognizing beginning inventories of lamb meat, B_{LM}, and ending inventories, E_{LM}, are held primarily for transactions purposes and can be treated as exogenous, the differential equation for lamb meat market clearing can be written:

(19) $$M_{LM}(dlnM_{LM}) = X_{LM}(dlnX_{LM}) + C_{LM}(dlnC_{LM}) + E_{LM}(dlnE_{LM}) - Q_{LM}(dlnQ_{LM}) - B_{LM}(dlnB_{LM}).$$

Defining sheep meat beginning and ending inventories similarly gives the market clearing for sheep meat:

$$(20)\qquad M_{SM}(dlnM_{SM}) = X_{SM}(dlnX_{SM}) + C_{SM}(dlnC_{SM}) + E_{SM}(dlnE_{SM}) - Q_{SM}(dlnQ_{SM}) - B_{SM}(dlnB_{SM}).$$

The remaining tasks to complete the model are to define the supplies of animals available and to clear the animal markets. In these relations, current period outcomes reflect earlier decisions due to gestation time plus the time needed for a lamb to mature. Lambs available for slaughter, Q_L, depend on the incoming inventory of lambs and ewes, $E_{S,-1}$, and the effect of the disease eradication programme on lamb supply, dln(ω). In its differential form this equation appears as:

$$(21)\qquad dlnQ_L = \eta_{LE} dlnE_{S,-1} + dln(\omega)$$

where η_{LE} is the elasticity of lamb supply with respect to initial lamb and ewe inventories and is non-negative. The change in initial lamb and ewe inventories is de-composed into the change in lamb inventory, $dlnE_{L,-1}$, and the change in ewe inventory, $dlnE_{E,-1}$, according to:

$$(22)\qquad dlnE_{S,-1} = (E_{L,-1}/E_{S,-1})dlnE_{L,-1} + (E_{E,-1}/E_{S,-1})dlnE_{E,-1}.$$

Behavioural equations are used to explain current period ending inventories of lambs and ewes. Inventories of rams, E_R, are generally constant and small so are treated as exogenous. The decision to retain or to slaughter is based on the expected earnings from retaining the animal compared to the current market price for a slaughter animal (Rosen, 1987). For lambs, future earnings are the expected discounted stream sales returns on lambs born plus the discounted sales value of a ewe for slaughter, P_L^*. Ending inventories also depend on the beginning inventory of lambs, $E_{L,-1}$. Differentiating the ending inventory equation for lambs yields:

$$(23)\qquad dlnE_L = \eta_{LL}[dlnP_L^* - dlnP_L] + \eta_{L,-1} dlnE_{L,-1}$$

where η_{LL} gives the elasticity of ending lamb inventories with respect to the expected future return relative to the current market price. Stabilizing expectations requires that this elasticity is non-negative, indicating that as the expected future return rises relative to the current price, more lambs will be retained. The parameter $\eta_{L,-1}$ is non-negative less than 1, and indicates the speed of lamb inventory adjustment over time. The expected return to holding a ewe, P_S^*, is the discounted value of future lambs plus the market price of a ewe over a one-year shorter time span than for the expected return to holding a lamb. Ending ewe inventories, E_E, depend on the expected return to retaining the animal relative to the market price and the beginning inventory of ewes, $E_{E,-1}$. In differential form this equation appears as:

$$(24)\qquad dlnE_E = \eta_{SS}[dlnP_S^* - dlnP_S] + \eta_{S,-1} dlnE_{E,-1}$$

where η_{SS} is the elasticity of ending ewe inventories with respect to the expected relative return from holding a ewe and is non-negative with stabilizing expectations. The parameter

$\eta_{S,-1}$ governs the speed of adjustment and needs to lie between 0 and 1 to preserve the model's stability.

Completing the model requires closure of the animal markets. Trade in live animals is small. Imports of live lambs by the USA, M_L, and exports of lambs, X_L, are treated as exogenous. The quantity of exports is another variable that can be altered under disease eradication as added market opportunities appear. Also included in the market clearing is any death loss of lambs, L_L. Since this loss depends on external factors like predators and weather it is modeled as exogenous. Live sheep imports by the USA, M_S, and US sheep exports, X_S, are small relative to world trade and treated as exogenous. Sheep exports have the potential to expand in the absence of scrapie, OPP and OJD in the USA. The market clearing for sheep also includes an exogenous death loss L_S. Finally, sheep are being removed due to the current scrapie programme, V_S. Elimination of scrapie in the USA adds to the sheep inventory by this amount, $d\ln V_S < 0$. This same effect is captured in the lamb market via the shift in ω, $d\ln(\omega) > 0$. Thus, the differential of the market clearing for lambs is:

$$(25) \qquad M_L(d\ln M_L) = E_L(d\ln E_L) + D_L(d\ln D_L) + X_L(d\ln X_L) - Q_L(d\ln Q_L) + L_L d\ln L_L.$$

The differential of the market clearing for sheep is:

$$(26) \qquad M_S(d\ln M_S) = E_E(d\ln E_E) + E_R(d\ln E_R) + D_S(d\ln D_S) + X_S(d\ln X_S) - E_{R,-1}(d\ln E_{R,-1}) - E_{E,-1}(d\ln E_{E,-1}) + L_S(d\ln L_S) + V_S(d\ln V_S).$$

Empirical Model Construction

The empirical model follows the structure of the conceptual model discussed above and relies on three sets of data. One set of data consists of supply, disappearance and price data for lambs, sheep, lamb meat and sheep meat. The second set of data consists of unit cost shares in animal slaughter. The final set of data is the various elasticities.

Most supply, disappearance and price data can be obtained or calculated from the reports of various agencies of the USDA as reported in 2003. Prices for lambs and ewes for slaughter at San Angelo, Texas are reported in the *Livestock, Dairy, and Poultry Outlook* published by the USDA Economic Research Service (ERS). The price of lamb meat, East Coast, is also from the *Livestock, Dairy, and Poultry Outlook.* Livestock slaughter for lambs and sheep, along with meat production and slaughter weights, come from the *Livestock Slaughter Annual* by the USDA National Agricultural Statistical Service (NASS). Data on imports and exports of meat and animals can be obtained from the *Trade Data Web* compiled by the US International Trade Commission (USITC). Import value data from the USITC data are used to calculate the unit value for sheep meat. Sheep and lamb inventory data are found in USDA NASS *Agricultural Statistics.* No data from meat stocks were found. Given this information lamb meat and sheep meat disappearance are calculated as the residual. For calibration it was assumed that 10% of the US population would not consume lamb and sheep meat due to the presence of scrapie and OJD. This assumption allows calculation of national *per capita* disappearance and calibration of the *per capita* demand. For animals death loss is treated as the residual of beginning inventories, ending inventories, births and trade.

Unit revenue shares are derived from the supply, disappearance and price data. Those data indicate the ratio of animals killed to meat production. Using the prices of animals and meat, the unit revenue shares from animals are calculated. Other exogenous mobile inputs are assumed to have a revenue share of 20%. Combined, those two revenue shares leave a residual for sector-specific factors. Indexing the rental rate at 100 determines the per-unit use of sector-specific factors.

The elasticities used in the model come from various sources. The own-price elasticity of demand is the value estimated by Paarlberg and Lee (2001) of −0.4371. The cross-price elasticity with sheep meat is set at a value that is smaller in absolute value to satisfy theoretical demand restrictions, 0.1. The cross-price elasticity of lamb meat in the demand for sheep meat is treated as symmetric, thereby, implying no income effect. The own-price elasticity of demand for sheep meat is assumed to be larger in absolute value, but still very inelastic, −0.2.[1] The substitution elasticities used in the model are taken from two papers by MacDonald and Ollinger (2000; 2001) that estimate factor substitution elasticities in hog slaughter. Their estimates indicate considerable factor substitution in hog slaughter. The substitution elasticity of animals with respect to the sector-specific factor is 1.493. The elasticity of substitution of the exogenous factor with respect to the specific factor is 2.663. The elasticity of lambs with respect to lagged ewe inventory is estimated using time series data to be 0.900894. The elasticities for ending inventories of lambs and ewes with respect to current prices are estimated to be −0.034079 and −0.422108, respectively. The elasticities on the lagged inventory terms are 0.970353 and 0.980749. The excess supply elasticities for lamb meat and sheep meat are assumed to be 10.0. The very elastic value means that the USA is modeled as nearly a "small country" price taker on world markets. This is plausible as the share of global trade moving to the USA is relatively small, with the USA taking around 7.4% of world imports in lamb and sheep meat (UN-FAO, 2003).

Simulations

As indicated above, the potential impact of eradicating scrapie, OPP and OJD from the US sheep population is modeled by introducing inventory, processing cost, trade and programme shocks. The total inventory shocks are derived from summing the production impacts of the three diseases on mature sheep. For scrapie, the prevalence in the US sheep population is estimated at 0.2% with a 100% mortality rate (APHIS, 2003a). OPP is much more widely spread as measured by a seroprevalence of 24.2% (APHIS, 2003b), but only 2 to 10% of sheep seropositive for OPP are believed to develop clinical signs (Bulgin, 1990). Because clinical cases of OPP also carry a 100% mortality rate, the reduction in mature sheep inventory due to OPP is calculated as ranging between 0.5 and 2.4%. OJD is estimated to be present on 4.7% of sheep operations in the USA (APHIS, 2004), causing an estimated 4% increase in mortality in the affected flocks (Topp and Bailey, 2001). The reduction in national sheep inventories due to OJD then is calculated to be 0.2%. The lagged inventory shocks introduced into the model for eradication of all three diseases are for a 3.0% increase due to reductions in mortality. Because the literature is mixed in reporting whether these diseases cause production losses other than increased mortality, no additional impacts on production are simulated.

[1] Other estimates of the own-price elasticity of demand are −0.62 (Byrne *et al.*, 1993) and −1.09 (Schroeder *et al.*, 2001). Sensitivity analysis of this parameter yielded little variation in simulation results.

The presence of scrapie in the US sheep population has led to a standard industry practice of charging producers for disposal of offal in landfills. The eradication of scrapie would be expected to allow the return of some uses of the sheep and lamb offal such that the landfill charge would be eliminated. Estimates of these charges were provided by slaughter establishments at $3 per carcass. Therefore, shocks reducing other processing costs for both lamb meat and sheep meat by 10% are included in the simulations.

Appropriate trade shocks were the most difficult to determine. For many years, an estimate based on industry-gathered market intelligence reports has set the cost of scrapie for the USA equal to a value of estimated potential live sheep exports of $20 million annually (Parker, C., personal communication, 2003). A review of export statistics of major sheep-exporting countries shows the quantity of live breeding sheep exported during the 1990s averaging 44,448 and 2,843 annually, for Australia and New Zealand, respectively (Figs 16.2 and 16.3). For Australia, the total value of breeding sheep exports during the 1990s was $29.2 million. While the sheep exports of these countries do not show any apparent trend, New Zealand's exports of ovine germplasm show more than a doubling of both ovine embryo and ovine semen exports since 1995, with Australia the dominant importer (Fig. 16.4). Given this information, it was decided to range the trade shocks to the model from 0 to 10% increases for sheep meat, live sheep, lamb meat and live lamb exports.

Finally, incomplete data indicate 4,472 sheep were purchased for indemnification under the Scrapie Eradication Program during Fiscal Year 2003. Therefore, a shock of adding 6,000 mature sheep to inventories to reflect ending of the indemnification programme was included in the simulations.

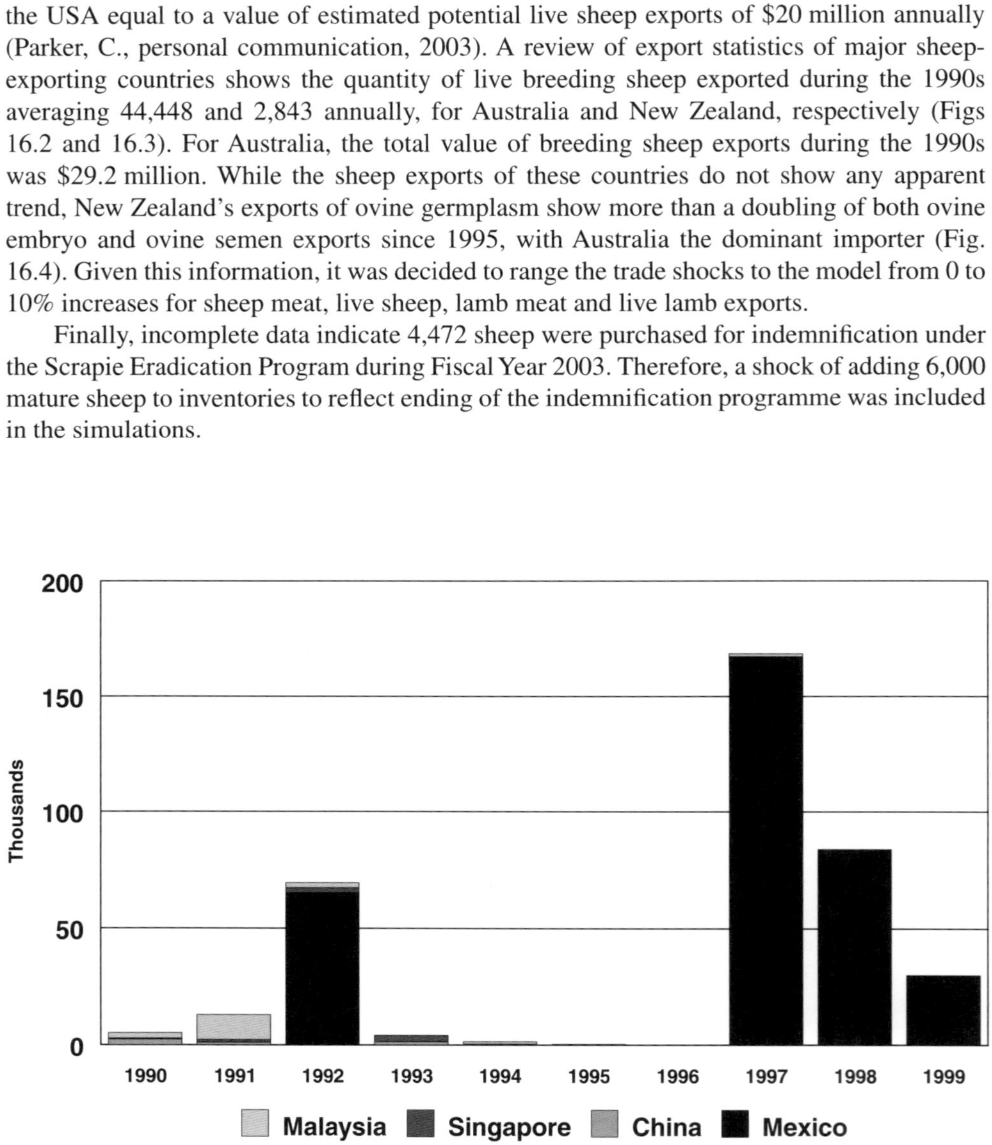

Fig. 16.2. Australian Breeding Sheep Exports by Destination.

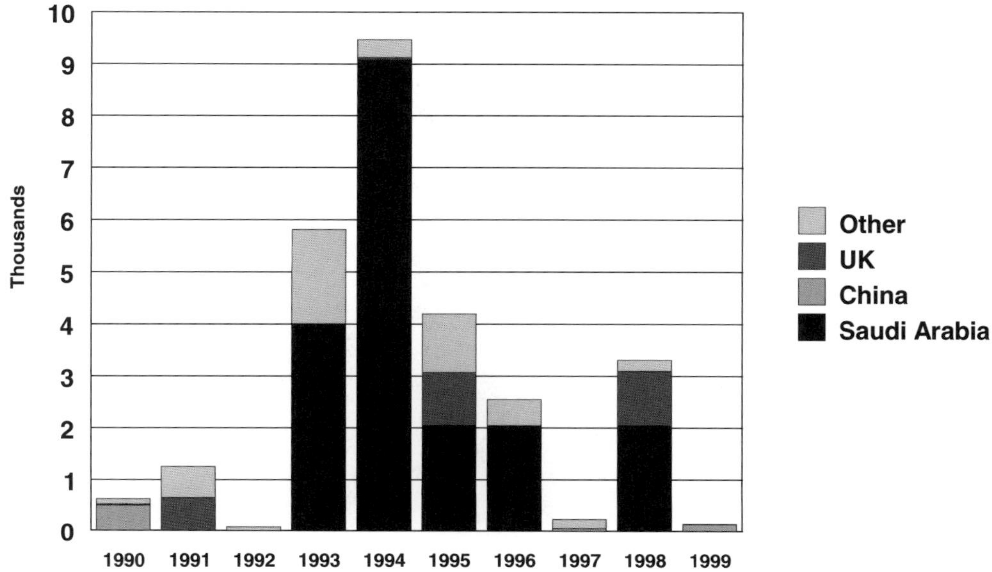

Fig. 16.3. New Zealand Breeding Sheep Exports by Destination.

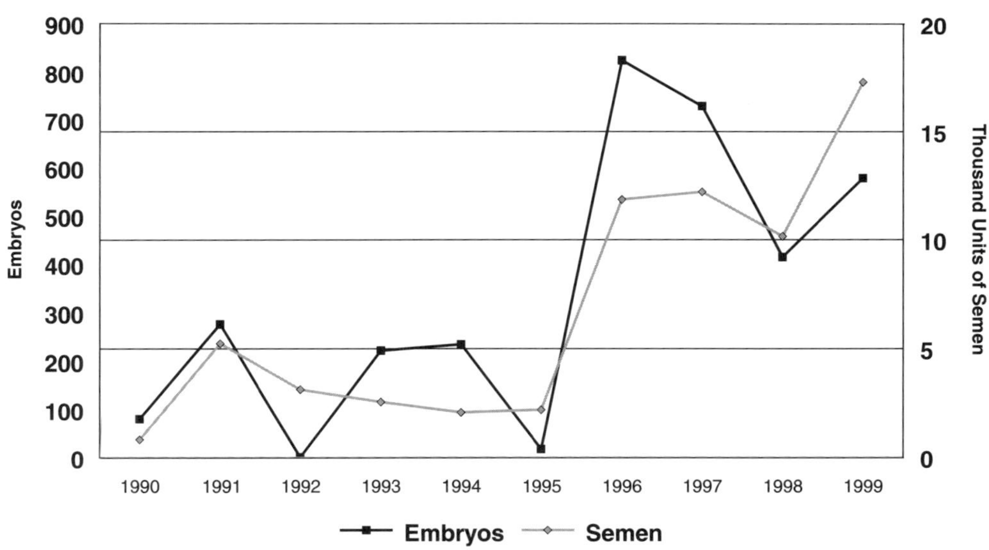

Fig. 16.4. New Zealand Ovine Germplasm Exports.

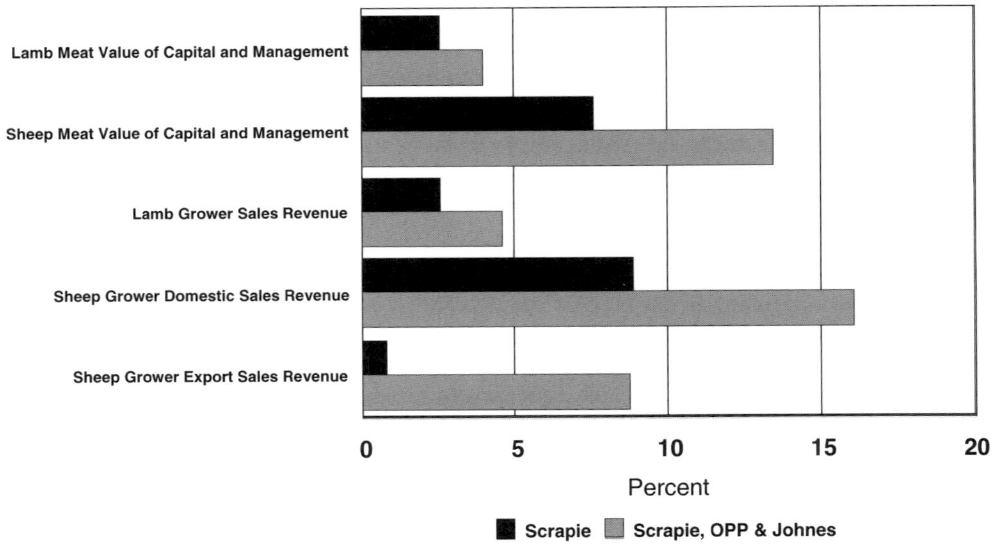

Fig. 16.5. Annual Percentage Changes in Returns to USA Producers and Processors.

Results

Figure 16.5 shows the changes in producer and processor returns due to shocks reducing other lamb meat and sheep meat processing costs by 10%, increasing beginning inventories of mature sheep by 0.2%, and increasing the supply of sheep by 6,000 in response to ending the Scrapie Eradication Program. In essence, this first scenario looks at the impact of scrapie alone without introducing changes in trade. Returns to the value of capital and management in the lamb meat and sheep meat sectors increase by 2.6 and 7.6%, respectively, as both outputs and prices increase. Sales revenues for lamb growers rise by 2.6% with prices increasing 2.4%. Sales revenues for domestic sheep go up by 8.9%, while sales revenues on exported sheep increase by 0.8%.

Table 16.1 records the changes in revenues from the base for this scenario indicating annual revenue increases of $1.8 million and $0.1 million in the value of capital and management involved in processing lambs and sheep, respectively. Lamb grower revenues increase by $7.4 million annually, while revenues on sheep sales in domestic markets increase by $1.1 million and revenues on sheep sales in export markets increase by $0.2 million. Consumer welfare also increases by $0.2 million.

Figure 16.5 also shows the changes in producer and processor returns due to eliminating scrapie, OPP and OJD which, it is believed, would allow the USA to participate more fully in international trade of sheep meat, live sheep, lamb meat and live lambs. Shocks reduce other lamb meat and sheep meat processing costs by 10% and increase supply by the estimated 6,000 sheep claimed by the Scrapie Eradication Program. Beginning inventories of mature sheep are increased by 3% and trade is increased by 10%. Table 16.2 records the changes in revenues from the base for this scenario indicating annual revenue increases of $2.8 million

Table 16.1. Annual Producer and Processor Revenues, Base Solution and Domestic Scrapie Simulation (Million US$).

	Base	Simulation	Change
Lamb Meat Value of Capital and Management	71.3	73.1	1.8
Sheep Meat Value of Capital and Management	1.1	1.2	0.1
Lamb Grower Revenue on Sales	278.5	285.9	7.4
Sheep Grower Revenue on Domestic Sales for Meat	11.5	12.6	1.1
Sheep Grower Revenue on Export Sales	31.9	32.1	0.2
Consumer Surplus	580.1	580.3	0.2

Table 16.2. Annual Producer and Processor Revenues, Base Solution and Full Trade Gains (Million US$).

	Base	Simulation	Change
Lamb Meat Value of Capital and Management	71.3	74.1	2.8
Sheep Meat Value of Capital and Management	1.1	1.3	0.2
Lamb Grower Revenue on Sales	278.5	291.3	12.8
Sheep Grower Revenue on Domestic Sales for Meat	11.5	13.4	1.9
Sheep Grower Revenue on Export Sales	31.9	34.7	2.8
Consumer Surplus	580.1	581.8	1.7

and $0.2 million in the value of capital and management involved in processing lambs and sheep, respectively. Lamb grower revenues increase by $12.8 million annually, while revenues on sheep sales in domestic markets increase by $1.9 million and revenues on sheep sales in export markets increase by $2.8 million. Consumer welfare increases by $1.7 million, or less than 1%.

In summary, the eradication of scrapie, OPP and OJD from the US sheep population is estimated to benefit the sheep, lamb, sheep meat and lamb meat sectors as well as US consumers of sheep meat and lamb meat. The eradication of scrapie alone is estimated to offer US sheep and lamb producers and processors increases in revenues of $10.8 million annually, while the eradication of all three diseases allows the capture of additional benefits, particularly in export markets. Under this latter scenario, revenues to US sheep and lamb producers and processors are estimated to increase by $20.5 million annually while consumer welfare increases by $1.7 million annually. These estimates of the benefits of eradicating disease must of course be weighed against the costs of the eradication programmes.

References

APHIS (2003a) Phase II: Scrapie: Ovine Slaughter Surveillance Study 2002-2003. US Department of Agriculture, Animal and Plant Health Inspection Service, Veterinary Services, Centers for Epidemiology and Animal Health, National Animal Health Monitoring System, Fort Collins, Colorado, #N419.0104.

APHIS (2003b) Ovine Progressive Pneumonia: Awareness, Management, and Seroprevalence. US Department of Agriculture, Animal and Plant Health Inspection Service, National Animal Health Monitoring System, Fort Collins, Colorado, #N414.1203.

APHIS (2004) Ovine Johne's Disease: Awareness, Management, and Seroprevalence. US Department of Agriculture, Animal and Plant Health Inspection Service, National Animal Health Monitoring System, Fort Collins, Colorado.

Bulgin, M.S. (1990) Ovine progressive pneumonia, caprine arthritis-encephalitis, and related lentiviral diseases of sheep and goats. *Veterinary Clinics of North America: Food Animal Practice* 6(3) 691-704.

Byrne, P.J., Capps, Jr, O. and Williams, G.W. (1993) US demand for lamb: the other red meat. *Journal of Food Distribution Research* 24, 158-166.

Cutlip *et al.,* (1992) Seroprevalence of ovine progressive pneumonia virus in sheep in the United States as assessed by analyses of voluntarily submitted samples. *American Journal of Veterinary Research,* 53(6), 976-979.

Howarth, J.A. (1932) Paratuberculous enteritis in sheep caused by an acid-fast organism. *Journal of the American Veterinary Medical Association* 81, 383-387.

Kimberling, C.V. (1997) Scrapie: get the facts. *American Sheep Industry Association Update.*

MacDonald, J.M. and Ollinger, M.E. (2000) Scale economies and consolidation in hog slaughter. *American Journal of Agricultural Economics* 82, 241-227.

MacDonald, J.M. and Ollinger, M.E. (2001) Scale economies and consolidation in hog slaughter: a reply. *American Journal of Agricultural Economics* 83, 1084-1086.

Paarlberg, P.L. and Lee, J.G. (2001) US trade policy on lamb meat: who gets fleeced? *American Journal of Agricultural Economics* 83, 196-208.

Rosen, S. (1987) Dynamic animal economics. *American Journal of Agricultural Economics* 69, 547-557.

Sanyal, K.K. and R.W. Jones. (1982) "The Theory of Trade in Middle Products", *American Economic Review.* 72; 16-31.

Schroeder, T.C., Jerrick, R.J., Jones, R. and Spaeth, C. (2001) US lamb demand. *Sheep and Goat Research Journal* 17, 14-17.

Topp, V. and Bailey, G. (2001) Ovine Johne's Disease: evaluation of control and eradication strategies. *ABARE Report to Agriculture, Fisheries and Forestry* – Canberra, Australia.

UN-FAO (2003) FAOSTAT. United Nations Food and Agriculture Organization, Rome, Italy. [Accessed 2003.] Available from http://www.fao.org.

US Federal Register (2000) "Declaration of Emergency Because of Scrapie in the United States", Vol. 65, No. 53., Notices. March 17. p. 14521.

Chapter 17

Understanding Broader Economic Effects of Livestock Insurance and Health Management: Impacts of Disease Outbreak on Allied Industries

James Pritchett, Dawn Thilmany and Kamina Johnson

Introduction

The impacts of an animal health outbreak can be quite costly and far-reaching. Thus, the use of livestock insurance products may be important for allied sectors to understand and introduce into their marketing considerations for strategic planning. This chapter discusses a conceptual framework that quantifies potential economy-wide losses due to an animal disease outbreak. Central to the framework are technical relationships and economic relationships. Technical relationships embody the growth, development and slaughter of livestock as well as relationships underlying the fabrication of meat products. Economic relationships link the stages of the meat-marketing channel by allocating meat products to their highest-value form and customer segment. Understanding the ripple effects and persistence of outbreak shocks may better inform those developing insurance on effects of these products (i.e. changing incentives) on upstream and downstream sectors.

Background

Media and governmental responses to Bovine Spongiform Encephalopathy (BSE) discoveries in the USA and Canada reinforce existing concerns among consumers, meat producers and allied industries of an outbreak in the USA. Immediate impacts of an animal disease outbreak include an initial reduction in the productive capacity of the animal products industry and a subsequent reduction in the supply of meat products and derived demand for livestock production inputs. At the same time, disease outbreaks may reduce the demand for meat and meat products. Risk management tools may be used to mitigate the economic loss from a disease outbreak; therefore the design of these tools, their implementation and the incentives created for the insured may be of interest to all participants in the meat-marketing sector. The magnitude of the economic losses will depend on a myriad set of factors including the absolute and spatial size of the outbreak, the geography of the outbreak and the strategy used to combat the outbreak.

To appropriately model a wide variety of animal disease impacts at different levels of the livestock marketing channel, a system of economic relationships is needed that accounts for the interdependencies and degrees of response (i.e. elasticities) among the various production, marketing and consumer sectors of the economy. For example, allied animal product industries, such as meat packers, processors and retailers, bear an initial loss in the supply of meat products as well as increased costs of finding and certifying safe food supplies while facing demand reductions and a possible persistent loss of consumer confidence. The size of economic effects at that level will not be easy to assess. An accurate assessment of losses due to animal disease is useful for policymakers who may weigh these potential losses against the cost of disease prevention and mitigation strategies – especially if long-term marketing strategies and incentives to meat producers are influenced by insurance programmes.

The chapter begins with a literature review of animal disease impact analysis, focusing on the broader economic impacts to allied industries and consumers. Attention is then turned to a synthesis of economic impact modelling that focuses on animal disease outbreaks.

Literature Review

Review of previous economic research concerning animal disease outbreaks is a starting point for developing an understanding of appropriate modelling approaches. Literature on the economics of animal disease, likely consumer response (both degree and persistence due to lack of confidence in animal products), and methods to analyze and simulate indirect and imputed costs are reviewed in this section to motivate the discussion of impact modelling techniques.

Economics of Animal Disease

Animal diseases are an example of an invasive species, and Evans (2003) explores the economics of invasive species when discussing the six categories of economic impacts: production effects, market and price effects, trade effects, impacts on food security and nutrition, human health and the environment, and financial costs (UN-FAO, 2001). Evans (2003) defines production impacts as specific to the host (e.g. livestock) resulting in mortality or reduced efficiency. Such impacts are generally easy to identify, but may be difficult to quantify. In livestock, for example, delays in reproduction result in fewer offspring, which has long-term effects not easily measured in the present. Outbreaks of diseases can also impact prices and markets in the short-run, with broader, longer-term multiplier effects.

Several studies of Foot and Mouth Disease (FMD) estimate economic effects beyond a simple reduction in productive capacity. In particular, a recent FMD outbreak in the UK has been the subject of several economic impact analyses. A report by the UK's Department of Environment, Food and Rural Affairs in conjunction with the Department of Culture, Media and Sport (2002) suggests a £3.1 billion impact on agriculture and the food chain from FMD in 2001, of which £1.9 to £2.3 billion are attributed to agribusiness sectors beyond the farm gate. Since the UK has a relatively small livestock sector, it is not surprising that the impact to agriculture would be matched in magnitude by a decline in tourist expenditures (an additional £2.7 to £3.2 billion). By ranking affected sectors, Blake, Sinclair and Sugiyarto (2002) found that hotels and pubs were actually the biggest relative losers from the UK outbreak, followed by railway transport, road transport, milk products and slaughtering/meat processing, illustrating the potential for diverse sectoral impacts. Blake, Sinclair, Sugiyarto (2002) extended the discussion of broader impacts by noting that lost tourism represents an even greater impact on a region's GDP losses because the economic impact multiplier on tourism expenditures is much higher than that for farm level production.

A 2002 Canadian study of a potential FMD outbreak focused primarily on trade impacts, as Canada is a large net exporter of many livestock products (C$8 billion plus in annual exports) (Serecon Management Consulting, Inc., 2002). The total net impact of an outbreak would range from C$13.7 to C$45.9 billion, depending on whether the outbreak was small- (50 herds over six weeks) or large-scale (1500 herds over six months). Of the total impacts, more than half are attributed to trade losses, and given the livestock sector's significant role in Canadian agriculture, the long-term costs of an outbreak could represent as much as 80% of Canada's 2001 agriculture cash receipts.

Among animal disease studies focused on the USA, Ekboir (1999) estimates $13.5 billion of potential losses from an FMD outbreak in California. The estimates are generated using input-output analysis and include direct losses to livestock producers, disease control costs including depopulation, and indirect or imputed losses to businesses. The estimates also show that there is great sensitivity depending on how quickly the disease is transmitted, the particular depopulation policies imposed, and the speed with which depopulation occurs. Ekboir (1999) does assume meat export losses are a substantial portion of total economic impacts ($6 billion of the $13.5 total), with only $1.4 billion attributed to direct production losses. Given recent trade actions in response to the small US and Canadian BSE outbreaks, these estimates seem reasonable.

Paarlberg, Lee and Seitzinger (2002) estimate losses to the USA of an FMD outbreak assuming the magnitude was similar to that experienced by the UK. Total losses to US farm income are estimated at $14 billion including a 7% reduction in domestic consumer expenditures. The authors suggest a need to decompose these gross effects into the components borne by individual groups, including producers, affected by the disease and those that remain disease free. In their more recent work, Paarlberg, Lee and Seitzinger (2003) found that some producers may actually benefit from an outbreak, which somewhat offsets the decrease in producer surplus for those with quarantined cattle. The extent that individual groups bear animal disease losses is an important question, and the conceptual model presented in this paper distinguishes between losses felt at differing levels of the marketing channel that might otherwise be aggregated as total losses to producer surplus.

In short, there is evidence that outbreak shocks resonate through many industry sectors, so that risk management tools focused primarily on animal production aspects may also have far broader effects. Moreover, timing, geography, size and relative shares of impacted producer and consumer populations are all important variables to consider in estimating expected losses.

Consumer Demand Responses to Animal Health Outbreaks

The previous literature emphasized losses to producers as a result of an animal disease outbreak, but did little to examine how consumers' demand for agricultural products might change due to an outbreak. The following section examines consumer response to an animal disease outbreak focusing primarily on BSE.

Consumer response to BSE estimates is derived primarily from analyses of the recent UK outbreak (Ashworth and Mainland, 1995; Verbeke and Ward, 2001; Henson and Mazzocchi,

2002; Thompson and Tallard, 2003; and Pennings, Wansink and Meulenberg, 2002). These studies estimated the decrease in demand using various methods, including response to media and the timing of the event. Thompson and Tallard looked specifically at the long-term effects on consumption and demand, and found that European consumption (in volume terms) appears to have recovered, but with significantly lower prices, and current trends would suggest long-term demand reductions approaching 25% of the original value. The effects on distributors, retailers and food service sectors will, subsequently, be quite large.

Although *E. coli* is not an animal disease outbreak per se, consumer responses to this food safety issue might give insight on the extent of consumer response to disease outbreaks. Schroeder, Marsh and Mintert (2000) estimated that beef recalls in 1993 caused a 1.6% decline in demand. In more recent years, Schroeder, Marsh and Mintert (2000) estimate that the loss to the industry from recalls between 1993 and 2002 was over $1.5 billion. In short, these allied sector impacts from consumer response to an outbreak may overwhelm simple losses due to weak sales: subsequently, losses will multiply as the agricultural sectors considered expand.

Consumers are becoming increasingly responsive to food safety information, suggesting that animal disease outbreaks will have significant consequences in the food production industry on the demand as well as the supply side. Henson and Mazzocchi (2002) found that returns to equities in the beef, pet food and animal feed industries (and to a lesser extent dairy) were significantly hurt by the BSE scare in the UK during 1996, which was only partially offset by positive returns to other meat sectors. In terms of indirect impacts, several meat processors also recorded substantial losses in 1996, but dairy processors fully recovered by the end of the first year. The relatively large impacts on animal feed, pet food and processing businesses again suggest that indirect impacts to allied agribusiness sectors are an important element of any analysis.

The previous literature has illustrated several important points. First, the immediate impacts of an animal disease outbreak at the farm level are significant, but may be exceeded by indirect effects to businesses in the meat marketing channel and impacts on industries that are more loosely associated with agricultural production (e.g. tourism, rendering, veterinary services and supplies). Second, the impacts are not limited to suppliers as consumer demand might be altered in response to an outbreak. Third, outbreaks are intertemporal: that is, immediate impacts are felt first, but changes in productive capacity and consumer response may persist. Each of these make the design of an effective livestock insurance programme more complex, because markets are very integrated across product form and space, and any risk management product that changes price signals and incentives to producers will likely impact allied sectors.

Methods for Measuring Animal Health Impacts

The conceptual framework presented in this chapter builds on the previous literature and extends it by considering specific relationships that would characterize broader effects in the marketing channel. Central to the model are the production relationships that link growth and development stages of livestock with the slaughter and fabrication of meat and meat products. In addition, economic relationships link each production stage by allocating scarce resources according to price signals. Important inputs in the model are the exogenous production shocks following a disease outbreak, and an important output is livestock and meat product quantities and prices reflecting market-clearing conditions at each marketing channel stage. These linkages, and other important relationships, are illustrated for the meat product industry in Fig. 17.1, and a brief description of elements of Fig. 17.1 follow.

Seed Stock Stage

The seed stock stage is the initial production stage of the meat marketing chain. In this stage livestock inputs are conceived, gestated and grown until ready for a feeding stage. Seed stock producers are relatively dispersed geographically in the USA, with more regional concentration in the poultry sector relative to cattle and hogs. In pork production, the seed stock stage is within the farrowing to weaning phase; in beef cattle the seed stock stage includes cow/calf producers and yearling stocker cattle.

If an animal disease outbreak occurs in this stage of production alone, supplies to the next stage may be interrupted and reproductive cycles delayed. Disease interruptions are of two types: the stock of breeding animals might be reduced, constituting an overall reduction in productive capacity, with restocking times varying by species; or alternatively, the flow of feeding animals may be delayed or interrupted.

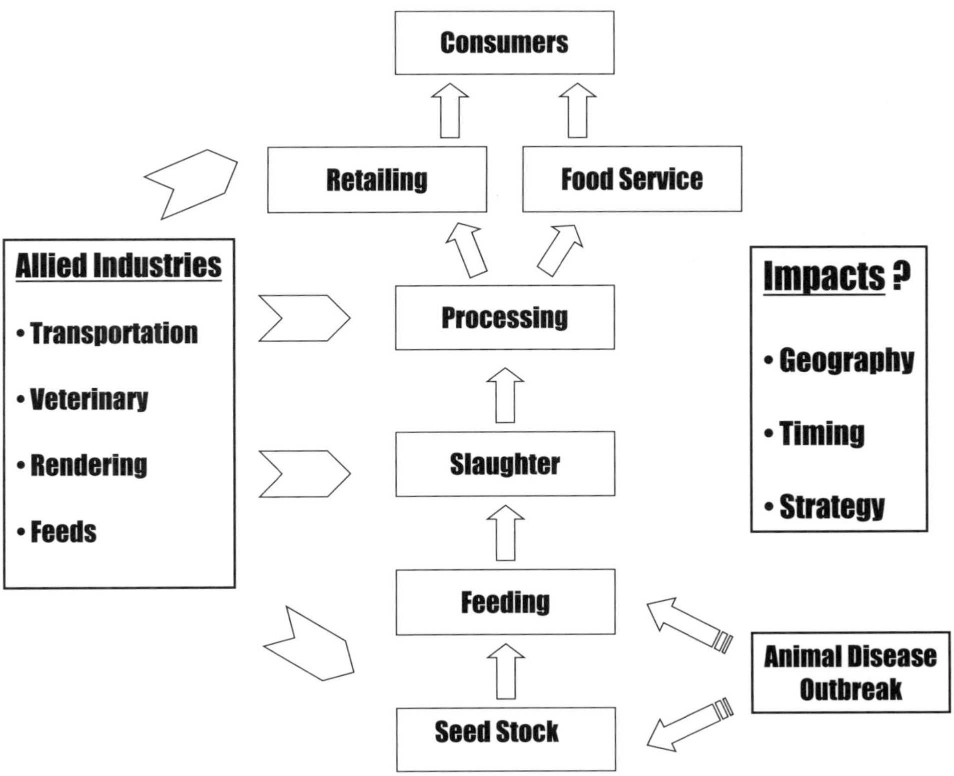

Fig. 17.1. Proposed Model of the Meat Products Industry.

Feeding Stage

Once meat animals have gained sufficient weight, they enter into a feeding stage that prepares them for slaughter. While the seed stock stage may be geographically diverse for some species, the feeding stage tends to be more regionally concentrated. Concentration occurs as it is generally cheaper to ship livestock than feed because the weight of the animals is less than that of the feed consumed. Further, the technical aspects of livestock feeding reward specialization as capital costs and labour costs per unit decline with size. As a result, particular regions tend to specialize in feeding operations and livestock are transported to those regions. As an example, beef cattle production tends to be concentrated in the Great Plains of the USA, while production has shifted to the West for dairy cattle and to the Midwest and Southeast for hogs (Shields and Mathews, 2003).

Regional concentration of feeding operations has important implications when studying animal disease impacts. First, transport of livestock from one region to another may accelerate disease transmission. Animal disease control strategies include the quarantine and non-transport of livestock that disrupts flows of inputs from the seed stock stage to the feeding stage. Disruptions in the flow of seed stock to the feeding stage will be reflected in higher prices for feeder livestock, and regional differences in feeder prices may be greater than the cost to transport animals, indicating regional markets are no longer integrated.

Shocks or disruptions from animal disease can be quantified in the technical relationship that transforms feeder livestock into its slaughter stage. As an example, a technical relationship can be estimated that indicates how feeder cattle supplies are grown to slaughter cattle supplies depending on such variables as feed efficiency, average daily gain, the time of year, mortality rates, etc. When an animal disease outbreak occurs, important technical relationships such as feed efficiency are altered (either due to disease impacts to animal health or indirectly when feeding policies and protocols might change). As a result, feed efficiency can be used as a control variable in an animal disease outbreak that can be adjusted to mimic disease impacts. Disruptions in the flow and number of animals caused by a change in the control variables are reflected in fed cattle prices that link the feeder and slaughter stages of the marketing channel.

Slaughter Stage

In the slaughter stage, animals are harvested and fabricated into boxes of primal and subprimal cuts of meat. In the beef industry, cattle carcasses are generally shipped in boxes of subprimal cuts such as top rounds, tenderloin and sirloins, collectively known as boxed beef. Purveyors or processors buy these meat cuts and transform the raw products into a form meeting customer needs. Recent trends have led slaughterers to further process meats until they are ready to be marketed as case-ready meats in grocery stores and supermarkets.

Slaughter prices reflect the relative supply and demand for livestock carcasses, while wholesale prices reflect the demand for processed meat products. The value added (defined as sales minus cost of goods sold) for this stage of the marketing channel totalled more than $39 billion in year 2000 (US Census Bureau, 2002) (See Table 17.1).

Because slaughter facilities are concentrated in relatively few regional locations, animal disease control strategies that limit livestock transport may produce substantial increases in costs for these businesses. In terms of insurance products, the processors would likely have interest in their own risk management portfolios, since animal flow constraints put them at risk of being below "capacity", thereby hindering profitability.

Wholesalers

While some processing occurs at the slaughter stage, wholesale purveyors of meat products also provide ready-to-eat meals, case-ready meats and frozen products to retailers. Average wholesale beef prices are indicators of the scarcity of meat products that journey to the retail market. In 1997, wholesale meat trade had sales totaling more than $57 billion (see Table 17.2). Firms that comprise this stage of the marketing channel include Sysco Corporation, US Food Service and Performance Food Group. Because the stock in many of these firms is publicly traded, share price reductions relative to general market valuations may be used as a proxy for business losses associated with animal disease.

Retailing and Food Service

The retail meat trade represents the end point for the meat-marketing channel. The retailing sector can be divided into two segments: retail food establishments (i.e. grocery stores and supermarkets) and food service establishments (i.e. restaurants, distributors, institutions). Retail food establishments, including specialty food stores and meat markets, grossed more than $107 billion in sales in 1997 (US Census Bureau, 1997) (See Table 17.3). The largest

Table 17.1. Slaughter & Manufactured Product Value.

Sector	Value Added ($1,000)
Dog and Cat Food Manufacturing	$4,355,843
Animal Slaughter [a]	$10,059,286
Meat Processing	$10,142,789
Poultry Processing	$13,846,024
Rendering	$1,067,940
Total	**$39,471,882**

[a] Excluding Poultry

Table 17.2. Wholesale Meat Trade.

Commodity Line	Sales ($1,000)
Beef	$28,139,621
Veal	$1,229,479
Lamb	$1,179,114
Pork	$10,708,940
Frozen Meat	$16,043,971
Total	**$57,301,125**

Table 17.3. Retail Trade.

Establishment	**Meat Product Sales ($1,000)**
Grocery Store	$50,133,442
Supermarkets	$49,613,711
Convenience	$519,731
Specialty Food	$3,971,253
Meat Markets	$2,957,543
Total	**$107,195,680**

retailers in terms of sales include The Kroger Company, Albertson's Inc., Safeway Stores, and Wal-Mart Supercenters (Kaufman, 2002).

Retailers make meat-pricing decisions within the context of many other items sold to consumers. However, retailers are not immune to the effects of supply shortages, and may absorb some of the costs of locating and segregating safe food supplies. In addition, retailers will be the first to face the effect of consumers' concerns regarding animal disease, possibly facing a more stringent regulatory climate for labelling, product traceability or other certifications. As a result, retail meat prices are likely to reflect, to a certain extent, any scarcity of meats due to production shocks, as well as the consumer response to outbreaks (and the potential for increased regulatory compliance costs).

Additional Indirect Effects of Animal Disease Outbreaks

In addition to firms that handle meat products directly, allied businesses that provide inputs to the meat-marketing channel (feed, veterinary, transportation and rendering sectors) will find revenues disrupted as a result of a disease outbreak. So, once again, any behaviour influenced by livestock insurance will indirectly influence their resulting business activity. The following is a synopsis of the effects on these sectors:

Feeds

With a reduction in the supply of animals, smaller quantities of feed and feed supplements will be needed. In contrast, if animal disease leads to a reduction in feed efficiency (more feed needed for each pound of gain), additional feedstuffs will be bid away from other sectors to support livestock feeding. Furthermore, if livestock transportation is restricted, regional feed price relationships will be altered. Secondary effects will occur both for wholesalers of processed feeds and to the purchasers of feed ingredients (e.g. maize, soybean meal). The value of feed mixed on-site was $6 billion in 1997, and the value of prepared feeds purchased off-site was more than $2 billion (Table 17.4). According to the USDA's World Agricultural Supply and Demand Estimates, roughly 50% of maize supplies and more than 50% of soybean supplies are used in livestock feeds (World Agricultural Outlook Board, 2002).

Table 17.4. Wholesale Livestock and Feed.

Enterprise	Sales Value ($1,000)
Livestock	$23,963,277
Feed (mixed on-site)	$6,192,031
Prepared Feeds	$2,182,161
Hay, Straw, Alfalfa	$169,153
Total	**$32,506,622**

Veterinary Services and Medicines

An animal disease outbreak will necessitate an increase in veterinary services such as testing and vaccinations, but, if sufficient numbers of livestock die or are destroyed in terms of remediation, there may actually be a reduction in long-run demand for other veterinarian services.

Transportation

Relatively little research has focused on the transportation of livestock and meat products with the notable exception of the recent report by Shields and Mathews (2003). The report outlines the principal geography of production by species, discusses the economics of livestock transport, and outlines the geographic distribution of each production stage. As noted by the authors, livestock kept in a single location may be shipped to another as they enter another productive stage, and shipped to yet another location for slaughter and processing. As discussed previously, livestock movement is often dictated by the relative cost of feed to the cost of transport, often because shipping livestock is an efficient use of feed and forage supplies that vary by region and by season.

Certainly, a reduction in live animal transportation will occur with an animal disease outbreak either due to restrictions on the transport of livestock or due to livestock mortality. If livestock movements were halted to reduce the transmission of a disease outbreak, one would expect higher costs for producers who are unable to move livestock to be marketed. These costs might include production losses due to overcrowding, extra feed costs, increased management and overhead costs, increased slaughtering costs at smaller, regional processing plants and increased veterinary costs (Shields and Mathews, 2003).

Following discussion of animal movements, it is clear that incorporating geography and a spatial dimension into an animal disease outbreak model is important for quantifying potential losses due to an outbreak, and for measuring the benefits of various control measures. While little data exist on livestock movements, the USDA Economic Research Service has developed a database of 2001 interstate livestock travel based on shipment data and animal health certificates issued by veterinarians, as tabulated by state departments of agriculture. This database may be useful in representing the geographic impact of animal disease outbreaks. Essential information includes an estimate of the reduction in distance traveled, the revenues generated per loaded mile and an assumption about the duration of the animal disease event.

Rendering

The rendering industry will also be impacted by a disease outbreak in several different ways. In the first case, disposal of diseased animals may represent a challenge for the industry depending on the size of the outbreak, the associated mortality rate and the nature of the outbreak. If diseased animals may be used in traditional rendered products, total revenues may increase or decrease depending on how rendered product prices change relative to the increase in supply. However, if diseased animals cannot be used in rendered products, or if transportation restrictions stop the flow of rendered inputs, rendered product outputs could fall. The ultimate impact is an empirical question depending on the initial assumptions of a disease outbreak.

From Sectors into Modelling – Summary

The previous discussion outlines important sectors of the meat products industries as well as the economic and the productive relationships interlinking various sectors. Since the development and use of livestock insurance will directly impact production stages, we can focus on shocks at that level, and then determine how insurance products influence upstream and downstream industries. As illustrated in Fig. 17.1, animal disease effects are principally felt at two production stages, the seed stock stage and the cattle feeding stage. Shocks to production are of two types, a reduction in the flow of animals and decreased productive capacity of the breeding livestock population. Ripples from these shocks are felt through the rest of the supply chain and the allied industries, but the degree of transmission is dependent on unique relationships linking the animal products sector.

As an example, suppose that an outbreak of FMD occurs at the cattle feeding stage. The output of this production stage, fed cattle, depends on technical relationships between the number of cattle placed on feed, the feed efficiency of cattle, the average daily gain of cattle, the days on feed and death loss to name but a few. An FMD outbreak will decrease feed efficiency and increase death loss either through outright mortality or slaughter due to loss in economic viability. In a simulation exercise, the technical relationship between feed efficiency and the fed cattle output can be altered to approximate the impact of FMD on fed cattle supplies. The shift in fed cattle supplies will subsequently impact fed cattle prices, wholesale meat prices and retail meat prices according to the estimated economic relationships in the system.

If technical coefficients such as feed efficiency and mortality rates are control variables used to simulate an animal disease outbreak, it becomes important to identify the criteria by which these variables will be altered. It is clear that the spread of animal disease in the USA will be affected by geographic production stages, the movement of livestock within and across these regions, and imports of live animals from foreign countries. Therefore, specific geography, timing and strategy scenarios will be evaluated to broaden the scope of results. Thus, stochastic disease outbreak elements that should be considered when estimating disease impacts include the following:

Geography – the location of the outbreak, size of the affected area, animal density and frequency of animal movements are important variables in determining economic loss.

Timing – the duration between the initial outbreak and pathogen recognition/response by animal health officials will be important determinants of economic loss. Losses vary with the time between introduction and response.

Strategy – the strategy employed to contain and respond to the outbreak will influence the degree of economic loss. In this case, insurance is one way to promote early communication by producers that increases the probability of effective control, but may decrease incentives to manage disease aggressively.

Broader Economic Impact Animal Disease Models

The previous literature is a heuristic discussion of complex disease impacts; subsequent discussion focuses on modelling the economic magnitude of impacts. In Fig. 17.2, a matrix is presented that describes various modelling approaches. The rows of the matrix represent the level of analysis ranging from sector level models to the national level. The columns summarize selected characteristics including common research objectives, typical methods to assess related objectives, policy instruments used to manage impacts, and potential research opportunities to improve methods of analysis and their subsequent estimates.

Characteristics of complete meat sector studies are found in the second row of Fig. 17.2. Typically, these studies are concerned with the industry losses from an animal disease event and subsequent regulations. Economic impacts are quantified with simulation models based on previous estimates of demand and supply response (i.e. elasticities), market clearing conditions, and technical relationships capturing the transformation of live animals to meat products. The previously cited Department of Environment, Food and Rural Affairs and Department of Culture, Media and Sport (2002) study in the UK suggests a £3.1 billion impact on agriculture and the food chain from FMD in 2001, of which £1.9 to £2.3 billion are attributed to agribusiness sectors beyond the farm gate. In Canada, Leroy and Klein (2003) chronicle the substantial short-term costs to federal and provincial agencies of responding to a single case of BSE, including the cost of indemnifying cattle feeders against large business losses.

Research suggests that the structure of the meat products sector is changing with the advent of more value-added or ready-to-eat meals, which will impact how disease shocks to commodity prices and food cost changes are transmitted through the marketing channel. Paul and MacDonald (2003) find that disembodied technical change, likely the result of value-added or ready-to-eat meals, has reduced the demand for agriculture inputs relative to other marketing inputs resulting in weaker impacts of farm level shocks on food prices. However, improving quality and real price declines of agricultural inputs has encouraged greater use of these inputs. When computing economic losses, sector analysis of animal disease impacts should account for these trends. This is an example of how agribusiness research might be integrated into subsequent analyses of economic impacts from animal disease events.

Recent models of antibiotic removal from livestock feed have used a sector approach. As an example, Hayes *et al.* (1999) estimated the likely economic effects of a subtherapeutic ban of antibiotics on the US pork industry by combining technical assumptions in pork production with economic relationships in a meat sector model developed by Buhr (1993). This type of modelling allows for both stock adjustments in livestock breeding herds as well as the flow effects associated with exogenous shocks. Buhr and Kim (1997) use a similar approach in their examination of dynamic adjustment in vertically linked markets.

Dynamic models of the meat market sector provide a venue for examining both the technical and economic impacts of animal disease. Additional insight may be gained as epidemiological models are integrated with sector modelling, as these sector models explicitly consider market structure.

Scope of Analysis	Research Objectives	Assessment Methods	Policy Instruments	Research Opportunity
Sector	Industry Losses	Simulation Efficiency Estimation	Traceability Certification	Post-Harvest Models Dynamic Models Epidemiological Links Market Structure, Distribution
Regional	Welfare Impact Industry-Specific Loss Inadvertent Loss	Input-Output Models Computible General Equilibrium Models	Travel Restrictions Compensation Prescribed Cull	Economic Geography Linking Epidemiological and Economic Mitigation and Prevention Costs
National and International	Welfare Impact Distribution of Loss	Partial Equilibrium Computible General Equilibrium Models	Regionalization Rapid Response Plans National ID Tariffs/Non-Tariff Barriers Restrictions	Economic Geography Distribution of Impacts

Adapted from Pritchett, Thilmany and Rosenstiel (2004).

Fig. 17.2. Matrix of Modelling Approaches with Broader Economic Effects as a Focus.

Regional Level Effects: Impacts Across All Industries

Economic losses in an animal disease outbreak are not limited to the meat sector and its allied agribusinesses. Rather, entire regions may suffer from the outbreak due to multiplier effects associated with lost sales and wages, and because disease mitigation strategies such as travel restrictions or consumer response also alter non-meat business activity (See row 3 in Fig. 17.2).

As mentioned previously, the impact of the 2001 FMD outbreak in Britain totalled £3.1 billion on agriculture and the food chain, but regional impacts were higher. Blake, Sinclair and Sugiyarto (2002) extended the discussion of broader impacts by noting that lost tourism represents an even greater impact on a region's GDP losses because the economic impact multiplier on tourism expenditures is much higher than for farm level production. Similarly, Ekboir (1999) estimates $13.5 billion of potential losses to an FMD outbreak in California. These estimates are generated using input-output analysis and include direct losses to livestock producers, disease control costs including depopulation and indirect or imputed losses to businesses. Input-output models are often used to generate regional economic impacts, and quantification relies on multipliers, which may be imperfect, but could be refined with analyses done at other levels of this matrix. For example, attention to potential travel restrictions, the size and distribution of relevant firms (i.e. economic geography) and the costs of disease prevention/mitigation strategies are opportunities to supplement baseline analysis.

National- and International-Level Effects

Due to the spillover effects of animal disease, disease mitigation strategies and the economic size and distribution of impacts are often computed at the national level. National studies include both domestic effects and the extent to which trade flows are altered as a result of the disease outbreak. In many instances, national data are readily available; yet, these data may not be easily disaggregated to consider sub-national disease management policy tools such as regionalization.

In addition to previously cited studies of national/international impacts, Jin, Skripnitchenko and Koo (2004) consider the ex ante effects of the US outbreak of BSE. The authors develop scenarios around decreased domestic consumption ranging from 5% to 20%, and decreased export scenarios ranging from 50% to 100%. Using previously published price elasticities, the authors' simulation results indicate the price of beef could decrease by 15%, while the price of beef substitutes would increase by about 3% as consumers switch consumption to pork and poultry. The price of fed cattle is expected to decrease 13.5% and the price of feeder cattle would decrease about 16%. The authors' base simulation findings on a single case of BSE, and do not consider potential productivity shocks due to culling, or the increased costs of new disease prevention measures.

Future Directions

Understanding how an animal disease event will impact the animal products marketing channel is a complex, multidisciplinary problem. Moreover, connecting the role of livestock insurance programmes to the potential frequency and magnitude of such events is even more difficult. Still, an accurate assessment of losses due to animal disease is useful for policymakers who may weigh these potential losses against the public costs of insurance that may partially manage these losses. It is clear that models that provide the most comprehensive assessment of potential losses are most useful to decision makers (US-GAO, 2002), and this chapter discusses the broader economic impacts of animal disease outbreaks, and also synthesizes approaches for quantifying economic impacts.

Several future directions exist for animal disease studies at each of the various market levels, which can subsequently feed better baseline data to broader sector, regional and national analyses. Similar to the empirical data needs for insurance product development, analysis of animal disease impacts requires a system of economic relationships that accounts for the interdependencies and degree of response (i.e. elasticities) among the various sectors of the economy, and which allows for sensitivity analysis of the magnitude and incidence of the initial animal health shock. Interdisciplinary work should encourage the merger of sophisticated epidemiological models used to trace the growth and demise of disease outbreaks and economic models that capture the technological and economic relationships linking stages, potential structural change and performance of the marketing channel.

As policymakers explore strategic responses to managing the risk associated with animal disease, the distributions of losses, policy costs and programme benefits become particularly important, as Paarlberg, Lee and Seitzinger (2003) conclude in their work. An important factor in the development of insurance or policy for animal health is that some individuals actually fare better after an animal disease outbreak, such as producers who are not quarantined, or consumers who are uninfluenced by animal disease outbreaks (and who are able to buy at lower prices), an issue for future researchers to consider.

The spatial dimension of animal disease also deserves additional attention. Too often data limitations prevent analysis of spatial economics when evaluating outbreak scenarios. The National Animal ID system allows for space to be added as a dimension for analysis, but the location, geographic distribution and movement of animals must be linked to probability assessment and economic data (e.g. market prices) to show the full effects.

References

Ashworth, S.W. and Mainland, D.D. (1995) The economic impact of BSE on the UK beef industry. *Outlook on Agriculture* 24, 151-54

Blake, A., Sinclair, M.T. and Sugiyarto, G. (2002) The economy-wide effects of foot and mouth disease in the UK economy. Working Paper, Nottingham University Business School, Nottingham, UK.

Buhr, B.L. (1993) A quarterly econometric simulation model of the US livestock and meat sector. Staff Paper P93-12, Department of Agricultural and Applied Economics, University of Minnesota, Minneapolis.

Buhr, B.L. and Kim, H. (1997) Dynamic adjustment in vertically linked markets: the case of the US beef industry. *American Journal of Agricultural Economics* 79, 126-138.

Department of Environment, Food and Rural Affairs and Department of Culture, Media and Sport (2002) *Economic Cost of Foot and Mouth Disease in the UK.* London, UK.

Ekboir, J.M. (1999) *Potential impact of foot-and-mouth disease in California.* Agricultural Issues Center, Division of Agriculture and Natural Resources, University of California, Davis.

Evans, E.A. (2003) Economic dimensions of invasive species. *Choices,* Second Quarter, 5-9.

Hayes, D.J., Jensen, H.H., Backstrom, L. and Fabiosa, J. (1999) Economic impact of a ban on the use of over-the-counter antibiotics. Staff Report 99 SR 90, Iowa State University, Ames, Iowa.

Henson, S. and Mazzocchi, M. (2002) Impact of Bovine Spongiform Encephalopathy on agribusiness in the United Kingdom: results of an event study of equity prices. *American Journal of Agricultural Economics* 84, 370-86.

Jin, H., Skripnitchenko, A and Koo, W. (2004) The effects of the BSE outbreak in the US on the beef and cattle industry. Special Report 03-4. Center for Agricultural Policy and Trade Studies. Department of Agribusiness and Applied Economics. North Dakota State University, Fargo, North Dakota.

Kaufman, P. (2002) Food Retailing. In: Harris, J.M., Kaufman, P.R., Martinez, S.W. (coordinator) and Price, C. *US Food Marketing System 2002.* US Department of Agriculture, Economic Research Service. Agriculture Economics Report-811, pp. 21-33.

Leroy, D. and Klein, K. (2003) Apocalypse cow: the effect of BSE on Canada's beef industry. *Western Economics Forum* 2, 2: 20-27.

Paarlberg, P.L., Lee, J.G. and Seitzinger, A.H. (2002) Potential revenue impact of Foot-and-Mouth Disease in the US. *Journal of the American Veterinary Medical Association* 220, 988-992.

Paarlberg, P.L., Lee, J.G. and Seitzinger, A.H. (2003) Measuring welfare impacts of an FMD outbreak in the US. *Journal of Agriculture and Applied Economics,* 35, 53-65.

Paul, C.M. and MacDonald, J. (2003) Tracing the effects of agricultural commodity prices and food costs. *American Journal of Agricultural Economics* 85, 663-646.

Pennings, M.E., Wansink, B. and Meulenberg, M.T.G. (2002) A note on modeling consumer reactions to a crisis: the case of the mad-cow disease. *International Journal of Research in Marketing* 19, 91-100.

Pritchett, J., Thilmany, D. and Rosenstiel, K. (2004) Animal disease economic impacts: a survey of literature and typology of research approaches. Working Paper, Department of Agricultural and Resource Economics, Colorado State University, Fort Collins, Colorado.

Schroeder, T., Marsh, T. and Mintert, J. (2000) Beef Demand Determinants: a research summary. Working Paper, Department of Agricultural Economics, Kansas State University, Manhattan.

Serecon Management Consulting, Inc. (2002) Economic impacts of a potential outbreak of foot and mouth disease in Canada. Report for the Canadian Animal Health Coalition, Calgary, Alberta.

Shields, D. and Mathews, K. (2003) Interstate livestock movements. US Department of Agriculture, Economic Research Service, Livestock, Poultry and Dairy Report LDP-M-108-01.

Thompson, W. and Tallard, G. (2003) Consumption versus demand: recovery after BSE? *EuroChoices* Spring, 24-25.

UN-FAO (2001) The state of food and agriculture 2001. United Nations Food and Agriculture Organization, Rome, Italy. [Accessed 2004.] Available from http://www.fao.org/docrep/003/x9800e/x9800e14.htm.

US Census Bureau (1997) Economic Census. Wholesale Trade Subject Series. Washington, D.C.

US Census Bureau (2002) Annual Survey of Manufacturers, Washington, D.C.

US General Accounting Office (2002) Invasive species: clearer focus and greater commitment needed to effectively manage the problem. Report GAO-02-1, Washington, D.C.

Verbeke, W. and Ward, R.W. (2001) A fresh meat almost ideal demand system incorporating negative TV press and advertising impact. *Agricultural Economics* 23, 359-374.

World Agricultural Outlook Board (2002) *World Agricultural Supply and Demand Estimates.* US Department of Agriculture, Washington, D.C.

Chapter 18

US Livestock Industry's Views on Livestock Disease Insurance

Jennifer L. Grannis, John W. Green and Megan L. Bruch

Introduction

In 2002, a research project supported by the US Department of Agriculture's (USDA) Risk Management Agency (RMA) and conducted by the USDA Animal and Plant Health Inspection Service (APHIS) investigated the potential for livestock disease insurance to improve animal disease risk management. Areas of particular importance to determining the feasibility of livestock disease insurance included livestock disease prevalence, animal concentrations and distributions, regulations related to animal disease, international approaches to livestock disease risk management, existing insurance products, and producer interest in livestock disease risk management tools, especially insurance. This chapter reports on producer interest in insurance.

Major livestock industry groups including the National Cattlemen's Beef Association (NCBA), National Milk Producers' Federation (NMPF), Delmarva Poultry, Inc. (DPI), American Sheep Industry Association (ASIA) and the National Pork Board (NPB) were contacted in the summer of 2002. Each of these industry groups participated in listening sessions with APHIS professionals to share their views about livestock disease insurance. In addition to the listening session, an informal discussion worksheet was provided to each group, though not all groups completed and returned the worksheets. Some industry groups completed the worksheets during the listening sessions and others were collected later from the group's membership. The results of these informal discussions cannot be considered representative of the opinion of the entire livestock industry.

The worksheet and discussion were designed to determine the importance of livestock disease as a peril to livestock production, current practices to manage disease perils, and perceptions about government and individual responsibilities in managing disease risk. Past research has provided some insights into risk management perceptions and practices by livestock producers. Boggess, Anaman and Hanson (1985) found that diseases and pests were a very important source of risk in livestock production for farmers in Florida and Alabama. Hall *et al.* (2003) found that beef producers felt that on-farm disease prevention was the most significant strategy that could be employed to reduce the risk to animal health. Twice-weekly herd inspections were also found to be important part of a health management strategy. Patrick *et al.* (2000) found that 70% of pork producers reported that disease in hogs had a relatively large effect on income. Pork producers also rated isolating new breeding stock, using an all-in/all-out system and the routine use of antibiotics and vaccines as relatively effective practices to manage on-farm disease risk.

The information from the listening sessions was used to guide research into particular insurance products and was incorporated into recommendations regarding insurance products with the best probability for industry acceptance. In general, the producer groups were favourably disposed to the idea of livestock disease insurance, but the cost of insurance and the ability for insurance to complement government indemnity programmes were significant concerns to these producers. The listening sessions indicated that disease is not the most important peril faced by livestock producers, but disease is a peril that can cause significant losses. Producers are also not regularly incorporating disease risk management into their day-to-day production practices. While producers had similar concerns, varying industry characteristics indicated that multiple types of insurance products could be necessary to reduce risk in different industry segments and for specific species.

For instance, poultry producers raising broilers under contract were interested in revenue insurance that covers disease as a peril. Beef producers, conversely, were opposed to revenue insurance because of potential market distortions (i.e. excess production). Beef producers believed that a more general policy (perhaps an all-farm policy) could include riders that would address their risk management needs. Pork producers desired coverage against the losses of catastrophic diseases, especially the consequential losses that might result from quarantine and trade interruptions. Beef producers concurred with the assessment that quarantine and trade losses from disrupted markets and decreased prices were perils that required insurance coverage. Sheep and dairy producers were interested in disease insurance, but were not specific about policy characteristics. The following sections summarize findings of the listening sessions held with each industry.

Beef Cattle

Four beef producers and two members of the NCBA staff were consulted at NCBA's 2002 summer meeting in Reno, Nevada. Producers specified potential revenue losses in export markets as the biggest risk of livestock disease restrictions, although concern was also stated for lost domestic market opportunities. Producers were not interested in insurance covering currently identified and managed production (endemic) diseases. However, insurance coverage was favoured in the event of the emergence of a new disease or from increased virulence associated with endemic diseases.

Producers were concerned about losses resulting from disease problems on farms adjacent to their operations or in neighboring countries (e.g. Canada and Mexico) and how insurance products might be structured to minimize risk resulting from these scenarios. Producers stated that no industry-wide biosecurity programme exists because the cost of implementation is high; this situation is not expected to change in the near future. Regionalization or movement restrictions could prevent cattle from reaching slaughter plants and insurance protection was desired against these losses. Market valuation for payment of indemnity was also a concern. The lack of a futures market for some segments of the cattle industry, the length of the futures contracts that do exist, and the significant independent variation between cash and futures market (or basis risk) were all specified as important components of effective policies.

Industry participants felt that insurance policies need to address three categories of animals: animals infected by disease, animals not infected by disease but impacted by the presence of diseased animals (exposed or quarantined animals), and depopulated animals. It was also desired that insurance products be kept simple and focused on specific risks in targeted species. Because of the great diversity in cattle production practices, policy riders to address issues specific to each production segment (cow-calf, feeder cattle, etc.) were favoured by producers. Participants believed that producers should pay for risk protection and believed that if government subsidized policy premiums, overproduction and moral hazard problems would result. Mid-size and smaller producers would not support mandatory insurance and biosecurity management practices as a prerequisite to obtaining bank credit approval. Participants agreed that commercial lending requirements are not the responsibility of APHIS or any other USDA agency.

Dairy Cattle

The Animal Health Committee of the NMPF participated in a listening session with APHIS in October 2002 in St Louis. In addition to discussion at the meeting, four producers provided written comments describing industry perils and the desired structure of risk management tools. The listening session participants did not express clear opinions about insurance products for dairy operations. They were more interested in learning about specific options that might become available. The NMPF members indicated interest in livestock disease insurance during the listening session, though written responses showed some hesitancy to embrace insurance. Again, neither the listening session nor the written survey responses should be considered representative of all dairy producers.

Producers were asked to prioritize the five most important perils facing their operations. Table 18.1 lists the perils included, the number of times each was mentioned and the weighted average rank of each peril. Economic viability, including concern about low milk prices and high input prices, topped the list. Mycotoxins in feed were also listed as an important peril. Natural disasters were important but less important than disease.

Producers were also asked to rate the significance of the top perils based on a scale of strongly disagree (1) to strongly agree (5). The perils and the ratings are presented in Table 18.2 for economic perils, mycotoxins and natural disasters. Disease as a peril was not rated highly by these producers. Producers agreed that the top perils have all caused significant economic losses to their operations. Economic perils and natural disasters do not occur

Table 18.1. Most Important Perils Facing Dairy Operations.

Peril	Number of Times Listed*	Weighted Average Rank**
1 = Most Important and 5 = Least Important		
Economic Viability (Milk and Input Prices)	6	2.17
Mycotoxins in Feed	3	3
Natural Disasters (Weather/Drought/Fire)	3	3.33
Disease	2	2.5

* A single respondent listed some perils more than once.
** Weighted Average Rank =(Σ(rank × number of times listed in rank)) / total number of times listed)

regularly but problems with mycotoxins occur at least annually. Producers did not agree that better management practices could reduce costs associated with these perils. Respondents did not believe that government should be responsible for covering losses caused by mycotoxins but agreed that government should intervene when losses occur from economic perils or natural disasters. Commercial insurance is not available for these perils, except for some protection from natural disasters such as fire and wind.

Producers were also asked to list diseases of greatest concern to their operations. Bovine viral diarrhoea (BVD), FMD, Johne's disease, mastitis and mycotoxins were each listed by at least two respondents. Tuberculosis, coccidiosis, foreign animal diseases, foot problems, metabolic diseases, pneumonia and salmonella were each listed once. These diseases cause a variety of losses to dairy operations, including reduced output, loss of markets and revenue, and downtime. Quarantine, disinfection costs, restart costs and lameness were stated as sources of losses caused by disease.

Producers were also asked to rate statements about disease incidence and preparedness (Table 18.3). Dairy producers are concerned that a catastrophic or foreign animal disease will occur in the USA and do not think that government indemnity programmes provide sufficient protection against losses from these events. Producers do not have business plans in the event of a major disease outbreak, although they do practice biosecurity. The respondents wanted a choice of disease insurance products for use in managing disease risk.

Table 18.2. Significance of Top Perils Facing Dairy Production.

Statement/Question	Average Rating *1 = Strongly Disagree and 5 = Strongly Agree*		
	Economic	Mycotoxins	Natural Disasters
1. This peril has caused significant economic losses in this industry.	4.3	5	4.3
2. This peril occurs on a consistent basis (annually, semi-annually, every 5 years).	Not consistent	Semiannually or annually	Not consistent
3. I have records of losses caused by this peril on my operation.	5	1.5	5
4. Management practices can affect the costs associated with this peril.	1.7	2.5	1.3
5. Government should be responsible for covering losses caused by this peril.	3	2	3.7
6. I can obtain commercial insurance for this peril.	1.67	1	4
7. Do you buy commercial insurance for this peril?	No	No	Yes (Wind/Fire) No (Drought/ Weather)
8. When this peril occurs, losses are:	Moderate-Severe	Moderate-Severe	Severe

Table 18.3. Dairy Producer Sentiment.

1 = Strongly Disagree and 5 = Strongly Agree	Rating	
	Average	Range
1. I am very concerned that a catastrophic or foreign animal disease will occur in the USA.	4.75	4-5
2. I have a business plan that includes business interruption, cleaning, disinfecting and labour costs that I will follow in the event of a major disease event.	1.75	1-4
3. Government indemnity programmes provide sufficient protection against losses from catastrophic disease events.	2	1-5
4. I currently incorporate management practices to reduce the impact of disease on my livestock.	3.25	1-5
5. I think that other producers would manipulate conditions to increase insurance payments should disease infect their livestock.	2.5	1-4
6. My operation practices biosecurity to protect against disease (endemic or catastrophic) introduction in my livestock.	4.5	3-5
7. I want a choice of livestock insurance products to reduce my disease risk.	4.5	3-5

Poultry

The listening session for poultry producers was conducted in August 2002 in Salisbury, Maryland with DPI. Eleven industry members holding contracts to produce broilers attended the session. Growers are directly affected by disease in their managed flocks and indirectly affected by disease infections in chicks from contracted operations (e.g. replacement chicks). Producers stated a desire for revenue insurance that included coverage for losses from disease. Concern was expressed about the need for additional inspections to assure compliance with insurance policy requirements; increased traffic in their facilities would, it was believed, compromise biosecurity practices.

Six poultry producers also provided written comments, including the ranking of perils to their operations. Table 18.4 presents the number of times each peril was listed and the weighted average rank. Natural disasters were listed by seven producers with a weighted

Table 18.4. Most Important Perils Facing Poultry Producers.

Peril	Number of Times Listed*	Weighted Average Rank**
1 = Most Important and 5 = Least Important		
Natural Disasters	7	3
Diseases	6	2
Power/Mechanical Failure	5	2.4
Downtime	4	3.5
Market Losses	3	3.33

* A single respondent listed some perils more than once.
** Weighted Average Rank =(Σ(rank × number of times listed in rank)) / total number of times listed).

average rank of three. Disease perils were ranked highest when listed. Power or mechanical failures were listed by five producers with the second highest ranking. Downtime and market losses were also listed as important perils.

Producers also ranked the significance of important perils (Table 18.5). Respondents agreed that natural disasters, disease and power or mechanical failures had all caused significant economic losses. They agreed that government should not be responsible for losses caused by power or mechanical failures. Commercial insurance is not available for these perils and losses were considered to be moderate to severe when they did occur.

When asked to name the diseases of most concern, avian influenza and laryngotracheitis were specified. Respiratory illnesses, coccidiosis, Newcastle disease, *Mycoplasma gallisepticum, Mycoplasma synoviae,* dermatitis and *E. coli* were also listed. Downtime, cleaning and disinfection costs, quarantine and disposal costs were the most significant losses caused by these diseases. Other losses of concern included production losses, lost markets and income, labour costs and the value of depopulated flocks.

The operators were asked to rate statements about disease incidence and preparedness (Table 18.6). Producers are concerned that a catastrophic or foreign animal disease will occur in the USA and do not believe that government indemnity programmes provide sufficient protection against losses. Poultry growers do not have emergency business plans in the event of a major disease incident. Producers practice some biosecurity to protect against disease introduction but would like a menu of livestock insurance for use in managing disease risk.

Table 18.5. Significance of Top Perils Facing Poultry Operations.

Statement/Question	Average Rating *1 = Strongly Disagree and 5 = Strongly Agree*		
	Natural Disasters	Diseases	Power/ Mechanical Failure
1 This peril has caused significant economic losses in this industry.	3.7	4.4	5
2. This peril occurs on a consistent basis (annually, semi-annually, every 5 years).		Semi-annually	
3. I have records of losses caused by this peril on my operation.	1.4	1.6	2
4. Management practices can affect the costs associated with this peril.	2	3.2	2.25
5. Government should be responsible for covering losses caused by this peril.	3.1	3.2	1.5
6. I can obtain commercial insurance for this peril.	2	1	1
7. Do you buy commercial insurance for this peril?	No	No	No
8. When this peril occurs, losses are:	Moderate-Severe	Moderate-Severe	Moderate-Severe

Table 18.6. Poultry Producer Sentiment.

1 = Strongly Disagree and 5 = Strongly Agree	Rating	
	Average	Range
1. I am very concerned that a catastrophic or foreign animal disease will occur in the USA.	4.33	3-5
2. I have a business plan that includes business interruption, cleaning, disinfecting and labour costs that I will follow in the event of a major disease event.	2.17	1-3
3. Government indemnity programmes provide sufficient protection against losses from catastrophic disease events.	1.4	1-2
4. I currently incorporate management practices to reduce the impact of disease on my livestock.	4.5	4-5
5. I think that other producers would manipulate conditions to increase insurance payments should disease infect their livestock.	1.67	1-3
6. My operation practices biosecurity to protect against disease (endemic or catastrophic) introduction in my livestock.	4.33	3-5
7. I want a choice of livestock insurance products to reduce my disease risk.	4.33	1-5

Sheep

Members of the ASIA represented the sheep industry at a listening session in October 2002 in St Louis, Missouri. Attendees participated in a discussion and four producers provided written comments. Table 18.7 presents the perils specified by sheep producers, the number of times each peril was listed and the weighted average ranking of each peril. Drought and weather, predators, and market and industry viability were each listed three times. Drought and weather received the highest ranking. Other perils include terrorism, grain prices, disease, insurance cost, social costs and international market factors.

Producers also rated the significance of the most important perils (Table 18.8). Each of the perils was listed consistently among producers and was thought to cause significant economic losses. Management practices were believed to affect costs caused by perils. Producers believed government should not be responsible for covering losses associated with these specified perils and insurance was not available to cover the moderate to severe losses resulting from these perils.

Table 18.7. Most Important Perils Facing Sheep Production.

Peril	Number of Times Listed*	Weighted Average Rank**
1 = Most Important and 5 = Least Important		
Drought/Weather	3	1.33
Predators	3	2.33
Market and Industry Viability	3	3.33

* A single respondent listed some perils more than once.
** Weighted Average Rank =(Σ(rank × number of times listed in rank)) / total number of times listed).

Sheep producers also listed diseases that were of greatest concern to their operations. Chlamydia, foot rot, internal parasites, ovine progressive pneumonia and scrapie were each listed twice. Johne's disease, lamb pneumonia/shipping fever and mastitis were each listed once. Losses caused by disease include decreased production, death, smaller domestic and export markets, and prevention and treatment costs.

Statements about disease incidence and preparedness were rated by producers and are summarized in Table 18.9. Producers do not have business plans in the event of a major

Table 18.8. Significance of Top Perils Facing Sheep Production.

Statement/Question	Average Rating *1 = Strongly Disagree and 5 = Strongly Agree*		
	Drought/Weather	Predators	Market/Industry Viability
1. This peril has caused significant economic losses in this industry.	5	5	4
2. This peril occurs on a consistent basis (annually, semi-annually, every 5 years).	5	4	4.33
3. I have records of losses caused by this peril on my operation.	3.67	2.67	3
4. Management practices can affect the costs associated with this peril.	3.33	4.67	2
5. Government should be responsible for covering losses caused by this peril.	2.67	2.67	3
6. I can obtain commercial insurance for this peril.	1.33	1	1
7. Do you buy commercial insurance for this peril?	No	No	No
8. When this peril occurs, losses are:	Severe	Moderate	Moderate-Severe

Table 18.9. Sheep Producer Sentiment.

1 = Strongly Disagree and 5 = Strongly Agree	Rating	
	Average	Range
1. I am very concerned that a catastrophic or foreign animal disease will occur in the USA.	3.75	2-5
2. I have a business plan that includes business interruption, cleaning, disinfecting and labour costs that I will follow in the event of a major disease event.	1.25	1-2
3. Government indemnity programmes provide sufficient protection against losses from catastrophic disease events.	2.25	1-5
4. I currently incorporate management practices to reduce the impact of disease on my livestock.	4.75	4-5
5. I think that other producers would manipulate conditions to increase insurance payments should disease infect their livestock.	3.5	2-5
6. My operation practices biosecurity to protect against disease (endemic or catastrophic) introduction in my livestock.	3.25	2-5
7. I want a choice of livestock insurance products to reduce my disease risk.	3.25	2-5

disease outbreak. They do, however, incorporate management practices to reduce the impact of disease into their production strategies.

Swine

Members of the NPB participated in a listening session in June 2002 in Des Moines, Iowa. The producers represented a range of production sizes from large, integrated to small, family-run operations. In addition to discussion, written comments were provided by eleven producers.

Smaller producers were more supportive of products that insured against losses from production diseases. All producers were interested in coverage for catastrophic, foreign or zoonotic diseases. Insurance covering losses from business interruption as a result of infected, quarantined or depopulated animals was also desired. Producers would like to see market price protection in the event of foreign animal disease and revenue insurance protection for uncontracted production losses.

Producers were asked to list the five most important perils facing their operation in order of importance (Table 18.10). Domestic diseases, including porcine reproductive and respiratory syndrome (PRRS), were listed by ten producers, resulting in a rank of 2.3. Low margins, market price and/or revenue variability was listed by nine producers but received a rank of only 1.9. Seven producers listed foreign animal diseases as a peril. Employee quality and turnover, environmental regulation and animal welfare regulation were also listed.

Producers were also asked to rate the significance of the most important perils (Table 18.11). Producers agreed that both domestic diseases and low margins have caused significant economic losses and occur frequently. The respondents do not believe that government should be responsible for losses caused by domestic diseases or low margins but do agree that government should cover losses caused by foreign diseases. Producers are not able to obtain insurance to protect against these losses.

Producers were also asked to list the most important diseases of concern to their operations. Nine operators listed PRRS and six operators listed foreign animal disease (FAD). Other diseases were listed once or twice, including FMD. These diseases cause significant production losses because of mortality and culling, herd condemnation, and decreases in feed

Table 18.10. Important Perils Facing Swine Operations.

Peril	Number of Times Listed*	Weighted Average Rank**
1 = Most important and 5 = Least important		
PRRS/Domestic Diseases	10	2.3
Low margins, market price, revenue	9	1.9
Foreign Animal Diseases	7	2.7
Employee quality/turnover	4	3.5
Environmental regulation	3	4
Animal welfare regulation	3	4

* A single respondent listed some perils more than once.
** Weighted Average Rank =(Σ(rank × number of times listed in rank)) / total number of times listed).

conversion efficiency and average daily gains. Market losses caused by decreased consumer confidence and export market restrictions were also major concerns. Other costs resulted from quarantines, downtime, cleanup expenses, treatment expenses, herd rebuilding and availability of credit.

Producers were also asked about disease incidence and preparedness (Table 18.12). Producers were concerned that a catastrophic or foreign animal disease will occur in the USA and they did not believe that government indemnity programmes would provide sufficient

Table 18.11. Significance of Top Perils Facing Swine Production.

Statement/Question	Average Rating *1 = Strongly Disagree and 5 = Strongly Agree*		
	Domestic Diseases	Low Margins	Foreign Diseases
1. This peril has caused significant economic losses in this industry.	4.8	4.1	1.7
2. This peril occurs on a consistent basis (annually, semi-annually, every 5 years).	4.5	4.4	1.3
3. I have records of losses caused by this peril on my operation.	4.2	4.3	1
4. Management practices can affect the costs associated with this peril.	3.3	2.8	2.4
5. Government should be responsible for covering losses caused by this peril.	2	2.2	4.7
6. I can obtain commercial insurance for this peril.	1.1	1.1	1.7
7. Do you buy commercial insurance for this peril?	No	No	No
8. When this peril occurs, losses are:	Moderate-Severe	Moderate-Severe	Severe

Table 18.12. Swine Producer Sentiment.

1 = Strongly Disagree and 5 = Strongly Agree	Rating	
	Average	Range
1. I am very concerned that a catastrophic or foreign animal disease will occur in the USA.	4.27	1-5
2. I have a business plan that includes business interruption, cleaning, disinfecting and labour costs that I will follow in the event of a major disease event.	2.50	1-4
3. Government indemnity programmes provide sufficient protection against losses from catastrophic disease events.	1.64	1-4
4. I currently incorporate management practices to reduce the impact of disease on my livestock.	4.55	3-5
5. I think that other producers would manipulate conditions to increase insurance payments should disease infect their livestock.	3.41	3-4
6. My operation practices biosecurity to protect against disease (endemic or catastrophic) introduction in my livestock.	4.64	4-5
7. I want a choice of livestock insurance products to reduce my disease risk.	3.55	1-5

protection from losses. They practice biosecurity to prevent the introduction of disease and they incorporate management practices to reduce the impact of disease introduction.

Conclusions

Contact with producer groups indicated that they are interested in seeing livestock disease insurance products developed for the USA. Some producer groups were interested in better understanding the potential for disease and then purchasing products available to manage that risk. Interest in livestock disease insurance or an interest in more education about risk management products do not indicate an estimate of potential demand or a desire to purchase livestock disease insurance. Instead, the overall impression from these listening sessions was that livestock disease insurance was an interesting option for livestock disease risk management. The availability of such a risk management tool would enable interested producers to purchase the product and those who were less interested could rely on government indemnity and self-insurance. If livestock disease insurance were developed, each livestock industry group would want a product tailored to the specific production practices of their industry. There are significant differences in the methods used to produce livestock and animal products that include nutritional requirements, length of feeding, animal husbandry, breeding methods and disease protection. Insurance products need to reflect the varying disease risk in alternative methods and stages of livestock production.

References

Boggess, W.G., Anaman, K.A. and Hanson, G.D. (1985) Importance, Causes and Management Responses to Farm Risks: Evidence from Florida and Alabama. *Southern Journal of Agricultural Economics*, 17, 105-115.

Hall, D.C., Knight, T.O., Coble, K.H., Baquet, A.E. and Patrick, G.F. (2003) Analysis of Beef Producers' Risk Management Perceptions and Desire for Further Risk Management Education. *Review of Agricultural Economics*, 25, 430-448.

Patrick, G.F., Baquet, A.E., Coble, K.H. and Knight, T.O. (2000) Hog Risk Management Survey: Summary and Preliminary Analysis. Staff Paper #00-9, Department of Agricultural Economics, Purdue University, West Lafayette, Indiana.

Chapter 19

Modelling the Impact of Compulsory Foot and Mouth Disease Insurance in the European Union

Torbjörn Jansson, Bo Norell and Ewa Rabinowicz

Introduction

In this chapter, we compare two ways of financing the programmes to combat Foot and Mouth Disease (FMD) in the European Union (EU) and use a simulation model to determine the welfare and production implications of the two systems. The two systems analysed are (1) a purely tax-financed system, where all costs for preventive measures and combating FMD outbreaks are financed by the member state governments and partly reimbursed by the EU, and (2) a compulsory insurance scheme where all costs are converted into regionally differentiated insurance premiums that are paid solely by the producers.

The first of the analysed systems represents a stylised version of the policy that is presently applied in the EU. There is a common policy for combating infectious animal diseases. The stamping-out strategy has been followed with culling and disposal of infected livestock and the preventive killing of animals that have been in contact with infected herds or that were on contiguous farms. Under the present policy, national governments bear almost all the costs connected to the outbreak. Part of those costs are reimbursed from the EU budget providing that certain measures are taken to control the outbreak. We disregard in this analysis, for the sake of simplicity, that some insurance schemes are already present in different member states and assume the entire cost is carried by national governments and the EU, i.e. paid by taxpayers. This simplification can be justified if it is argued that national insurance schemes seem to be of minor importance.

The second system is an exploration of alternatives to the present policy. The choice of this particular option can be motivated as follows. Although government involvement in the *eradication* of an infectious disease can be justified on efficiency grounds, because freedom from infectious diseases is a public good, the same does not apply to the government responsibility for *financing* the outbreak clean up. The present system implies that manageable risks are not reflected in the cost of production. The result is that too much risky production is generated and allocation of production between regions and countries is affected as well. Howe and Whittaker (1997) have argued that the present policy acts as a free insurance of last resort. In addition, the fact that part of eradication costs are reimbursed from the EU budget creates a substantial redistribution between member states since the risks of outbreak vary considerably among those countries.

Liability assignment could constitute an alternative option for financing eradication. However, this solution is not suitable in the case of FMD. (Causation is not always clear and the liable party may not be able to pay the damages.) Private, first-party insurance could be an alternative to liabilities but only compulsory insurance would be efficient, as uninsured farmers would otherwise lack the incentives to limit the disease. As insurance companies show little interest in sectors where basic statistical data are not available, an exigency fund seems to be a possible option. Such a fund would in reality operate as a compulsory insurance scheme. A Veterinary Fund already exists in the EU.[1] Accordingly, this chapter will analyse some implications of the introduction of a compulsory insurance scheme assuming that full cost of eradication is included in insurance premiums, i.e. paid by the producer. It should be pointed out, however, that the question of how best to design alternative risk-financing instruments for covering costs of eradication of epizootic diseases is a very complicated one, and this paper does not aspire to resolve it. The ambition is only to illustrate implications of one particular option that can be deemed reasonable. For more discussion of alternative financing schemes see Van Asseldonk *et al.* (2003 and this volume).

This chapter has four sections: In the first section, we describe the scenarios analysed, and the assumptions underlying the analysis are carefully explained. In the second section, a brief overview of relevant parts of the CAPRI modelling system is given. In the third section, the results of the modelling exercise are presented, and in the last section we conclude with some policy implications.

Background for Constructing the Scenarios

In order to explore the implications of a compulsory insurance scheme, the premiums have to be established. In a normal insurance system, such premiums depend on past development. In fact, one of the important conditions for insurability of risk is the availability of reliable information on the probabilities and the magnitudes of future outbreaks. However, lack of sufficient data and the fact that the FMD situation worldwide has deteriorated significantly during recent years (Ryan, 2001), makes historical data potentially even less reliable predictors of the future. This implies that calculations of the premium cannot be based on only statistically sound estimates. Instead, a hypothetical case has been developed, in cooperation with experts in veterinary medicine, based on the experiences from the analysis carried out in Rabinowicz *et al.* (2002). Based on costs of eradication in this scenario, the insurance premiums were calculated for Sweden. The premiums in all regions in the EU were, subsequently, based on the cost estimates for Sweden modified with respect to risk.

The study (Rabinowicz *et al.*, 2002)[2] assumed one outbreak of the disease per 30 years, each affecting 500 herds, or equivalently, 1% of livestock being affected every 30 years. That

[1] However, it is not a fund in the proper sense of the word, merely a line in the EU budget. Total spending on eradication of infectious diseases has vastly exceeded allocations.

[2] The main purpose in this Swedish study was to compare costs and benefits of stamping out, i.e. the present eradication strategy, with a preventive vaccination. The stamping-out strategy turned out to be much less costly for Swedish society under a wide range of assumptions. It is reasonable to believe that the same applies for the rest of the EU. Hence, the calculations in this chapter are based on an assumption that FMD will continue to be eradicated by stamping out.

Table 19.1. Assumed annual costs for FMD in Sweden.

Livestock category	Annual cost per head (Euro)
Dairy cows	1.23
Heifers, bulls and steers	0.80
Calves	0.39
Suckler cows	0.83
Sows	0.36
Pigs for fattening	0.17
Ewes and goats	0.26

Source: Rabinowicz *et al.* (2002) and veterinary expertise.

assumption was arrived at after consulting veterinarian experts, i.e. people with knowledge and understanding of current and future probabilities at the Swedish Board of Agriculture and the Swedish National Veterinary Institute (SVA). The cost per animal included full repayment to the farmer of the value of culled animals and reimbursement of business interruption costs. Costs for the government included cost for destruction of the animals and disinfecting facilities. Costs for other sectors (slaughterhouses, traders and tourist facilities) were assumed to be 30% of the direct cost for agriculture. The costs were annualized using an interest rate of 4%. Table 19.1 shows the annual cost for FMD outbreaks in Sweden resulting from the assumptions above. Those costs were used as a basis for computing the costs for FMD in all regions of the EU.

Constructing Scenarios for the EU Level

Specific features of FMD disease can provide some guidance for constructing scenarios at the EU level. FMD is an epizootic, acute, infectious, viral animal disease. The virus is present in several varieties and it affects most animals in agricultural production, such as cattle, pigs and sheep. Airborne spread of the disease can occur and the animals can also pick up the virus from direct contact, as well as from contaminated foodstuffs.

The magnitude of an outbreak is influenced by a number of factors. The most important are the number of susceptible animals, livestock density, the types of livestock that get the infections first, transport of animals and movement of people in the area, breeding method, climatic conditions (moisture, wind) and delay between outbreak and measures taken. The determining factors in limiting the outbreak are prompt identification and an effective surveillance programme. In this study we only consider stocking density, because it is the most easily measured factor. Accordingly, the premiums in the hypothetical insurance analysed in this chapter will be based only on this factor. Moreover, if a common compulsory insurance scheme were to be introduced, it would have to be based on widely available data. With increasing understanding about the relevant factors influencing risk, more detailed criteria can be used.

Table 19.2. Simulated scenarios by density and probability assumptions.

	Probability of infection is 1% of the herd every 30 years in Sweden	Probability of infection is 1% of the herd every 30 years in the EU
Premium depends linearly on stocking density	FMD1	FMD3
Premium depends on squared stocking density	FMD2	FMD4
Premium is identical in all regions in the EU	FMD5	

There is no precise information or knowledge at the EU level about how livestock density affects the magnitude and distribution of FMD. Although more detailed studies at micro-level are available, more complex relationships cannot easily be used for an approximation of the relationship between risk and two parameters. Hence, the calculations are based on alternative assumptions and should merely be seen as an illustration of the effects of different risk scenarios.

To probe the range of possible effects of introducing compulsory FMD insurance for livestock, five scenarios were constructed. In all scenarios, premiums depend on stocking density in order to capture the impact of density on the probability of spreading. The scenarios differ with respect to assumptions about *spreading of infection* and *infection probability.* All the scenarios share the assumptions that spreading of the disease depends only on stocking density, and that the composition of the herd (different types of livestock) does not influence the spread.[3] Furthermore, it was assumed that each animal should bear its own cost, i.e. costs of eradication of the disease are not pooled between different livestock categories but covered within each category. This implies that the premium is lower for a pig than for a dairy cow.

The scenarios can be inscribed in a two-by-two matrix (Table 19.2), where the rows represent different assumptions about how infection spreads and the columns represent different assumptions about infection probability. The method by which insurance premiums were adjusted depending on assumptions is discussed in detail below.

As argued above, stocking density affects the spread of the disease. Unfortunately, no data are available to *estimate* the interdependence between disease spread and stocking density. Instead, three different forms of functional relationships were used, all based on the quadratic form:

(1) $$p_{ri} = a_i + b_i q_r + c_i q_r^2$$

[3] It is well known that different livestock categories have a different impact on how the disease is spread. If available, more detailed information on such relationships could be incorporated in the calculations.

where the premium p_{ri} for animal species i in region r depends on stocking density q_r and the three parameters a, b and c. The stocking density q_r is the total number of FMD-susceptible animals per square kilometre of land in region r, i.e. it is animal density and not herd density. This is a rather flexible definition that could be used to approximate, within a limited range, other functional forms. As we base all scenarios on the cost for Sweden from Rabinowicz *et al.* (2002), we only have a single piece of information, sufficient for determining only one of the parameters. It would indeed be desirable to obtain some information also on how the cost changes when stocking density changes.

Three different functions used were obtained by using, in turn, each of the parameters alone, thus obtaining for (a>0, b=c=0) a constant premium, identical in all regions, for (b>0, a=c=0) a premium that depends linearly on stocking density with intercept zero, and for (c>0, a=b=0) a premium that depends only on the squared stocking density. These relationships can be motivated, for example, by assuming that the probability that a new infection starts is the same for all animals, and then varying the principles for spreading this new infection to surrounding farms.

The simplest possibility is that the premium is the same in all regions. This could be the case if the disease is spread from each infected animal to a constant number of other animals, regardless of the number of animals in the region (e.g. every animal would have to pay for the stamping out of itself and for 100 "neighbouring animals"). This is not a very realistic representation of how the disease spreads, as the number of farms affected would also depend on stocking density.

The next, more complex (from a mathematical point of view), possibility is linear dependence, arising when a circle is drawn around the infected farm and the *radius* of the circle does not depend on stocking density. Then the number of animals within the circle will be proportional to stocking density, as will the cost *per animal.*

Yet another possibility is that by drawing a circle around an infected farm, enclosing all animals to be destroyed or quarantined, the *area* of that circle is proportional to stocking density. In this case, the number of animals in the area, and hence the cost per animal, will depend on the square of the stocking density. That could be a reasonable case if we imagine that the spreading velocity of the disease is dependent on stocking density. It is easy to think of good motivations for cubic and exponential forms as well, by assuming that the radius of the circle depends linearly (or some higher order expression) on stocking density, for example by the hypothesis that the number of transports to and from farms in the area is an important means of spreading the disease, and that the number of transports is proportional to stocking density. Due to limited space and scarce empirical information, we limit the study to the quadratic form given above.

As pointed out above, the *probability for infection,* i.e. the frequency with which FMD outbreaks occur (columns in Table 19.2), was a second parameter that was varied while constructing the scenarios. Two assumptions were considered: (1) the expert judgement that 1% of the animals infected every 30 years applies only to Sweden on average, and (2) the judgement applies to the entire EU. Those two assumptions do, in combination with the different spreading functions, imply very different annual costs for FMD, because most Swedish regions have a below-EU average stocking density.

Assumption (1) is actually the assumption that each year each animal species i in Sweden, on average, costs k_i euro per head per year, with the amounts k_i corresponding to the figures in Table 19.1. Now, this amount has to equal the sum of premium payments in all Swedish regions divided by the total number of animals in that livestock category. With the notation introduced above, letting x_{ri} represent the number of animals of species i in region r, the relationship can be written as:

$$(2) \qquad k_i = \sum_r \left(p_{ri}\, x_{ri}\right) / \sum_r x_{ri} \qquad (r \in SE)$$

The sums are taken over the set of all regions in Sweden ($r \in SE$). If the expression for p_{ri}, according to equation (1), is substituted into equation (2), the non-zero parameter (a, b or c) can be solved for in each of the functional relationships that have been described above. Inserting this parameter back into equation (1) allows for the calculation of the value of the premium in different regions. This procedure can be repeated with assumption (2) by taking the sums over all regions in the EU instead of only regions in Sweden. This gives different values for k_i and as a result, different levels of premiums.

In the case where the functional relationship only contains the constant a, assumptions (1) and (2) are equivalent. In contrast, if the function is linear (b>0), the average premium per head in the EU under assumption (2) will be exactly as in Table 19.1, whereas it will be about seven times higher under assumption (1), when the figures in Table 19.1 are assumed to hold for Sweden. This is, of course, because the average stocking density is seven times higher in the EU than in Sweden. If we use only the quadratic term in the function for p, i.e. (c>0), the average premium per animal in the EU will be also be precisely the amount in Table 19.1 under assumption (2), but under assumption (1) it will be much higher (the average will instead hold for Sweden). In the latter case, a few densely stocked regions in the EU pay the major share of the sum of premiums in the EU. The regional premium amounts are presented in somewhat greater detail in the next section.

Each of the five scenarios was implemented for the model base year, which is the average of 2000, 2001 and 2002, and the results can be roughly interpreted[4] as answering the question "what would have been the situation in 2001 if a compulsory insurance programme had been in place?" Each scenario is compared to a reference scenario, which is simply the base year where *the taxpayers have to pay the insurance premium instead of the producer.* As the model is a partial one, this does not influence the model solution, but only the welfare analysis of the results. It should be kept in mind that, because we use a different set of assumptions about how the risk relates to certain measurable parameters, the cost for FMD in the reference scenario will be different for each insurance scenario analysed. Accordingly, there is no common reference scenario but each insurance scenario is compared with a reference scenario of its own.

Regional Distribution of Animals and Premiums

The regional premium amounts are endogenous to the model in the sense that if the number of animals in a region is reduced, the density is reduced and consequently the premium is reduced (except for FMD5).[5] Hence, the premium amounts are to be considered as model

---

[4] The model is a comparative static, hence the result should not be interpreted as the outcome of any specific year, but as the long-term equilibrium.

[5] The adjustment of the premium amount is done by iteratively changing the premium and solving the model. In that way, the premium amount is considered fixed in the economic optimization problem of the producer.

results as much as model inputs. We present them in this section, and leave the results section for the presentation of effects on welfare, financing and production.

Figure 19.1 shows the stocking density at a regional level in Europe in the reference scenario. The darker colour is associated with higher stocking densities. The highest densities are found in Noord-Brabant in the Netherlands, with more than 2000 animals per square kilometre on average, whereas in many regions, white on the map, the density is less than one. In the GINI diagram on the left hand side of the map, the regions are ordered from the

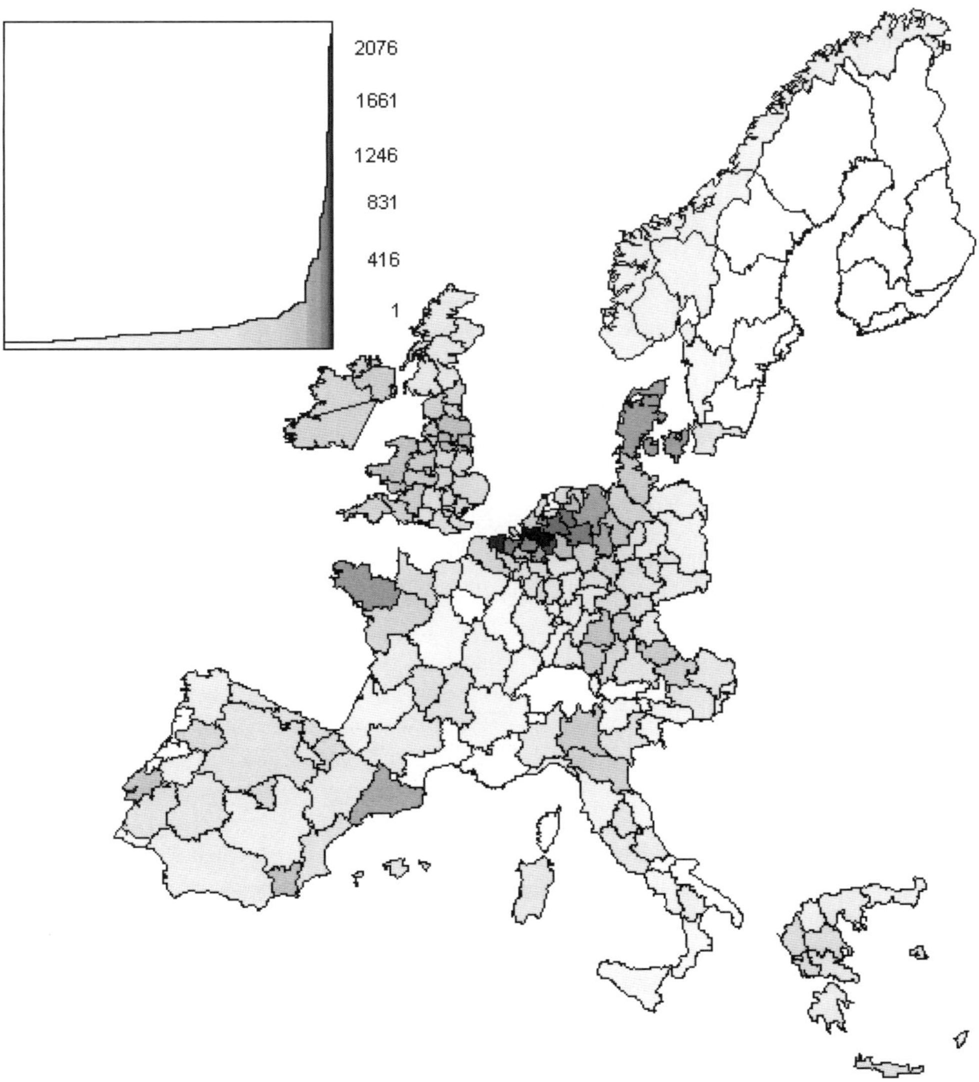

Fig. 19.1. Stocking densities of the EU in the base year. Animals per square kilometre.

smallest density (left) to the highest, with a pillar whose height and colour is proportional to density. The strongly convex shape of the upper contour in the GINI diagram indicates the uneven distribution of the animals over the land area.

Table 19.3 shows how the premium amounts for different livestock categories vary in the scenarios. Clearly, the second scenario leads to prohibitive insurance costs in a few regions, and rather high average costs. In the other scenarios, however, the average EU premiums are not very high, at least not compared to the direct payments that apply in many cases (suckler cows, bulls and steers, and slaughter premiums). A small number of regions have significantly higher risks with this risk model.

The following two maps illustrate the premium distributions. In scenario 5, the distribution is trivial, it is flat, so this map is omitted. Furthermore, the choice of risk level, i.e. the difference between scenarios 1 and 3 and between 2 and 4 is only a matter of scaling. As the colours of the maps are automatically fitted to the span of the data, the maps for scenario 1 would be identical to that of scenario 3, and that of scenario 2 would be identical to that of scenario 4. Consequently, the maps for 1 and 2 are omitted. The premium amounts are different for different animal species. To conserve space, maps for dairy cow premiums are used as examples, and the reader is referred to Table 19.1 to get an idea of the distribution of the premiums for the other livestock categories.

In the GINI-diagram of Fig. 19.2, showing the spatial distributions of the insurance premiums for dairy cows in scenario 3, it can be seen that in a few regions, the premiums are much higher than average. Those regions are darker coloured on the map. In most regions in scenario 3, the insurance premium will be much less than one euro per animal each year, and in the most densely stocked region, Noord-Brabant, it will be about 11 euro per head. In scenario 1, where the premium function is calibrated to a certain infection frequency for Sweden, the premiums are about seven times higher in all regions than in scenario 3, which is still not very expensive in any region except for the regions with the highest stocking densities (about 72 euro per dairy cow in Noord-Brabant, see also Table 19.3). As the premium is proportional to stocking density in this scenario, the difference in premiums is a direct effect of the fact that in a few regions, husbandry is very intensive in terms of animal density.

Table 19.3. Premium costs for FMD.

Livestock category	FMD1	FMD2	FMD3	FMD4
Dairy cows	$0 \leq 8 \leq 72$	$0 \leq 75 \leq 2346$	$0 \leq 1 \leq 11$	$0 \leq 1 \leq 35$
Heifers, bulls and steers	$2 \leq 5 \leq 44$	$0 \leq 27 \leq 1389$	$0 \leq 1 \leq 8$	$0 \leq 1 \leq 26$
Calves	$1 \leq 3 \leq 21$	$0 \leq 15 \leq 651$	$0 \leq 0 \leq 315$	$0 \leq 0 \leq 9$
Suckler cows	$2 \leq 3 \leq 36$	$0 \leq 10 \leq 978$	$0 \leq 1 \leq 11$	$0 \leq 1 \leq 50$
Sows	$1 \leq 3 \leq 13$	$0 \leq 23 \leq 322$	$0 \leq 0 \leq 2$	$0 \leq 0 \leq 3$
Pigs for fattening	$0 \leq 1 \leq 6$	$0 \leq 8 \leq 149$	$0 \leq 0 \leq 1$	$0 \leq 0 \leq 1$
Ewes and goats	$1 \leq 1 \leq 17$	$0 \leq 3 \leq 610$	$0 \leq 0 \leq 4$	$0 \leq 0 \leq 32$

First number is the lowest premium in any region, the middle number is the EU average, and the last number is the highest value in any region. Amounts in euro per head per year.

Source: Own calculations and simulations with CAPRI.

The distribution of premiums over regions is much more askew when stocking densities enter the risk function with their squared term, as in scenarios 2 and 4. This is evident in Fig. 19.3. A ratio of 1:2000 for stocking density translates into a 1:4,000,000 ratio in premiums if only the quadratic term is considered. In those scenarios, then, a few densely stocked regions are responsible for a major share of the costs for FMD. In Noord-Brabant, the premium per dairy cow amounts to about 35 euro per head per year in scenario 4, whereas it is close to zero in almost all other regions. In scenario 2, where the square risk function is calibrated to

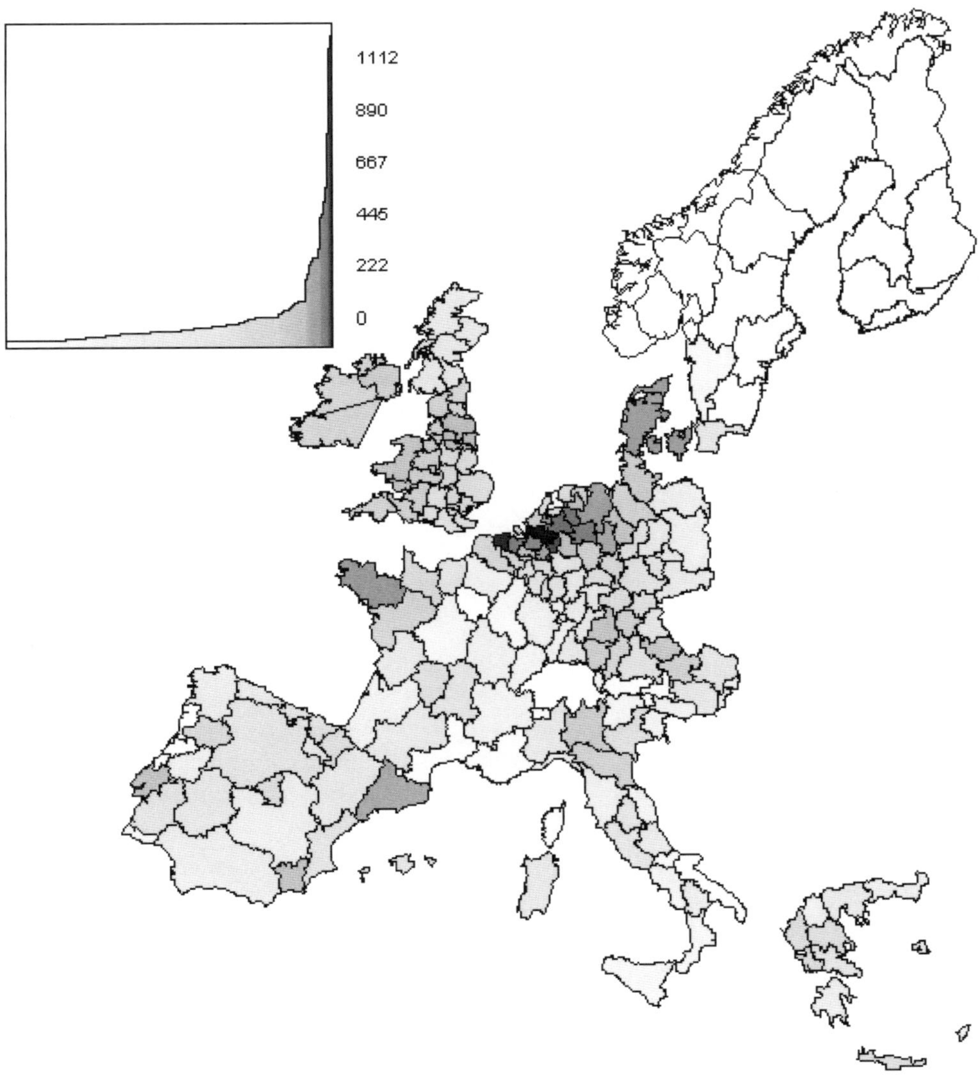

Fig. 19.2. Distribution of the insurance premium for dairy cows in scenario 3. Premium in euro cents.

the assumption that 1% of the animals in Sweden are infected every 30 years, this implies prohibitive risk premiums in the most densely stocked regions, with insurance premiums of up to 1000 euro per dairy cow.

It would be very useful to assess the soundness of calculated levels of premiums by comparing them with other studies. Unfortunately, we are not aware of any such studies at the EU level. Van Asseldonk *et al.* (2003) conduct Monte Carlo-based simulations for the Netherlands subdivided into two regions. It seems (after minor recalculations) that the level of premiums the authors arrive at for a milk cow is 5.50 euro for the region with a lower

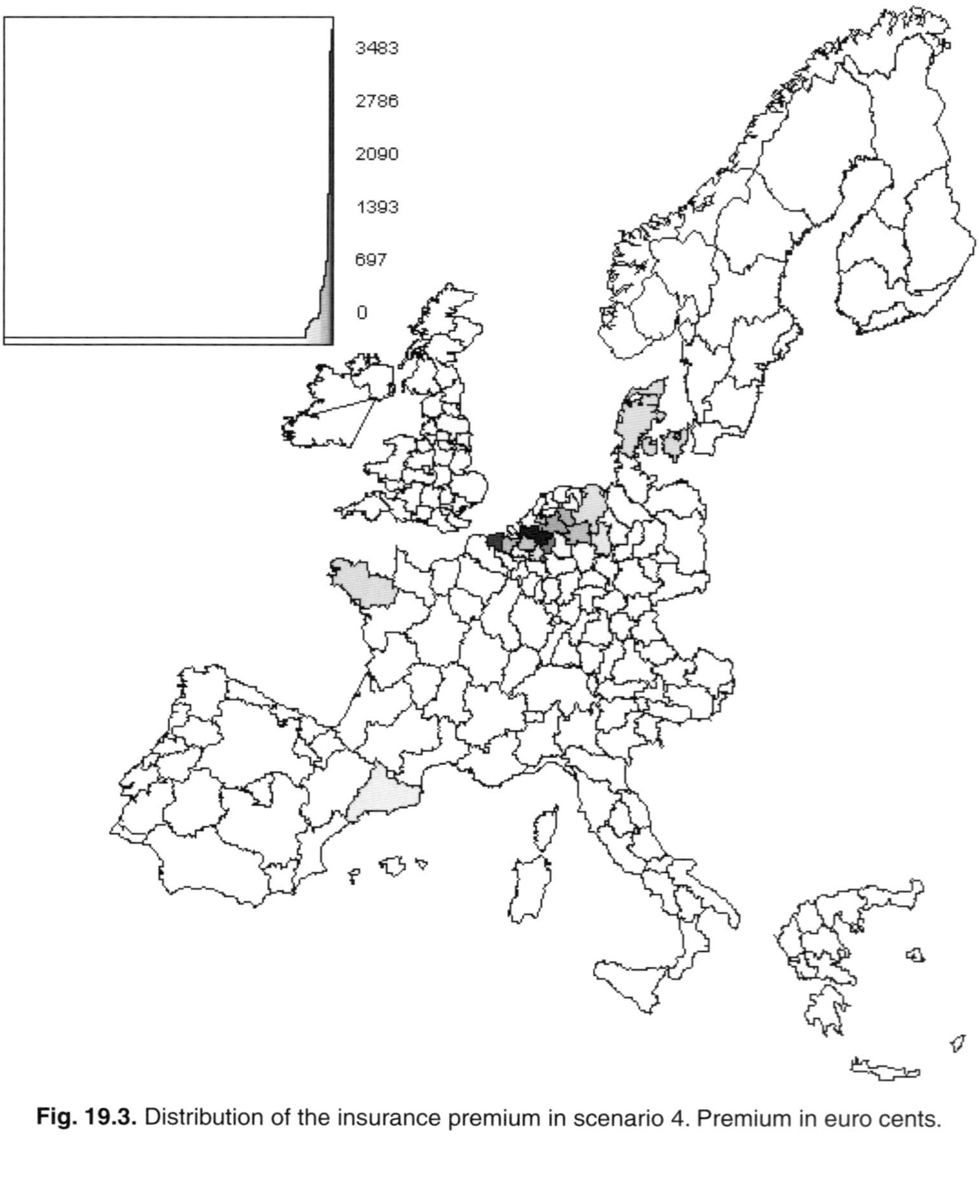

Fig. 19.3. Distribution of the insurance premium in scenario 4. Premium in euro cents.

livestock density and 56 euro for the region with a higher density. The latter figure is in the same order of magnitude as our figure of 70 euro under the linear assumption. It should be noted that our model contains more than two regions in the Netherlands. Hence, 70 euro corresponds to a higher livestock density than in the Van Asseldonk *et al.* (2003) study.

The CAPRI Modelling System

The introduction of compulsory FMD insurance was analysed using the CAPRI modelling system. CAPRI (i.e. Common Agricultural Policy Regional Impact Analysis) is being developed at the Institute for Agricultural Policy, Sociology and Market Research at the University of Bonn, Germany. Most of the materials in this chapter are based on the overview in Wieck *et al.* (2004). However, as the system is constantly evolving, some new material has been added. An extensive documentation can be found in Britz (2004), and on the website of the CAP-STRAT (2004) project. The CAPRI modelling system is designed as a projection and simulation tool for the agricultural sector of the EU based on:

- A *physical consistency framework,* building upon balances for agricultural area, young animals and feed requirements for animals as well as nutrient requirements for crops, realised as constraints in the regional supply models. The market model ensures closed fat and protein balances for processed dairy products.
- *Economic accounting principles* according to the definition of the Economic Accounts for Agriculture (EAA). The model covers all outputs and inputs included in the national EAAs, with revenues and costs broken down consistently to regions and production activities.
- A *detailed policy description.* The regional supply models capture all relevant payment schemes with their respective ceilings as well as set-aside obligations and sales quotas. The market side covers tariffs, TRQs, intervention purchases and subsidised exports.
- *Behavioural functions* and allocation steering are strictly in *line with micro-economic theory.* Functional forms are chosen to be globally well behaved, allowing for a consistent welfare analysis.

The model distinguishes a supply and a market module, which are iteratively coupled. The *supply module* consists of about 200 aggregate programming models at NUTS 2[6] level, working with exogenous prices during each iteration. After being solved, the results of these models – crop areas, herd sizes, input/output coefficients, etc. – are aggregated into member state-level models, which are then calibrated to these results by using techniques borrowed from Positive Mathematical Programming (see Heckelei and Wolff, 2003). Next, prices on young animals are determined by linking these member state models into a non-spatial EU model with market balances for young animals. Afterwards, supply and feed demand functions of the *market module* are calibrated to the results from the supply module. Solving the market model delivers producer prices at the member state level, which drives the next

[6] NUTS is a system of administrative regions used by Eurostat, the official statistics agency of the EU. NUTS 0 is the member states of the union, and NUTS 1, 2, 3 and so on represent regionalizations of finer geographical resolution.

iteration. Similarly, in between iterations, premiums for activities are adjusted if ceilings are overshot according to the rules of Common Market organizations.

The supply models for annual crops and animals are based on the assumption of a two-stage decision process. In the *first stage,* producers determine optimal variable input coefficients (nutrient needs for crops and animals, seed, plant protection, energy, pharmaceutical inputs, etc.) per hectare or head for given yields exogenously determined by trend analysis. Nutrient requirements enter as constraints in the supply models, whereas all other variable inputs together with their prices define the so-called accounting costs. The preceding reflects the calculation of gross margins in farm management. In the *second stage,* the non-linear aggregate programming models define the profit maximizing crop mix and animal numbers simultaneously with a cost minimizing feed and fertilizer mix. Availability of grass and arable land restrict production possibilities, with the crop mix influenced by set-aside obligations and the two-tier quota system for sugar beets. A cost-minimized feed mix covers animal requirements (energy, protein, etc.), whereas either organic or purchased fertilizers meets crop nutrient need. Fodder (grass, straw, fodder maize, root crops, silage and milk from suckler cows) is assumed to be non-tradable across regions and hence links animal processes to crop production and regional land availability. All other outputs and inputs can be sold and purchased at fixed prices, with milk bounded by quotas.

The use of a *mathematical programming approach* allows direct embedding of compensation payments, set-aside obligations, voluntary set-aside and sales quotas, as well as capturing important relations between agricultural production activities. Likewise, environmental indicators, such as fertilizer nutrient (N,P,K) balances and output of gases linked to global warming, are integrated in the system.

The *market module* breaks the world down into 12 country aggregates[7] and the EU member states, each featuring systems of supply, human consumption, feed and processing functions. The parameters of these functions are derived from elasticities of other studies and modelling systems, and calibrated to projected quantities and prices in the simulation year, where the choice of the *functional form* (normalized quadratic for feed and supply, Generalised Leontief Expenditure function for human consumption) and *further restrictions* (homogeneity of degree zero in prices, symmetry, correct curvature) ensure regularity. Accordingly, the demand system allows for the calculation of welfare changes for the consumers. The *processing stage of dairy products* for the EU member states comprises balancing equations for fat and protein ensuring that processed products use up exactly the amount of fat and protein comprised in the raw milk. Production of processed dairy products is then driven by the difference between the dairy product's market price and the value of its fat and protein content, based on a normalized quadratic profit function. Lastly, prices of raw milk are equal to its fat and protein content valued with fat and protein prices.

Policy instruments in the market module include bilateral tariffs (ad valorem and specific), Producer/Consumer Subsidy Equivalent price wedges (PSE/CSE) and important bilateral agreements[8], as well as globally or bilaterally allocated Tariff Rate Quotas (TRQs) for the EU

---

[7] EU, East European Candidate countries, Mediterranean countries, USA, Canada, Australia and New Zealand, Free trade-developing countries, High-tariff traders (e.g. Japan), India, China, African-Carribean-Pacific countries, Rest of the World.

[8] Including Double Zero Agreements with Central and Eastern European Countries and certain bilateral sugar quotas.

and the 12 country aggregates. Additionally, intervention sales and subsidized exports under World Trade Organisation (WTO) commitment restrictions are explicitly modelled.

The *Armington assumption,* that imported and domestically produced goods are imperfect substitutes, drives the composition of demand from domestic sales and the different import origins depending on price relations, and thus determines *bilateral trade flows* (e.g. Armington 1969). The model comprises a two-stage Armington system: the top level determines the composition of total demand from imports and domestic sales, the lower stage determines the import shares from different origins. Product markets are hence directly linked by import flows and prices, where observed in the base year. Accordingly, no uniform world market price is found in the system. In the current analysis, the upper stage substitution elasticities for meat were all set to 4.0 except beef, where it was set to 2.0, and the lower stage substitution elasticities were set to 6.0. This implies that most products are good substitutes regardless of origin, but that imports from different sources outside the EU are more perfect substitutes to one another than are imports in general to domestic products.

Results

Though the premiums differ strongly between scenarios, there is little effect on production except in scenario 2, where production is all but wiped out in some Dutch regions. Even in scenario 4, production is seriously affected only in West-Flaandern and Noord-Brabant in the Netherlands, where the sheep and goat insurance premium of about 30 euro causes decreases of 65% and 34%, respectively, in production of those animals, and where the number of suckler cows is reduced by about 6% in response to an insurance premium of 47 and 50 euro, respectively. Note that the insurance premiums for suckler cows could be compared to the direct payments from the Common Agricultural Policy of about 150 and 80 euro per head for those regions. Qualitatively, the model slightly reduces production compared to the reference scenarios where the premiums are high. Total animal production decreases, which in turn cause the prices of the outputs to rise, and prices of intermediate inputs like fodder and cereals to fall. The rising output prices counteract the cost of the insurance premium, so that in some regions with low premiums, production may increase.

An overview of herd size changes between scenarios is given in Table 19.4. In that table it can be seen that the decrease is as high as 90% for several livestock categories in the second scenario. That means that the model has run against a technical bound. The results for those regions are thus almost useless, indicating that the model is unsuited to simulate insurance premiums of 1000 euro and more. Furthermore, the results show that in all scenarios except for scenario 2, the total effect in the EU is small, always less than 0.5%. This indicates that if stocking density really is a major determinant of infection risk, then a small number of regions would be affected by a policy.

The most interesting effects can be observed in the welfare analysis. Even though the quantitative effects on production are small in all scenarios, the principle that the one causing the costs pays the costs results in reallocation among the member states. Currently, the costs of an outbreak of FMD are paid partly by the member state and partly by the EU.[9] Member

[9] In the simulations, it has been assumed that member states pay 40% of the costs with the remaining 60% being paid from the EU budget. Member states have been assumed to contribute to the EU budget in proportion to their Gross National Income.

Table 19.4. Herd sizes in scenarios compared to reference scenario measured in percentages.

Livestock Category	FMD1	FMD2	FMD3	FMD4	FMD5
Dairy cows	–2 < 0 < 1	–70 < 0 < 9	–0.5 < 0 < 0	–1 < 0 < 0	0 < 0 < 0
Heifers, bulls and steers	–4 < 0 < 2	–90 < 0 < 9	–1 < 0 < 1	–4 < 0 < 1	0 < 0 < 1
Calves	–4 < 0 < 1*	–90 < –1 < 33	–1 < 0 < 2	–3 < 0 < 3	–1 < 0 < 1
Suckler cows	–5 < 0 < 0	–80 < –2 < 4	–1 < 0 < 0	–6 < 0 < 0	0 < 0 < 0
Sows	–3 < 0 < 4	–70 < –3 < 5	0 < 0 < 0*	–1 < 0 < 0	0 < 0 < 0
Pigs for fattening	–5 < 0 < 1	–90 < –4 < 10	–1 < 0 < 1	–1 < 0 < 1	0 < 0 < 0
Ewes and goats	–35 < 0 < 3	–90 < 0 < 14	–8 < 0 < 0	–65 < 0 < 3	0 < 0 < 0

First number is the lowest premium in any region, the middle number is the EU average, and the last number is the highest value in any region. Amounts in euro per head per year.

Source: Simulations with CAPRI.

states with low stocking density currently pay a larger share of the total European costs for FMD than what is actually associated with the producers in that country, whereas the opposite is true for regions with high stocking densities. Introducing the compulsory insurance also provides incentives to reduce the total costs for FMD. In the simulations this is only possible by reallocating production. In reality, on-farm management practices strongly influence the risk of infection and spreading of the disease, letting us assume that the potential reduction of the FMD risk is greater than the modelling results indicate. Introducing the compulsory insurance indirectly shifts the burden of the costs from taxpayers to producers. The reduction of the tax burden that results in the model if the cost for the insurance scheme is shifted from the taxpayer to the producer is shown in Table 19.5.

Budget outlays in the model consist, apart from the cost for combating FMD, mainly of direct payments, export subsidies and intervention purchases. Those other costs of the Common Agricultural Policy (CAP) remain largely unchanged in all the scenarios, because production is very little influenced and because most direct payments have ceilings that stabilize the total amount paid even when production fluctuates. Therefore, the changes in tax burden reported in Table 19.5 are almost completely due to the changed financing of programmes to combat FMD. And as the insurance premium is assumed to correspond precisely to the true cost of FMD outbreaks, *those numbers are also the entire cost of FMD outbreaks per year in the baseline for the member state* under the different infection risk and spreading assumptions. It can, as expected, be seen that member states with the highest costs for FMD outbreaks are the larger member states: Germany, France, Italy and UK, as well as those states with high stocking densities: the Netherlands and Belgium.

Furthermore, we see that the more strongly the risk is assumed to depend on the stocking density, the higher the costs of FMD in densely stocked member states rise. This is especially obvious if the cost of FMD in the Netherlands is compared between scenarios 3, 4 and 5. In scenario 5, where the premium does not depend on stocking density at all, the annual cost for FMD is seven million euro. If a linear relationship is assumed, then the annual cost turns out to be twice as high, or 14 million euro, and if the quadratic relationship is assumed, the

Table 19.5. Taxes saved with compulsory insurance compared to completely tax-financed programmes to combat FMD.

	FMD1	FMD2	FMD3	FMD4	FMD5
EU total	768	7941	121	119	121
Belgium & Lux.	49	832	7	12	4
Denmark	37	316	5	4	4
Germany	162	1491	25	22	27
Austria	15	130	2	2	3
Netherlands	95	1748	14	24	7
France	125	1117	20	17	22
Portugal	10	77	2	1	2
Spain	62	455	10	8	13
Greece	12	90	2	2	3
Italy	78	697	12	11	15
Ireland	17	85	3	2	3
Finland	7	73	1	1	2
Sweden	13	132	2	2	3
UK	86	696	15	12	15

Million euro per year.

cost is 24 million euro per year. For a member state with lower average stocking density, the converse is observed. Note that in scenarios 3, 4 and 5 the premium function was set up so that the total cost for the EU was the same in all three scenarios, so the effects just mentioned are pure redistributions between member states. The reader may object that the total tax savings for the EU turn out lower in scenario 4 than in scenarios 3 and 5. This is due to changes in agricultural production in scenario 4, when the premiums are rather high for some animal species (see Table 19.3).

In most cases, agricultural income is reduced when the cost for combating FMD is shifted to the industry. Calculations of the change in income are shown in Table 19.6. However, agricultural income does not decrease in all member states in all scenarios, but in some cases there are actually minor income increases. That is because the increased cost leads to slightly higher prices, and in regions with low stocking densities – and therefore low premiums – price increases are sufficient to offset the negative effect of the premium. Of course, this effect can only occur if the premium depends in some way on stocking density, and is stronger in the quadratic scenarios 2 and 4.

If the taxpayers' gains in Table 19.5 are compared to the producer losses in Table 19.6, it is clear that in scenarios 3, 4 and 5, where the premiums are low, the insurance is merely a redistribution of money between taxpayers and producers. All taxpayers in the EU together save about 120 million euro and the producers lose about the same amount in all three scenarios. It is important to notice that even if there is no dependence of the risk upon animal

Table 19.6. Gain of agricultural income* with compulsory insurance compared to tax financed programmes to combat FMD.

	FMD1	FMD2	FMD3	FMD4	FMD5
EU total	–419	–556	–120	–124	–111
Belgium & Lux.	–64	–434	–11	–21	–3
Denmark	–51	–268	–9	–6	–5
Germany	–50	221	–17	–12	–20
Austria	3	114	0	1	–2
Netherlands	–151	–1616	–23	–47	–6
France	–49	108	–24	–23	–22
Portugal	–3	59	–3	–2	–4
Spain	–24	485	–19	–11	–26
Greece	–8	66	–7	–6	–8
Italy	22	451	–2	–2	–2
Ireland	–15	42	–4	–1	–5
Finland	2	45	–1	–1	–2
Sweden	4	82	0	0	–2
UK	–33	91	–1	6	–3

* Gross value added plus direct payments minus insurance premiums.
Million euro per year. Negative number = loss of income.

stocking density, as in scenario 5, the shift of responsibility for the costs from taxpayers to producers is enough to cause some redistribution between member states.

In scenarios 1 and 2, where the premium amounts are much higher, the incentive to shift animals to regions with lower premiums is strong and leads to a cost saving. This manifests itself in Tables 19.5 and 19.6 in that the taxpayers gain much more than the producers lose when the insurance is introduced. In the extreme scenario 2, with high costs and a squared term in the risk function, this has a radical effect: the taxpayers save almost eight billion euro, while the producers lose only 556 million. That, perhaps surprisingly small producer loss, is due to the price effect: animal output prices rise significantly, and the inputs get significantly cheaper. That also implies strong effects on the consumption side, where the higher meat prices induce a welfare loss for the consumer that is not offset by lower prices for crop outputs, and also losses for the processing industries when processed volumes are reduced. In the other scenarios, production is little influenced, so price changes are in the range of less than plus or minus one percent.

If *agricultural income, money metric* for consumers, and *processing industry profits* are added and *taxpayer costs subtracted,* a measure of overall welfare is obtained. Table 19.7 shows the difference in that measure between the insurance and the reference scenarios. In each scenario, there is an overall welfare gain, although very small. In scenarios 3, 4 and 5 it is 37, 38 and 26 million euro per year, respectively. Only in the extreme scenario 2 is the

Table 19.7. Welfare* gain with compulsory insurance compared to completely tax-financed programmes to combat FMD.

	FMD1	FMD2	FMD3	FMD4	FMD5
EU total	28	3357	37	38	26
Belgium & Lux.	–24	172	–2	–8	2
Denmark	–21	–24	–3	–2	–1
Germany	52	824	20	24	15
Austria	9	150	2	2	0
Netherlands	–69	–68	–8	–22	2
France	29	580	10	13	6
Portugal	–1	34	–1	0	–1
Spain	–6	407	–5	1	–11
Greece	–6	60	–3	–2	–4
Italy	42	590	12	14	8
Ireland	–3	90	–2	–1	–3
Finland	6	78	1	1	0
Sweden	11	133	3	3	1
UK	9	331	14	17	12

* Welfare is defined as money metric for consumers plus agricultural income plus processing industry profits minus taxpayer outlays.

Million euro per year.

difference significant at 3.4 billion euro. Only Denmark and the Netherlands lose overall welfare in that scenario. More interesting is the redistribution between member states, which in some cases is greater than the overall welfare change. Germany, Austria, France, Italy, Finland, Sweden and UK show net welfare gains in all scenarios, whereas Denmark, the Netherlands, Greece and Ireland lose in all scenarios except in scenario 2. For the three remaining member states, Belgium (with Luxembourg), Portugal, and Spain, the net effect depends on the choice of functional form for the risk function.

Summary and Conclusions

This chapter compares the impact of a hypothetical compulsory insurance scheme where the premiums are related to some of the manageable risks in production (stocking density and number of animals) with the present EU policy. Impact on production, welfare and distribution between regions and member states is analysed under different risk scenarios.

According to the analysis in the chapter, the introduction of a compulsory insurance scheme would improve social welfare, since the most risky production would decrease or reallocate to regions with lower risks (lower density). This result is consistent with the economic theory. Our analysis has ignored the transaction costs involved in collecting premiums (or levies). If such costs are significant, the welfare gain will be reduced.

The impact is, *in general,* quite small but it should be observed that we analyse only one disease. It does not make much sense to limit compulsory insurance to one disease. Infectious diseases (e.g. swine fever) with risks related to stocking density should be covered by such a scheme if applied. In such a case, the impact would be bigger. However, even when only one disease is analysed, the impact on the most affected regions, i.e. those with highest livestock density, is not negligible. The most significant impact of the insurance scheme would be a substantial redistribution of welfare and budgetary flows between the member states.

There are major uncertainties involved in estimation of the total annual cost of FMD. However, our results are not critically dependent on the correct level of the total costs, as the objective is to analyse welfare gains stemming from a system that distributes the costs in relation to livestock density.

The analysis in this chapter is based on very simplified assumptions and the results, accordingly, should be treated with caution. Nevertheless, we believe that the results are solid enough to contribute to the discussion on how the future policy for the eradication of infectious diseases should be formulated, in particular how the costs should be paid. The alternative risk-financing instrument explored in this chapter would result, if introduced, in a welfare gain. This could constitute an argument for reform. However, in order for the gains to materialize, the premiums need to be correctly related to the risk factors. The results of model experiments indicate that the welfare gains are strongly dependent on how risks are connected to production. Our analysis does not aspire to have provided a reliable way of making the connection – rather it indicates a strong need for further research in this area.

Such a further research seems, furthermore, motivated by the fact that the risk of outbreaks of infectious livestock diseases is likely to increase in the future, especially in view of the recent and upcoming enlargements of the EU.

The alternative risk-financing instrument explored in this chapter implied that the producer paid the full costs of eradicating FMD. Such an alternative is efficient if the producer is the only one who can manage risks connected to the disease, or if the producer is the one who can avoid them at lowest costs. Producers can do a lot to affect risk. By the choice of the location, but also by conduct on the farm, producers are able to mitigate the risk. However, the cost of the outbreak also significantly depends on the conduct of the public authorities in the member states where the outbreak takes place. Prompt identification and an effective surveillance programme are crucial for limiting the number of animals that have to be culled. However, the present system, where a large part of the costs are paid by a common budget, is not efficient either as it limits incentives both for farmers and for national governments to apply costly precautionary measures.

References

Armington, P. (1969) A theory of demand for products distinguished by place of origin. International Monetary Fund Staff Papers 16, 159-178.

Britz, W. (ed.) (2004) *CAPRI Modeling System Documentation.* Final Report: Common Agricultural Policy Strategy for Regions, Agriculture, and Trade Project. QLTR-2000-00394; University of Bonn, Germany.

CAP-STRAT (2004) Project Web Site [Accessed 2004.] Available at http://www.agp.uni-bonn.de/agpo/rsrch/capstr/capstr_e.htm.

Heckelei, T. and Wolff, H. (2003) Estimation of constrained optimisation models for agricultural supply analysis based on generalised maximum entropy. *European Review of Agricultural Economics* 30, 27-50.

Howe, K. S. and Whittaker, J. M. (1997) Guiding decisions on methods and responsibilities for epidemic disease prevention and control: perspectives from environmental and insurance economics. *Proceedings of the Conference of the Society for Veterinary Epidemiology and Preventive Medicine,* pp. 223-235.

Raboinwicz, E., Kaspersson, L. and Norell, B. (2002) *Att bekämpa mul - och klövsjuka en ESO - rapport om ett brännbart ämne.* Government offices of Sweden, Department of Finance. Ds 2002:31 [Accessed 2005.] Available at http://www.sli.ln.se/pdf/ESO_rapport_200231.pdf.

Ryan, J. (2001) FMD, Risk and Europe. Paper presented at the International Conference on Prevention and Control of Foot and Mouth Disease, European Commission, Brussels. [Accessed 2004.] Available at http://www.warmwell.com/riskassessment.html

Van Asseldonk, M., Meuwissen, M. and Huirne, R. (2003) A risk financing model for livestock epidemics in the European Union. IRMA, Wageningen University, the Netherlands.

Wieck, C., Britz, W., Pérez Domínguez, I. and Jansson, T. (2004) *Auswirkungen der Mid-Term Review Beschlüsse auf die Landnutzung in den europäischen Regionen und Betrieben – eine quantitative Analyse mit dem CAPRI Modellsystem.* Schriften der Gesellschaft für Wirtschafts – und Sozialwissenschaften des Landbaues 39, pp. 87-96.

Chapter 20

Investigating the Feasibility of Livestock Disease Insurance: a Case Study in US Aquaculture

Keith H. Coble, Terry R. Hanson, Stephen H. Sempier, Saleem Shaik and J. Corey Miller

Introduction

The Agricultural Risk Protection Act of 2000 mandated that the US Department of Agriculture (USDA), through its Risk Management Agency (RMA), investigate the feasibility of offering risk management protection to livestock producers. Several projects have arisen from this initiative. Our project, the National Risk Management Feasibility Program for Aquaculture (NRMFPA), a partnership between RMA and Mississippi State University, is in the process of examining the feasibility of insurance and non-insurance tools for the aquaculture industry. Specifically, we are examining the four highest-valued aquaculture industries (as of 1999) in the USA: catfish, salmon, trout and baitfish (Miller *et al.*, 2002). At the time of writing the project is not completed, nor have we formulated our final recommendations to RMA. However, our experiences with this feasibility study provide some issues and lessons that are likely common to most attempts to provide livestock insurance. In this chapter we expound upon some common livestock insurance issues and use our aquaculture project to illustrate our themes. We lay out what we believe to be a logical approach to developing a livestock disease insurance product:

1. Investigate the risks and production practices of the industry under consideration, and conduct listening sessions with the industry to get preliminary indications of what risks are economically significant.
2. Evaluate the nature of the risks that the insurance provider (public or private) would consider insuring.
3. Develop draft underwriting language that defines the risks insured and the nature of the insurance coverage.
4. Collect actuarial data assessing the covered risks so that premium rating can be conducted.
5. Assess producer willingness to pay for the draft product.

In the next section a brief discussion of aquaculture producer interest follows a description of the differences between the design of private and public insurance. In the following section, the insurability conditions with aquaculture experience are detailed. The classification of the perils is presented next using aquaculture examples.

Determining Producer Risks and Their Economic Significance

An initial step in evaluating the feasibility of aquaculture insurance was to develop knowledge of aquaculture risks and the perceived economic impact of the risks. Background studies were conducted to summarize existing knowledge of these risks. The NRMFPA has since its inception informally sought such information through multiple workshops, conferences and meetings. Through these events, responses have been obtained from producers, industry experts and researchers from the four target aquaculture species. In the first workshop conducted by the NRMFPA, groups of producers, researchers and other individuals associated with specific aquaculture industries were specifically asked their opinion about risks facing aquaculture producers. Interestingly, the one peril common to the responses of each species group was disease (Miller *et al.,* 2002). Participants were asked to what degree each peril posed significant economic losses to producers. Diseases received the highest ranking among the perils listed for each species, as they represent the largest source of loss for most species, and the potential exists for these losses to be very severe. Furthermore, many producers are equally, if not more, concerned about the potential quarantines that could result from diseases, especially the "unknown" disease.

The responses NRMFPA have received from producers have been encouraging from an insurability standpoint in that their primary interest has been in catastrophic losses. This situation is obviously complicated by the fact that the primary peril of concern is disease, which can involve large variations in the application of insurability criteria. The likelihood of clearly identifying and measuring the magnitude of specific diseases within and across species is very small. However, strictly in terms of perils, the potential exists for insurance to at least be a part of a programme that provides producers with relief for truly random, catastrophic disease losses, their primary concern.

Approaches to Risk for Private and Publicly-Provided Insurance

Having identified perils that aquaculture producers perceive as economically significant, the next logical step is to consider insurance designs. Private aquaculture insurance has existed for a number of years, but has not been equally available for all species or production systems (Secretan, 2003). For example, relatively little private insurance has been offered for pond-based aquaculture, which we learned through investigations early in the life of our project. However, we soon discovered that governmentally provided insurance and privately underwritten insurance are not likely to operate or to be designed in the same fashion. Thus, there are limits to how much one can generalize from a policy underwritten by the private sector versus a government programme. In this section we describe some of the fundamental differences we have encountered.

The rationale for public sector (i.e. government) involvement is due to market failures. Stiglitz (2000) identifies six situations involving market failures. Specifically within the insurance context, public sector intervention could overcome market failures by providing insurance for correlated risks due to disease outbreaks and subsidies on premiums, improving the welfare of society (including producers).

In contrast to the publicly offered crop insurance programmes by RMA, private aquaculture insurance companies tend to offer individualized policies for heterogeneous risks that are serviced by specialized and skilled personnel. However, they will not offer such policies to small and risky operations where the premiums are not likely to cover the relatively high transactions costs. In the private sector, rates are often negotiated and thus are not subject to clear actuarial standards. Rate structures are based on many factors (management, farm surveys, etc.) and are subjectively derived. Current publicly offered insurance tends to be standardized policies for homogenous systems and risk levels. Every producer that meets the minimal standards is eligible to purchase publicly offered insurance from RMA that is serviced by a generalized service/sales force. Furthermore, the cost of delivery is relatively low. Rates are based on a few characteristics (crop type, production practices and geographic location) and objective data. Rates for the publicly offered insurance by RMA are formula-driven but are still held to an actuarial standard with provisions for subsidies and transactions costs. An integrated private-public approach would consist of maintaining standardized policies with the current public-private structure, an in-depth risk assessment prior to insuring to identify risk levels and insurability, establishing rates according to the probability and magnitude of losses based on objective and/or subjective data, and finally, assessments of losses conducted by experts.

Defining a Policy That Conforms to Insurance Principles

Since producers have indicated to us how they perceive aquaculture disease risks, we turn to the development of an aquaculture insurance design that conforms to the aforementioned actuarial principles as they apply to a governmentally supported policy. A key portion of our research is investigating the possible underwriting and actuarial components of potential aquaculture insurance policies. As we use the term in this report, underwriting involves designing policies and associated materials that determine the policy language. For example, defining the risks that will be insured or excluded is a part of underwriting, as well as defining the mechanisms by which liabilities and indemnities will be calculated. Another dimension of insurance design is actuarial development, which involves the statistical analysis of setting premium rates. While these are two different activities, they are interdependent and neither should be conducted in isolation from the other. For example, if one designs an insurance policy that covers a peril where data are difficult to obtain for rating purposes, the lack of data has an effect on the actuarial performance of the product. In this section we will discuss a number of issues confronted by our aquaculture project that may broadly apply to livestock insurance. We believe this is a fairly useful case study of the problems that will be encountered as one attempts to develop livestock insurance products for the first time.

We have organized this section in a consistent fashion, with Table 20.1 (and Table 5.1) outlining the criteria for insurability. We first address the issue of whether a determinable and measurable loss is possible. Following that, we consider the issue of accidental and unintentional losses as it applies to aquaculture. Third, we discuss determining whether sufficient data exist to establish accurate premium rates, which will be a major part of our discussion. Fourth, we consider the problem of risk classification for a new insurance design.

Table 20.1. Conditions for Insurability.

Producer Loss Exposure	Risk Characteristics	Insurability Conditions						Potential Risk Management Tool
		Determinable & Measurable Losses	Accidental & Unintentional Losses	Sufficient Information to Conduct Risk Classification	Sufficient Data to Establish Accurate Premium Rates	Losses Sufficiently Uncorrelated to Allow for Pooling	Economically Feasible Premium	
Production Losses	Completely controllable with proper management	Hard to tell if loss was caused by an insured peril or poor management	No	No	Perhaps in some cases, but research required	Yes	No	No justifiable public or private solution
	Outbreaks indemnified by APHIS	Usually	Usually	Perhaps, but may be very expensive	Perhaps in some cases, but research required	Often not	Often not	Wrap-around insurance for high-valued animals
	Widespread outbreaks not indemnified by APHIS	Usually	Usually	Perhaps, but may be very expensive	Perhaps in some cases, but research required	No	Often not	Government insurance
	Localized outbreaks not indemnified by APHIS	Usually	Usually	Perhaps, but may be very expensive	Perhaps in some cases, but research required	Yes	Perhaps	Private disease insurance or government insurance
	Endemic – persistent losses	Usually	No	Yes, high risk for everyone	Yes	No	No	Government exit assistance
Market Losses	Depopulated, lag in repopulating or rebuilding genetics	Yes	Depends on disease	Depends on disease	Difficult	Depends on disease	Depends on disease	Private business interruption insurance
	Missed marketing window due to quarantine	Yes	Yes	Yes	Difficult	Depends on extent of quarantine	Perhaps	Private business interruption insurance
	Localized, short-term lower prices	Yes, if prices are observable	Yes	Yes	Difficult	Yes	Perhaps	Private insurance
	Widespread, short-term lower prices	Yes, if prices are observable	Yes	Yes	Difficult	No	No	Government standing market loss programme
	Long-term loss of market due to endemic disease	Yes, if prices are observable	No	Yes, high risk for everyone	Yes	No	No	Government exit assistance

We next discuss whether losses are sufficiently uncorrelated to allow risk pooling. Finally, we consider how one determines if an economically feasible premium rate can be established.

Determinable and Measurable Losses

A viable insurance policy requires the equitable indemnification of losses such that the insurer can be reasonably satisfied that dishonest insureds are not cheating the policy, while at the same time individuals purchasing insurance feel as though they are treated fairly when a claim is made. Creating this equilibrium makes the capacity to objectively measure losses a central requirement for a viable insurance product. One of the more interesting aspects of our aquaculture insurance feasibility study is how challenging measuring losses is in some aquaculture production systems. Aquaculture production systems are not all equally conducive to loss measurement. A raceway system is quite transparent: the fish are visible and contained in specific areas where one could physically examine and enumerate either live or dead fish. Pond culture, on the other hand, is much more challenging due to the difficulty in actually observing the fish grown in 10- to 15-acre ponds, other than when they surface at feeding time.

In general two basic approaches exist to measuring an indemnity. One can either actually observe the loss; i.e. count dead fish, or one can know the expected production, count the remaining live fish, and measure the difference between the two. Either approach has precedence in the insurance industry, but both are particularly difficult in the context of aquaculture insurance. However, similar problems would likely exist if one were trying to insure cattle, where actually counting how many head are on a large ranch becomes problematic. In the course of our feasibility study, we have investigated various ways to enumerate both live and dead fish. In terms of measuring quantities of live fish, we have concluded an essential requirement is a regular detailed report of the fish inventory provided by the producer on at least a quarterly basis, so that one would have current records about the existing inventory. We also believe, in attempting to measure inventory in this fashion, that requiring auxiliary information such as feed records is also necessary. The producer would generate a record that could be validated by a third party, verifying that the amount fed was consistent with the number of animals that the producer claims to exist.

Another key reason to be concerned about determinable and measurable losses in any type of livestock production system is the general likelihood of physically moving live animals – as opposed to the virtual impossibility of moving a field of maize in traditional crop insurance. This situation leads us to believe that an aquaculture insurance policy must indemnify at the producer level and not insure subunits, i.e. individual ponds, pens or raceways within a production system, because of how easily an individual could move animals from one location to another. Many private livestock insurance policies require adjusters to observe lost animals to preclude intentional hiding of the remaining live animals. However, in many instances we have observed the potential difficulty in inventorying lost animals. For example, dead fish that sink to the bottom of a pond may not be measurable. Prolonged disease events may also occur that cause relatively small losses over a period of time, rather than a single large loss event, thus requiring an adjuster to return on a regular basis to observe how many animals are ultimately lost.

Accidental and Unintentional Losses

In our effort to evaluate the most feasible aquaculture insurance product, a key issue is avoiding the moral hazard problem of insuring intentional loss situations where producer behaviour may increase indemnities. Two primary techniques exist to mitigate this behaviour. One involves using deductibles or co-payments so that the insured absorbs a portion of losses. A deductible simply requires the producer to absorb the first tier of losses before the insurance subsumes the remaining losses, while a co-payment requires absorbing a percentage of all losses. The greater the concern about moral hazard, the more we generally expect to see a greater deductible or co-payment. A key aspect to avoiding moral hazard also includes the valuation of losses. One of the primary challenges is to make sure that losses will be assessed at a reasonable value per unit. However, an insurer always wants to be careful to make sure the inventory does not have an insured value greater than its economic value to the individual who owns the animals. In other words, an insurance policy should not value a lost fish at $1.00 per pound when the market price of live fish is only $0.70 per pound. Such a situation creates a very strong moral hazard incentive. Thus, another aspect of making sure one does not induce intentional losses is to be very careful that all deductibles are effective and not mitigated by some factor such as an overvalued animal price.

Another concern involves insuring losses that may be determined by producer management decisions. In our aquaculture project, we have investigated at length the management associated with various potential losses of aquaculture animals. Our primary objective in this research is identifying whether or not a particular disease or cause of loss can be mitigated or eliminated by proper management. Aquaculture policies should be defined in a manner that specifically excludes causes of loss that are preventable with proper management. Policies should explicitly require that producers use certain mitigating practices such as vaccinating fish or applying proper treatments once a disease breaks out.

Establishing Accurate Premium Rates

Accurately rating an insurance policy requires a great deal of statistical information. One needs a full probability distribution rather than an expected value to calculate the probability of loss and the expected indemnities. That is, one needs to know both the frequency of loss and the magnitude of loss. Failing to know either of these two components leaves one without the ability to estimate premium rates. As with most property and casualty lines of insurance, insuring aquaculture or most livestock would be a case of determining where a complete loss of all animals or varying degrees of partial loss could occur. This situation makes identifying the expected indemnities under these various scenarios even more imperative. In conducting actuarial analysis, one typically expects to examine objective data such as historical observations on production. Ideally one would like a large sample size so that strong statistical properties of the estimated premium rate could be obtained. Importantly, the nature of the risk that is being insured will determine the statistical accuracy of the premium rates. For example, in health insurance one hopes that a great deal of independence exists between losses so that, if one person gets sick this year, the probability of other individuals getting sick as well is not significantly affected. Alternatively, one can encounter the situation observed in crop insurance where losses are highly correlated with other losses because of drought and other severe weather conditions, or disease outbreaks that affect wide areas. In such a scenario, more years of experience are required to be comfortable with the validity of the statistical properties of the estimates that are made. In automobile insurance, for example, two or three years of data with a large sample of cross-sectional data might be deemed quite sufficient for rating because there is no systemic component to those losses. In crop insurance, RMA currently uses at least 20 years of data due to the systemic nature of crop losses.

Another complicating factor when one attempts to use objective data is how well the historical data reflect the current situation. Historically, in insuring animal diseases or various other forms of losses, there have been changes in technology associated with the expected losses. Possible changes in disease prevalence over time might make historical data irrelevant to the current situation. For example, power outages might be a great concern to an aquaculture producer, but if advances in electric delivery systems have significantly reduced the probability of a power outage, then the historical data may be largely irrelevant for the current situation.

Our project has been confronted with a significant challenge in that the type of objective historical data needed to rate aquaculture insurance products for any US species have not been collected. Therefore, the type of data just described simply do not exist. One could begin to collect such data, but several years of data collection would be needed before one could move forward with an insurance policy, and in many instances pressures exist to move forward more rapidly. Given the nature of this situation, our feasibility project has proposed approaching the collection of actuarial data from two standpoints. One approach attempts to elicit historical data from individuals where possible. In other words, it asks individuals to recall historical production and historical loss events so that we can have some sense of the frequency and magnitude of those losses. The second approach is the elicitation of subjective probabilities based on the current situation, which involves asking experts what they believe the frequency and magnitude of an insurance loss would be. In the private insurance sector, subjective assessments of loss risks are regularly used based on the expert opinion of actuaries who have a sense of the industry but have relatively little data on which to base their analysis. Our project has undertaken eliciting subjective probabilities from experts, because this allows one to generate actuarial information where objective data do not exist. Furthermore, this elicitation also allows one to get current estimates that would not reflect various changes in technologies and trends and the other complicating factors previously mentioned.

Once one defines the group of experts from whom subjective probabilities are elicited, one has to select an approach to collect this information. Two dominant approaches exist: one involves eliciting a consensus probability distribution by putting a group of experts together and asking them to discuss the risk until they reach the consensus. The alternative is composite forecasting of the probability distribution, where experts are kept apart and each independently develops his or her own estimate of the loss probability.

Finally, one has to decide what technique to use to capture the probability distribution from participants. There are two primary techniques when one is attempting to capture a full probability distribution. The first is the fractal approach, where individuals are given points of probability between zero and one and asked to assign a value to each. Conversely, one can give individuals ranges under what we call the histogram approach. One can give ranges of outcomes, and then ask experts to assign probability to those ranges.

The fractal approach is illustrated by two survey questions below that involve asking producers how likely they think certain perils will cause losses beyond "normal" production losses on their farm next year. The answers to these questions will allow us to generate the distribution from which probabilities and magnitudes of losses can be estimated. The specific example below is consistent with assuming a beta distribution and allows one to approximate the mean and standard deviation. Other methods make other distributional assumptions.

Example 1: Subjective elicitation question using the fractal approach

a) What do you expect your most likely catfish production (pounds/year) to be next year?
b) What do you expect your low catfish production (pounds/year) to be in the next ten years? That is, there would be only about a 10% chance that the production would go below this level.
c) What do you expect your high catfish production (pounds/year) to be in the next ten years? That is, there would be only about a 10% chance that the production would go above this level.

Example 2: Subjective elicitation question using the three-point approach

a) What are the chances you will lose between zero and ten percent of your total production next year due to these perils [provided to respondent in a handout]?
b) What are the chances you will lose between 10 and 20% of your total production next year due to these perils?
c) What are the chances you will lose between 20 and 30% of your total production next year due to these perils?
d) What are the chances you will lose between 30 and 50% of your total production next year due to these perils?
e) What are the chances you will lose between 50 and 100% of your total production next year due to these perils?

Risk Classification

Another key element intimately related to subjective probability elicitation is how well one can characterize different individual operations in terms of their relative riskiness and in a justifiable manner. In aquaculture production, a number of criteria may be conditioned on the production system. For example, consider salmon production in coastal waters. A much greater risk of disease may be present in one location than another as water quality, tides and other factors may be very different in one cove versus another (Forster, 2003). The ability to conduct risk classification to avoid adverse selection is crucial. In other words, if the probabilities that one has are applied broadly across a set of individuals who actually vary significantly in their riskiness, then the classic adverse selection problem may set in where the high-risk individuals agree to purchase the insurance and low-risk individuals opt out. Again, a lack of historical data constrains the ability to perform risk classification. We have attempted to overcome this problem in our aquaculture project by linking the responses to subjective probability questions to characteristics of individuals so that factors such as particular production practices or location would be related to the subjective probability in such a fashion that one could adjust rates based on those responses.

Losses Sufficiently Uncorrelated to Allow Risk Pooling

In our earlier discussion we noted that risk pooling is an issue extremely important to a private sector insurer because it determines how much capital must be retained in reserve for losses. In a government insurance design, risk pooling is much less of an issue because of the government's "deep pockets" that could likely withstand significant correlation of losses in the indemnity pool. In animal diseases, the correlation of losses across insured individuals has a great deal to do with the transmission of the particular disease. For example, if a disease tends to appear in isolated incidents and is not highly transmissible, then a much smaller number of correlated losses will occur than if this disease tends to have a high transmissibility to spread rapidly and control mechanisms are largely ineffective. A widespread set of losses could then occur in a region requiring greater reserves from the insurer. In this instance, we think disease spread models merit attention, where epidemiologists and other scientists familiar with disease transmission would actually attempt to quantify the transmissibility of a disease and bring together various research and knowledge about how a disease is transmitted, the probabilities of transmission, and any mitigation methods effective in providing an estimate of how likely one is to have correlated losses.

Economically Feasible Premiums

In the context of a feasibility study, considering whether an economically feasible premium for the product can exist is imperative. In other words, if the willingness of insurers to provide insurance requires a rate that is much higher than any producer is willing to pay, then a market will not arise. Economically feasible premiums are the reason our project is concerned with measuring the willingness to pay for insurance. Measuring willingness to pay is a challenge in the sense that in most instances one is asking a hypothetical question of producers, unless there is a pilot programme to determine the actual interest in the product. Obviously, willingness to pay is a function of policy-specific attributes and a number of individual attributes, including the risk aversion of the individual and his or her ability to manage this risk with other mechanisms. The approaches we have considered to measure willingness to pay include contingent valuation, which involves asking individuals about their willingness to pay for policies of various characteristics; and conjoint analysis, which has a greater ability to evaluate an individual's willingness to pay for various attributes but is also more demanding on participants, who are asked a number of questions.

Potential Aquaculture Insurance Policy Design

Having considered the insurability conditions described in the previous section, we have attempted to define the general framework of a policy for further evaluation. Several designs have been considered (Shaik *et al.*, 2003), but as a case study we will describe a named-peril policy covering aquaculture disease losses. The named-peril policy would provide insurance against loss of fish production due to mortality from a list of specific perils (including disease) and could contain the following elements: a) the sum insured, b) a producer-elected deductible percentage, c) liability, and d) the producer-reported and adjuster-verified fish (number/weight by fish size categories) lost to mortalities from a peril event. Due to the multiple-batch stocking that occurs in some aquaculture systems (catfish in ponds), the sum insured and the indemnity for the named-peril policy would rely on the fish lost (number/weight) and the inventories, respectively, evaluated by each fish size category. The value of the sum insured, liability and indemnity would be based on a predetermined quantity-price table per fish size category for a species.

Due to the production practices in aquaculture (especially pond culture) and the inability to accurately identify and measure the contents (fish) of the pond, the *sum insured* might be computed as the mean of the two most recent quarterly reported inventories.

The producer may also select the deductible percentage level from within the range allowed.[1] Based on this selection, the deductible will be equal to the sum insured multiplied by the deductible percentage. The deductible selection would in turn influence the premium rate charged for the policy, with a larger deductible resulting in a lower premium. Likewise a smaller sum insured would result in a lower premium. The liability is the maximum value the insurer is required to pay in the event of a complete loss and is computed as the sum insured minus the deductible value. In the named-peril aquaculture insurance policy premiums are calculated based on quarterly reported liability.

The named-peril policy design would provide an indemnity payment when fish are lost in a production unit due to an insured peril based on the value of the inventory of fish lost.[2] The indemnity is subject to a deductible and co-payments. Both deductibles and co-payments are introduced to address potential abuses of the policy. A co-payment is the percentage of the loss shared by the insured after meeting the deductible. Thus, the total indemnity paid to a producer is reduced by the co-payment percentage.

A producer does not receive an indemnity payment if the value of lost fish due to mortality is less than the deductible amount he or she selected. However, indemnity payments occur if the value of lost fish due to mortality exceeds the deductible amount. The indemnity reaches a maximum when the value of fish lost equals or exceeds the amount of the liability coverage minus the deductible.

Application to Aquaculture Diseases

The policy described in the previous section requires the specific perils to be defined. Thus, we have been forced to consider fish diseases on a case-by-case basis. As described in the next section, the answer as to whether a disease is insurable often varies by species, production system, or even region.

In this section we illustrate this process with a set of representative cases we have encountered. Because we are considering a governmentally provided insurance plan, we are assuming that correlation of losses is not a primary concern and that willingness to pay will be addressed in a later stage of either a willingness-to-pay study or through a pilot programme. We focus on aquaculture production losses occurring from diseases and do not consider marketing losses. Specifically, we look at the predominant aquaculture diseases for the aquaculture species of concern in this project and their "fit" with insurability criteria detailed

---

[1] Current crop yield deductibles range from 50% to 15% and price guarantees range from 50% to 100%. However, for aquaculture this range needs to be elicited from producers as well as their records.

[2] In crop yield or revenue insurance, indemnities are realized if current year yield or revenue is less than the liability guarantee.

above. Insurability criteria questions posed and the responses from veterinarians, researchers and producers of aquaculture species will be discussed when unique aquaculture instances occur. Finally, we will discuss categorizing aquaculture diseases as "potentially insurable", "definitely non-insurable" and "mixed conditions for insurability" and use example diseases for each species. These examples will provide an insight into the process of how a specific aquaculture disease could be placed in one of these insurability categories. In this section we again try to be consistent with Table 20.1 in presenting the criteria for insurability.

Earlier in this chapter we discussed determinable and measurable losses and issues concerning the measurement of fish losses in different production facilities, approaches to indemnification and the potential for fraudulent actions through moving fish around the farm, which is often a normal management action. In this section, we look at fish disease loss measurement from the viewpoint of determining if a species-specific disease is likely to be insurable. Key questions to be asked to determine the insurability of a disease loss are:

- Can one clearly determine whether or not the loss was caused by this disease alone?
- When more than one disease occurs simultaneously, can one clearly determine whether or not the loss was *primarily* caused by this disease?

The easiest case to categorize is when a disease is clearly the culprit and there are no other infections. This may be a rare event, as many diseases may be present and waiting to erupt under optimal conditions or when fish are stressed from another disease. In less clear cases, determining a primary cause of loss is difficult. For example, in catfish pond culture numerous bacterial diseases may simultaneously occur in the spring or autumn and together cause fish losses, but attributing the loss to only one disease is not easy. This scenario does not automatically disqualify a disease from being insurable. For example, if fish die from an outbreak of two diseases that are both insurable according to all other criteria, then it does not matter which of the two is the primary disease since the producer would have coverage for a loss from either disease.

From our insurability perspective, the main aquaculture disease measurability question to ask is:

- Is the disease characterized by chronic or acute losses?

Chronic losses occur over prolonged periods, two weeks to three months or more in duration, and may be characterized by small quantities of daily losses, but in the end sum to a large quantity of overall fish lost. Acute losses occur quickly, in a matter of hours or over a couple of days, and are characterized by large, easily observable quantities of dead fish. The former are particularly difficult to enumerate for pond aquaculture systems as a few dead fish found daily may be considered routine and not initially counted due to labour and time constraints. However, the greater enumeration problem arises in how to distinguish and count today's dead fish from yesterday's, when fish may remain floating or may have already sunk or been eaten by resident animals. Typically, large ponds are not seined to collect dead fish: instead they are left and quickly deteriorate. On the other hand, trout raceway systems may be able to measure daily for losses, but on large commercial facilities, the sheer number of raceways and total area under production may make counting daily losses difficult (Hinshaw *et al.*, 2004).

As stated earlier in this chapter, our primary objective in assessing if a loss is accidental and unintentional is to identify whether a particular fish disease can be managed. The necessary questions to pose are:

- Are the probabilities of getting these diseases conditioned on management?
- Do available control measures exist to *prevent* the disease outbreak?
- Do the available control measures *mitigate* the severity of the outbreak?

In the clear case of a fish disease being preventable or its severity reduced through proper management, this fish disease would not be insurable, as per row two in the production losses rows of Table 20.1. However, even these seemingly clear cases are often not so clear. In some catfish disease cases, specified best management practices (BMP) to contain a disease outbreak might be vague and refer to common management practices. While the BMP statements are good, such vague recommendations could lead to the false conclusion that a disease outbreak is management related and therefore not insurable, when in reality following the recommendations may not stop or reduce losses.

Commercial aquaculture production has a goal of profitability and routinely requires intensification to the point where additional management and equipment are required to maintain water quality. Alternatively, medicated feed treatments may be available to combat one disease but may cause another looming disease to increase in severity. Multiple diseases occurring simultaneously can be complex and may be difficult to prevent or reduce, even if one of the diseases is treatable. In the cases where management cannot prevent or reduce fish losses from a specific disease, this disease would be insurable.

Earlier in this chapter, risk classification was cited as a key element in determining whether groups of aquaculture operators could be characterized as having similar or different levels of risk in their operations. In our discussion of aquaculture diseases the pertinent questions are:

- Within a region can one visit two aquaculture operations and label one farm more prone to this disease than another farm?
- Between two regions can one go to two operations/farms and label one farm more prone to this disease than another farm?
- How contagious is the disease?
- Is the disease endemic or exotic?

As stated earlier there may be a much greater risk of disease in one location than another. The ability to conduct risk classification to avoid adverse selection is crucial. Causes of the differences in disease riskiness among operations within the same region may be due to environmental conditions (water characteristics, sources, tide patterns, weather, migratory bird routes), disease characteristics (historical presence, contagiousness, endemic/exotic), farm management (on-farm practices, regional harvest/transport practices) or other factors. Observable physical differences (e.g. water wells) between farms within a region and between regions will help determine the risk classification of a specific farm toward a specific disease that may come from surface water sources, and the appropriate premium for that farm.

The correlation of disease losses across insured individuals has much to do with the transmission of the particular disease, and the pertinent questions to pose are:

- Are losses largely independent of other producers (idiosyncratic) or likely to affect many insured producers simultaneously (systemic)?
- Are there factors that may affect the frequency of occurrence over time?
- Will future outbreaks be more probable after the first outbreak occurs?

As mentioned earlier in this chapter these are important considerations in assessing how much capital the insurer must keep in reserve. The impact of disease outbreaks is evaluated on both a farm and regional level. Diseases such as Spring Viremia of Carp (SVC), a reportable disease, may be considered idiosyncratic outbreaks since this exotic disease will appear on a farm only if infected fish are introduced to the farm (Stone, N., personal communication, Arkansas, 2004). This will impact the individual farm, and under current APHIS regulations the farm will be quarantined and all fish on the farm destroyed before the disease can be spread to other farms (Stone, N., personal communication, Arkansas, 2004). The full market value of fish destroyed due to SVC (or ISA for salmon) may not be covered as APHIS uses no set amounts for indemnification, so the possibility of "wrap-around" insurance to cover the remaining market value exists, as per row three of the production rows in Table 20.1. A systemic outbreak could arise from a disease that is dependent on weather conditions such as *Saprolegnia* in catfish pond systems, also known as winter fungus (Park, E., personal communication, Arkansas, 2004). All farms in the area experience an increased risk of contracting the disease if cooler weather descends on the region. Finally, when certain diseases occur on a farm it becomes predisposed to having another outbreak. This is the case with the parasite that causes whirling disease in trout raceway systems. This parasite can remain on the farm in its dormant spore phase for several years before conditions are favorable to infect fish again (Bruno and Ellis 1996). In addition, there is no management to effectively remove the dormant spores from the system; therefore, farms that experienced a case of whirling disease are more likely to have an outbreak in the future.

Examples of Classifying Diseases

The flow diagram in Fig. 20.1 presents a pragmatic approach to classifying diseases according to the primary criteria just described. This method of analysing each disease one question at a time leads to diseases being classified as "potentially insurable" or "non-insurable". In some cases there is overwhelming evidence to support a disease being placed into the "potentially insurable" or "non-insurable" category. There are several instances where there is not a "yes" or "no" answer to one of the questions in the flow diagram and there are strong arguments that a disease should be insurable and other points that suggest the same disease should not be insurable. This section provides examples of the diseases that best fit into the "non-insurable", "potentially insurable" and "mixed conditions for insurability" categories.

When the basic insurability conditions are not met no additional analysis is needed for a disease to classify it as "non-insurable", see Table 20.2 for examples. One clearly uninsurable disease is digenic trematodes on catfish farms. These parasites can cause high mortalities in smaller fish but not larger fish, yet can cause economic loss to the farmer since this parasite burrows into the skin of the fish making the fillets unmarketable (Tucker *et al.*, 2004). The trematodes require a specific snail and pelican species to be present to complete their lifecycle and thus continue to infect catfish. Farm management practices are now well established to treat and reduce the requisite snail population, effectively preventing trematode outbreaks

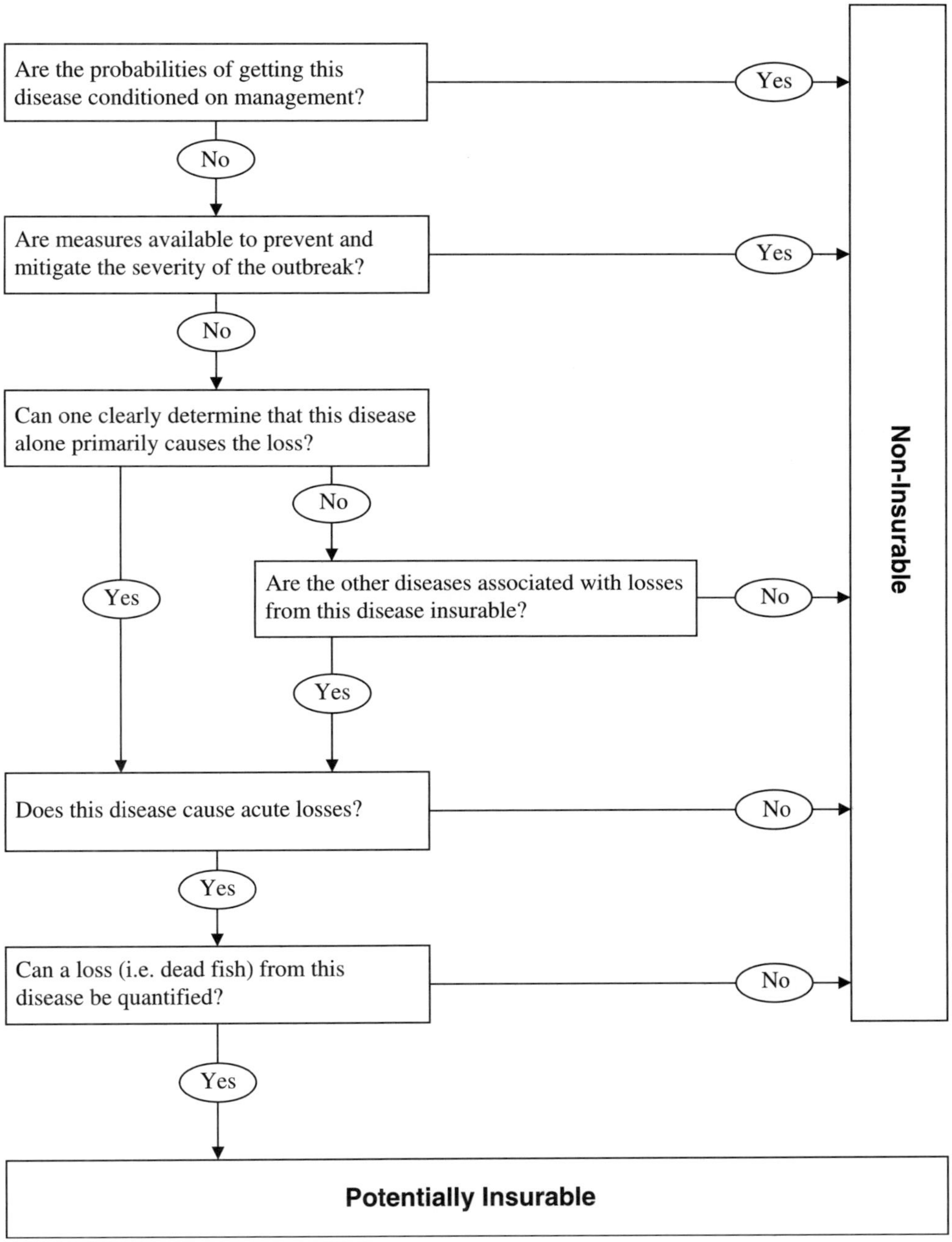

Fig. 20.1. Flow Diagram to Assess Diseases Using the Primary Insurability Criteria.

Table 20.2. Non-Insurable Diseases.

Insurability Conditions	Catfish	Trout	Baitfish
Disease and Description	Digenic Trematodes Parasite Causes a marketing problem more than a mortality issue	Infectious Pancreatic Necrosis Virus Fish that survive a disease outbreak are carriers of disease	*Saprolegnia* Fungus A secondary disease that stressed fish contract
Determinable Losses	Yes, but loss is not from mortality	Yes	Not primary cause of death
Measurable Losses	Acute if snail population increases rapidly	Can be acute	Chronic
Accidental and Unintentional Losses	Management can greatly reduce risk	Producers can use only certified disease-free eggs, use well water and disinfect eggs to reduce risk	Management can treat disease
Sufficient Information to Conduct Risk Classification	Yes, some farms are more prone to disease than others	Farmers that do not use certified disease-free fish are a higher risk	Since this occurs in winter there is an increased risk of an outbreak in the winter
Sufficient Data to Establish Accurate Premium Rates	Not available	Not available	Considered within normal threshold of loss
Losses Sufficiently Uncorrelated to Allow for Pooling	Uncorrelated since it depends on management practices of farm	Increased risk of transmission of disease since survivors can pass disease on to others	Can affect many producers at the same time since it is temperature dependent
Is this Disease Insurable?	No	No	No, does not cause deaths above normally acceptable range

on the catfish farm. Since this is purely a management-related disease it is not insurable. However, when the first trematode outbreak occurred there was no known treatment and at that time the peril would have been insurable. Thus, the insurability of specific perils can change over time and policy underwriters need to be aware of such developments to keep their policies current and avoid covering non-insurable perils.

There are also examples of non-insurable diseases for trout and baitfish culture. Infectious Pancreatic Necrosis (IPN) in trout raceway production is not insurable. As mentioned previously, management practices, such as purchasing certified disease-free eggs and fry and using well water, effectively prevent the disease from impacting a farm (Hinshaw, J.M., personal communication, North Carolina, 2004; Hinshaw *et al.,* 2004). Finally, a fungus known as *Saprolegnia* can affect baitfish. Although this disease cannot be prevented, several points argue against the insurability of this disease. Since an outbreak of this well-known fungus is a chronic, secondary infection, and can be mitigated through management, this disease is non-insurable for baitfish (Stone *et al.,* 1997; Engle and Stone, 2004; Park, E., personal communication, Arkansas, 2004).

Potentially insurable diseases (see Table 20.3) are also revealed following the questions in the flow diagram in Fig. 20.1. Viral Hemorrhagic Septicemia (VHS) easily falls into the potentially insurable category according to the criteria outlined in Fig. 20.1. The virulent form of VHS impacts salmonids and is exotic to the USA (Winton, J.R. personal communication, Washington, 2004). The disease is not conditioned on management and measures will not prevent or mitigate an outbreak if the disease did enter the USA (Kinnunen, R., personal communication, Michigan, 2004.; Fornshell, G., personal communication, Idaho, 2004). One can clearly determine the acute, quantifiable losses from VHS. All of the above factors contribute to make this disease insurable according to the initial criteria explained above.

Many diseases are not easy to classify as potentially insurable or non-insurable and fall into the "mixed conditions for insurability" category: see Table 20.4. Experts often disagree with regard to management, treatment and prevention of these diseases. Some regions of the USA are predisposed to contracting a disease because of local rearing practices or the natural presence of the disease in surface water, rendering it insurable in one region of the USA but non-insurable elsewhere.

One disease classified with "mixed conditions for insurability" due to regional differences in the USA is Infectious Haematopoietic Necrosis (IHN). This virus impacts trout production, can impact fish of any age and cause either chronic or acute losses depending on the water temperature (Bruno and Ellis 1996). Although losses can be acute or chronic, the greatest factor making this a "mixed condition for insurability" disease is regional management practices. Fish that survive exposure to an IHN outbreak are less susceptible to future outbreaks. Since the disease is common in Idaho, farmers intentionally expose young fish to the disease to cull the most susceptible individuals from the production system (Hinshaw, J.M., personal communication, North Carolina, 2004). This ensures no investment in growing fish that may later die of IHN before harvest. This practice makes the disease uninsurable since the fish are intentionally exposed to the disease and losses will result from this exposure. On the East Coast of the USA, however, IHN is exotic and farmers do not intentionally expose fish to the disease. In fact, there is a certification process that screens for IHN (Hinshaw, J.M., personal communication, North Carolina, 2004). If a farmer purchases only certified disease-free trout then effective management exists to prevent IHN from occurring on the farm. These factors tend to classify the disease as insurable for farmers on the East Coast.

Table 20.3. Potentially Insurable Diseases.

Insurability Conditions	Catfish	Trout	Baitfish
Disease and Description	Visceral Toxicosis of Catfish Idiopathic Disease First identified in 1999	Viral Haemorrhagic Septicaemia Virus Virulent strain not present in USA	Spring Viremia of Carp Virus OIE disease
Determinable Losses	Yes	Yes	Yes
Measurable Losses	Acute	Acute	Acute, due to farm being shut down and fish destroyed
Accidental and Unintentional Losses	No management can prevent or mitigate	No management can prevent or mitigate	Use of ground water, and screening of fish is only management
Sufficient Information to Conduct Risk Classification	Mostly found in the Mississippi delta region, but does occur elsewhere at lower frequency	Virulent strain is exotic to the USA	Outbreaks in NC, WI, IL and WA; not discovered elsewhere
Sufficient Data to Establish Accurate Premium Rates	Not available	Not available	Not available
Losses Sufficiently Uncorrelated to Allow for Pooling	Losses are independent of other producers	If disease entered the USA it could spread via wild fish and birds; initially it would likely be idiosyncratic	Farm that contracts virus will be shut down, therefore reducing risk to others
Is this Disease Insurable?	Yes	Yes	Yes

Table 20.4. Mixed Conditions for Insurability.

Insurability Conditions	Catfish	Trout
Disease and Description	Enteric Septicaemia of Catfish from *Edwardsiella ictaluri* "Hole in the Head Disease" Bacteria Mainly impacts fry and fingerlings	Infectious Haematopoietic Necrosis Virus Impacts fish of all ages
Determinable Losses	Often combined with two other diseases, hard to determine primary cause of loss	Yes, can determine primary cause of death
Measurable Losses	Acute for small fish: chronic for larger fish	Acute (above 10° C): chronic (below 10° C)
Accidental and Unintentional Losses	Some management can worsen loss; treatments for three diseases vary	A certification process is in place but no management exists once fish contract disease
Sufficient Information to Conduct Risk Classification	Cannot say one farm is more prone to this disease than another	In Idaho fish are intentionally exposed to the disease, but elsewhere producers try to keep this disease out of farm
Sufficient Data to Establish Accurate Premium Rates	Not available	Not available
Losses Sufficiently Uncorrelated to Allow for Pooling	Losses are independent of other producers	Future outbreaks may be more common since survivors are carriers of the virus and younger fish in production will be susceptible
Is this Disease Insurable?	Yes, since the other two diseases associated with ESC are also insurable	Yes, outside of Idaho: no, farms in Idaho

Another disease that can be considered a disease with "mixed conditions for insurability" is Enteric Septicaemia of Catfish (ESC). When catfish are diagnosed with ESC the fish often have infections from other diseases (Goodwin, A.E., personal communication, Arkansas, 2004; Hemstreet, W.G. personal communication, Alabama, 2004). This makes determining that the fish died primarily from ESC difficult. Alternative treatments are used when catfish have a suite of diseases including ESC (Camus, A.C., personal communication, Mississippi, 2004). These treatments are contradictory to the strategies for treating just ESC. Experts have mixed opinions on methods to treat a pond with multiple diseases so there are no clear management practices to reduce losses. The other diseases that are associated with ESC outbreaks are classified as insurable; therefore, since ESC meets all other criteria for insurability it tends to be more insurable than not insurable. In this case it is not absolutely necessary to say that the fish primarily died of ESC since all common diseases associated with ESC during the spring and autumn periods are also insurable.

Conclusions

Insuring livestock in general is a challenging exercise, and developing coverage for aquaculture production is as difficult as for any other species. Complications exist in the application of the standard insurability criteria, particularly with regard to the ability to measure and inventory aquaculture stocks in various production systems. However, we believe that our project will be able to establish a set of guidelines for the development of insurance policies for aquaculture. How well these guidelines will function in the creation of a successful policy will depend on numerous factors, not the least of which include the characteristics of the species under consideration and its associated production system.

As a case study, aquaculture can provide insight into how insurance policies should deal with diseases, one of the most prominent and problematic perils of concern to all livestock producers. The application of the insurability criteria is also key to providing coverage for diseases, and even those diseases that result in considerable losses to producers may be subject to exclusion from a policy. Diseases will differ in how well each will fit the criteria, and may not always be clearly classified as "insurable" or "non-insurable". Furthermore, the inclusion of certain diseases may change over time as production management may adapt to mitigate their effects. However, for catastrophic losses caused by disease outbreaks, insurance – perhaps in conjunction with other programmes – has substantial potential to provide aquaculture producers with risk protection.

References

Bruno, D.W. and Ellis, A.E. (1996) Salmonid disease management. In: Pennell, W. and Barton, B.A. (eds), *Principles of Salmonid Culture*. Elsevier, The Netherlands, pp. 759-832.

Engle, C. and Stone, N. (2004) Industry Profile: The Aquaculture of Baitfish. National Risk Management Feasibility Program for Aquaculture. Mississippi State University, Starkville, Mississippi. [Accessed 2005.] Available from http://www.agecon.msstate.edu/Aquaculture/pubs/Baitfish_Profile.pdf.

Forster, J. (2003) Industry Profile: The US Farmed Salmon Industry. National Risk Management Feasibility Program for Aquaculture. Mississippi State University, Starkville, Mississippi. [Accessed 2005.] Available from http://www.agecon.msstate.edu/Aquaculture/pubs/Salmon_Industry_Profile.pdf.

Hinshaw, J.M., Fornshell, G. and Kinnunen, R. (2004) A Profile of the Aquaculture of Trout in the United States. National Risk Management Feasibility Program for Aquaculture. Mississippi State University, Starkville, Mississippi. [Accessed 2005.] Available from http://www.agecon.msstate.edu/Aquaculture/pubs/Trout_Profile.pdf.

Miller, J.C., Lee, J.E., Hanson, T.R. and Coble, K.H. (2002) 2002 National Risk Management of Aquaculture Workshop Summary Report. Department of Agricultural Economics Information Report 2002-023, Mississippi State University, Starkville, Mississippi.

Secretan, P.A.D. (2003) The Availability of Aquaculture Crop (Stock Mortality) Insurance. National Risk Management Feasibility Program for Aquaculture, Mississippi State University, Starkville, Mississippi. [Accessed 2004.] Available from http://www.agecon.msstate.edu/aquaculture/pubs/availability.pdf.

Shaik, S, Coble, K.H., Miller, J.C. and Hanson, T.R. (2003) Aquaculture Insurance: Issues, Policy Designs, and Potential Research. Department of Agricultural Economics Information Report 2003-001, Mississippi State University, Starkville, Mississippi.

Stiglitz, J.E. (2000) *Economics of the Public Sector.* W.W. Norton and Company, New York.

Stone, N., Park, E., Dorman, L. and Thomforde, H. (1997) Baitfish culture in Arkansas: golden shiners, goldfish and fathead minnows. Cooperative Extension Program MP 386, University of Arkansas at Pine Bluff, Arkansas.

Tucker, C., Avery, J., Engle, C. and Goodwin, A. (2004) Industry Profile: Pond-Raised Channel Catfish. National Risk Management Feasibility Program for Aquaculture, Mississippi State University, Starkville, Mississippi. [Accessed 2005.] Available from http://www.agecon.msstate.edu/Aquaculture/pubs/Catfish_Industry_Profile.pdf.

Index

cabi-publishing.org/bookshop
Browse Read and Buy
ANIMAL & VETERINARY SCIENCES
BIODIVERSITY CROP PROTECTION
HUMAN HEALTH NATURAL RESOURCES
ENVIRONMENT PLANT SCIENCES
SOCIAL SCIENCES
CABI Publishing
A division of CAB International
Online BOOK SHOP
Search
Reading Room
Bargains
New Titles
Forthcoming
Order & Pay Online!
MasterCard
VISA
AMERICAN EXPRESS
Crop Pollination by Bees
Keith S. Delaplane and Daniel F. Mayer
A DICTIONARY OF Entomology
G Gordh and D H Headrick
Principles of CATTLE PRODUCTION
C.J.C. Phillips
Seeds
THE ECOLOGY OF REGENERATION IN PLANT COMMUNITIES
2nd EDITION
Edited by Michael Fenner
FULL DESCRIPTION
BUY THIS BOOK
BOOK OF THE MONTH
Tel: +44 (0)1491 832111 Fax: +44 (0)1491 829292